CPSS Power Electronics Series

This series comprises advanced textbooks, research monographs, professional books, and reference works covering different aspects of power electronics, such as Variable Frequency Power Supply, DC Power Supply, Magnetic Technology, New Energy Power Conversion, Electromagnetic Compatibility as well as Wireless Power Transfer Technology and Equipment. The series features leading Chinese scholars and researchers and publishes authored books as well as edited compilations. It aims to provide critical reviews of important subjects in the field, publish new discoveries and significant progress that has been made in development of applications and the advancement of principles, theories and designs, and report cutting-edge research and relevant technologies. The CPSS Power Electronics series has an editorial board with members from the China Power Supply Society and a consulting editor from Springer.

Readership: Research scientists in universities, research institutions and the industry, graduate students, and senior undergraduates.

More information about this series at http://www.springer.com/series/15422

Huafeng Xiao · Xiaobiao Wang

Transformerless Photovoltaic Grid-Connected Inverters

Springer

Huafeng Xiao
College of Electrical Engineering
Southeast University
Nanjing, China

Xiaobiao Wang
College of Electrical Engineering
Southeast University
Nanjing, China

ISSN 2520-8853 ISSN 2520-8861 (electronic)
CPSS Power Electronics Series
ISBN 978-981-15-8524-1 ISBN 978-981-15-8525-8 (eBook)
https://doi.org/10.1007/978-981-15-8525-8

This Springer imprint is published by the registered company Springer Nature Singapore Pte Ltd.
The registered company address is: 152 Beach Road, #21-01/04 Gateway East, Singapore 189721, Singapore

Contents

Chapter 1
Introduction

Abstract Photovoltaic power generation (PVPG) is an important part of renewable energy and is considered as the most potential renewable energy around the world. Major countries have invested huge amounts of money to compete in research and industrialization activities. Until 2018, global-installed solar PV capacity had reached 505 GW [1]. PVPG started developing a bit late in China, but it is developing significantly fast. According to the statistics, between 2010 and 2015 years [1], a miracle was created that worldwide cumulative installation capacity increased 100 times in 5 years. As a result, China has taken the first place in total PVPG accumulative installed capacity since 2015, as well as the ratio of China's PVPG installed capacity to total capacity had reached 7% at the end of 2018. It produced 4% of the total power generation amount during 2018. At the same time, China Electric Power Research Institute reported that the average cost of PVPG plants had been reduced by 45% in 2017 compared with 2012.

Keywords Photovoltaic power generation · China · First place · Cost down

1.1 Energy Challenges

Human beings experienced energy utilization styles from the age of firewood to coal, oil, gas, and electricity. While the total energy consumption is increasing constantly, the proportions of different energy are also changing predicted by the Joint Research Centre of European Commission in 2004 [2]. Every transition of the energy is accompanied by a huge leap in productivity, which has greatly promoted the development of economy and society.

At the same time, with the increasing consumption of energy, especially for fossil energy, the restriction of energy on economic and social development and the damage to the environment are becoming significantly apparent, such as urban smoke caused by soot pollution, automobile exhaust and industrial exhaust pollution, acid rain, the ozone layer damaged, soil loss, mining earthquake, marine pollution, and nuclear pollution, so far. Scientific observations show that the concentration of CO_2 in atmosphere has increased from 280 ppmv before the industrial revolution to the current

H. Xiao and X. Wang, *Transformerless Photovoltaic Grid-Connected Inverters*, CPSS Power Electronics Series, https://doi.org/10.1007/978-981-15-8525-8_1

379 ppmv, and the global average temperature has also increased by 0.74 °C in the past century.

A series of challenges in front of human beings force us to think about the solutions, that is, how to find a way out for human energy? The answer is "drive the world with sunlight!" The sun is the source of life on earth and also the source of energy, which can be continuously regenerated. The development of renewable energy can gradually meet the human energy needs, and is also an urgent need to protect the environment.

As a kind of renewable energy, solar energy can be directly converted and used. Calculations show that only 130,000 km^2 of photovoltaic (PV) panels need to be installed in China to meet the whole nation's energy consumption. According to China's annual building construction area of 2 billion square meters, China's urban construction area can reach 32,000 km^2, rural construction area can reach 21,000 km^2, and the effective install area for PV panels can reach 10,000 km^2 by adding the southern wall areas at the end of 2020. Therefore, the development potential of building-integrated PV (BIPV) is huge.

1.2 Utilizing Styles of Solar Energy

Since the birth of life on earth, it has been mainly survived by the thermal radiation provided by the sunshine. Humans have also understood that objects can be dried by the sunshine since ancient times, and treat that as a method of making food, such as salt making and salted fish. With the decrease of fossil fuels, solar energy is becoming an important part of human energy utilization, and develops quickly. In a broad sense, wind energy, chemical energy, and water energy on the earth all belong to solar energy. From a narrow sense, utilization of solar energy mainly includes photothermal, photochemical, photovoltaic, and other conversion forms.

1.2.1 Photothermal Conversion

The principle of photothermal conversion is to collect solar radiation energy, and then convert it into thermal energy by medium interaction [3]. There are four types of solar collectors so far; they are flat-plate collector, vacuum tube collector, ceramic solar collector, and focusing collector. According to medium temperatures, photothermal utilizations can be classified into low-temperature utilization (<200 °C), medium-temperature utilization (200–800°C), and high-temperature utilization (>800 °C), respectively. For example, low-temperature utilization mainly includes solar water heaters, solar dryers, solar air-conditioning refrigeration systems, and so on; solar stoves belong to medium-temperature utilization; high-temperature solar furnace is one example of high-temperature utilization.

Currently, the most successful commercialization form of photothermal conversion is solar water heater. Its working principle is to convert sunshine energy into thermal energy to heat water for needs in daily life and industrial applications.

1.2.2 Photochemical Conversion

Photochemical conversion is a process of absorbing sunshine energy for chemical reactions to obtain chemical energy [3], such as photosynthesis of plants.

In recent years, a lot of PVPG plants were built in western China, but electricity consumers are located in eastern China, so electricity transmission pressure is put on the high-voltage transmission grid, and has caused severe abandonment of PV electricity. In order to relieve this pressure, some cities have launched a plan named as "ten cities one thousands of hydrogen fuel cell vehicles". In this situation, use PV electricity to produce hydrogen by electrolyzing water in the places of PVPG plants, and then ship the produced hydrogen to different cities to fuel vehicles. Along with the fast development and cost down of PV electricity and electrolyzing water, the conversion between solar energy and hydrogen through photovoltaic process will be a profitable way to guarantee China's energy security and optimize the energy structure.

1.2.3 Fuel Using

A new engine fuel named as "solar fuel" can be made of water and carbon dioxide catalyzed by high-temperature energy from sunshine. The European Union has successfully implemented the full-process production of laboratory-scale renewable fuel around the world for the first time in June 2011, and the produced fuel fully complies with the European Union standards in terms of aircraft and automotive fuel. As a result, there is no need for any modification of aircraft and automobile with engines.

The designed "solar fuel" prototype mainly has two parts: the first part uses high-temperature energy generated by the concentrated solar light, supplemented by the metal oxide material additives invented by the ETH Zürich's intellectual property, water and carbon dioxide to synthesize mixture gas. The second part works based on the Fischer–Tropsch process, which converts the high-temperature mixture gas into "solar fuel" product that can be commercially applied to the market.

1.2.4 *Photovoltaic Conversion*

There are two ways to convert sunlight into electricity, that is, photovoltaic converting directly, or indirectly converting from sunshine to electricity through thermal energy.

1. Light to heat to electricity conversion

The operation process is that the thermal energy generated from solar radiation is used to generate electricity [4]. Commonly, a solar collector is used to convert the absorbed thermal energy into working medium steam, and then the steam drives a gas turbine to drive a generator to generate electricity. The first process is light to heat conversion, and the second process is heat to electricity conversion.

At present, the most promising solar thermal power generation systems can be roughly classified into three kinds: trough-shaped parabolic focusing system, central receiver or solar tower focusing system, and disc-shaped parabolic focusing system. The three forms that are technically and economically feasible are: 30–80 MW focused parabolic trough solar thermal power-generation technology (parabolic trough type); 30–200 MW point-focused central-receiving solar thermal power-generation technology (central receiving type); 7.5–25 kW point-focused parabolic disc solar thermal power-generation technology (parabolic disc type).

2. Light to electricity conversion

The basic operating principle is to directly convert solar radiant energy into electrical energy using the photovoltaic effect [5]. Its basic device is solar cells. The development history, technical characteristics, and application styles will be discussed in detail in the next section. It is the main application form of renewable energy technology and is the most successful form of solar energy utilization for commercialization and large-scale application.

1.3 Solar Cell

Solar cell is a device that converts sunlight into electricity based on the photovoltaic effect. The photovoltaic effect was first discovered by French scientist A.E. Becquerel in 1839. The first solar cell was manufactured by Charles Fritts in 1883. At the very beginning, Charles overlaid a very thin metal layer on the selenium semiconductor to form a semiconductor metal junction, but its conversion efficiency was only 1%, which was not practical. With the improvement of semiconductor physics and manufacturing technology, researchers from Bell Labs in the United States found that doping a certain amount of impurities in silicon would make it more sensitive to light, and produced the first single-crystal silicon solar cell with an efficiency of 6% in practical application value in 1954 [3–5]. Nowadays, solar cell technology has made great progress, and the efficiency of commercialized single-crystal silicon solar cells has reached 20%, as well as many types of commercialized batteries have been designed, such as cadmium telluride batteries, copper indium gallium selenium batteries, and amorphous silicon batteries. At the same time, novel solar cells with high efficiency are being developed. Advances of photovoltaic cell technology drive the cost of photovoltaic power generation down. The cost gradually approaches the level that can compete with conventional power generation and promotes the development of the photovoltaic industry.

There have been many types of solar cells so far [3–5]. According to structures: homojunction solar cells, heterojunction solar cells, Schottky solar cells, and liquid junction solar cells, and so on; while according to materials: silicon solar cells, sensitized nanocrystalline solar cells, organic compounds and inorganic compounds, semiconductor solar cells, thin-film solar cells, and so on. The following briefly introduces the mainstream solar cells with the highest acceptance in the market.

1.3.1 Monocrystalline Silicon Photovoltaic Cell

Monocrystalline silicon photovoltaic cells are a type of photovoltaic cell that was developed earlier and has the highest conversion efficiency. The conversion efficiency of monocrystalline silicon photovoltaic cells has reached an average of 16.5% in China, and the highest conversion efficiency recorded in the laboratory has exceeded 24.7% [3–5]. This photovoltaic cell generally uses high-purity monocrystalline silicon rods as raw materials, and the purity requires 99.9999%.

1.3.2 Polycrystalline Silicon Photovoltaic Cells

Polycrystalline silicon photovoltaic cells are made of polycrystalline silicon materials [3]. Because polycrystalline silicon materials are mostly produced in casting instead of the drawing process of the monocrystalline silicon, the production time is shortened significantly, and the manufacturing cost is greatly reduced as well. In addition, the monocrystalline silicon rod is cylindrical, and the produced wafers are round shape too; as a consequence, the utilization rate of the panel is low after forming the photovoltaic module. Compared with monocrystalline silicon photovoltaic cells, polycrystalline silicon photovoltaic cells appear to have a certain competitive advantage in the market.

1.3.3 *Amorphous Silicon Photovoltaic Cells*

Amorphous silicon photovoltaic cells are a new type of thin-film battery composed of amorphous silicon as a raw material [4]. Amorphous silicon is a kind of semiconductor with amorphous crystal structure. The thickness of amorphous silicon photovoltaic cells is only 1 μm, which is equivalent to 1/300 of monocrystalline silicon photovoltaic cells. Most importantly, its manufacturing process is greatly simplified, the silicon material consumption is low, and its unit power consumption is also greatly reduced.

1.3.4 *GaAs Photovoltaic Cells*

GaAs photovoltaic cells are III–V compound semiconductor photovoltaic cells [4]. Compared with silicon photovoltaic cells, GaAs photovoltaic cells have high photoelectric conversion efficiency. For instance, the theoretical efficiency of silicon photovoltaic cells is 23%. However, the conversion efficiency of single-junction GaAs photovoltaic cells has already reached 27%. It can be made as thin film and ultra-thin solar cells. For example, GaAs photovoltaic cells only need to be 5–10 μm in thickness to absorb 95% of sunlight, but silicon photovoltaic cells need to be thicker than 150 μm.

1.4 Photovoltaic Power-Generation System

PVPG usually consists of PV arrays, solar tracking equipment (optional), battery pack (optional), battery controller (optional), and inverter, which can supply AC and DC loads, as well as store excess energy in batteries or send it to the grid, as shown in Fig. 1.1.

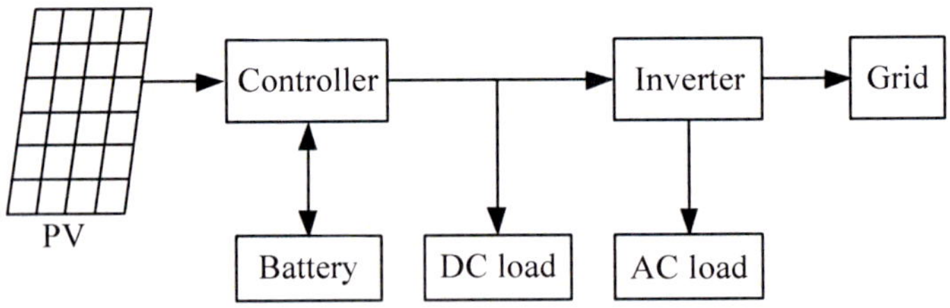

Fig. 1.1 Schematic diagram of PVPG

1.4.1 Off-Grid PVPG

Off-grid PVPG is also named as independent PVPG, which means that this system does not connect to the grid. It is mainly composed of solar arrays, controllers, and batteries. In addition, an AC inverter is also required if there are AC loads in the PVPG. Generally speaking, off-grid PVPGs include remote village power supply, household solar power supply, communication power source, street lights source, and so on.

1.4.2 On-Grid PVPG

On-grid PVPG is also called as grid-connected PVPG, and the DC power generated by solar array is directly delivered into the public grid. On-grid PVPG may integrate with batteries or not [5]. Obviously, the on-grid PVPG with batteries is dispatchable according to the needs; in addition, it can be treated as a backup power source for emergency loads when the grid fails.

On-grid PVPG without batteries is the most popular PVPG form. There are different circuit structures according to power ratings, battery connection methods, converter architectures, as shown in Fig. 1.2.

1. Centralized type

The centralized PVPG is the earliest structure in practice, as shown in Fig. 1.2a. Its capacity is generally tens to hundreds of kilowatts. The system structure connects a large number of photovoltaic arrays in series to a higher voltage level first, and then collects them to a high current output through diodes in parallel. Finally, collected DC electricity is fed into the grid through a centralized inverter. The main advantages of centralized PVPG are high efficiency and low cost. But there are also issues, such as low utilization of single photovoltaic arrays, poor ability to against local shadows, and inconvenient maintenance. The centralized PVPG is generally used in ground photovoltaic power farms with large capacity.

2. String and multi-string types

In order to avoid the mismatch loss of multiple strings in the centralized PVPG, string and multi-string structures have emerged on the market, as shown in Fig. 1.2b and c, respectively. Among them, each string has an independent maximum power point tracking (MPPT) controller. Generally, the string-type PVPG is installed on the roof of small power residential; the multi-string-type PVPG adopts a two-stage circuit, and the front stage is scalable to fit different positions of PV arrays on the building. However, these systems still have the following problems in practical applications: (1) Although the MPPT controller solves the mismatch loss between the strings, the battery modules in each string cannot be guaranteed to run at the maximum power point; (2) the system maintenance is still difficult.

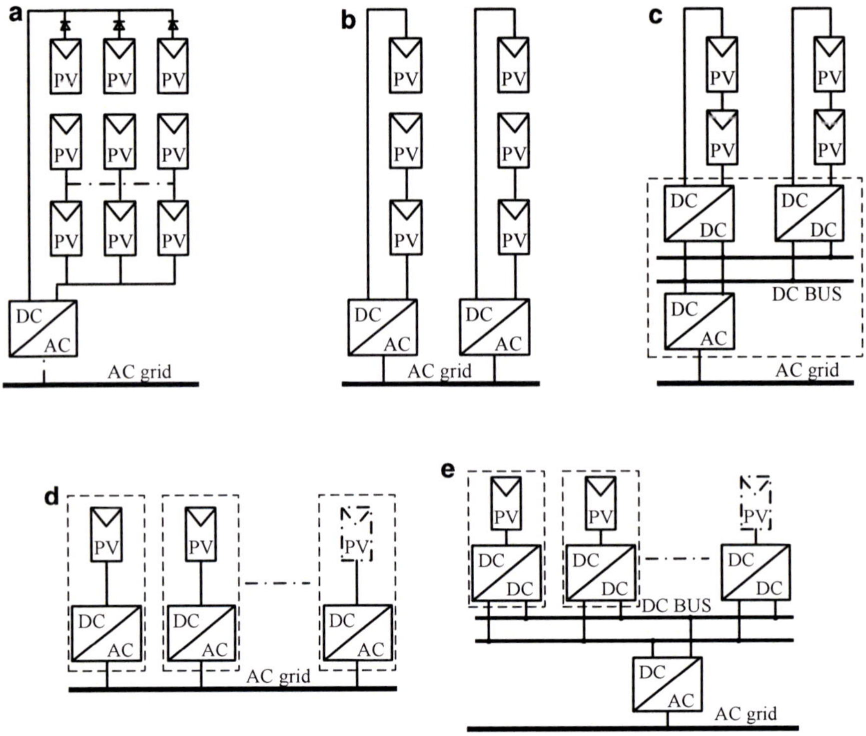

Fig. 1.2 Structures of typical PVPG systems, **a** Centralized type, **b** String type, **c** Multi-string type, **d** AC modular type, **e** DC modular type

3. AC module type

The AC module PVPG integrates a grid-connected inverter with MPPT function for each PV module, as shown in Fig. 1.2d. The advantages of this structure are that each PV module can be guaranteed to work at its maximum power point, and has strong ability to stand local shadow. It is convenient to scale up and maintain. In addition, the AC module PVPG supports plug and play. However, it still has the shortcomings of low conversion efficiency, complex system circuit, and high cost.

4. DC module type

The DC module PVPG combines the characteristics of multi-string and AC module structures, composed of multiple DC modules and a centralized inverter, as shown in Fig. 1.2e. The MPPT function is implemented in different DC converters for each PV module, and the centralized inverter connects DC bus and the AC grid.

With the development of PVPG technology, some new system structures have also appeared in recent years, such as improved string structure based on bypass

DC modules, improved string system based on power generation control circuits, cascaded DC module system, and AC battery cell system [3].

1.5 Maximum Power Point Tracking

The efficiency of PVPG system is the product of PV panel conversion efficiency, MPPT efficiency, and the inverter efficiency. Therefore, improving the MPPT efficiency is of great significance for increasing the system efficiency and reducing the cost. Under different radiation levels and PV cell temperatures, the maximum output power points of photovoltaic cells are different, as shown in Fig. 1.3. Figure 1.3a shows the current–voltage (I–U) curves of a PV module, and its power–voltage (P–U) curves are shown in Fig. 1.3b, where S represents the radiation level, and I, U, and P are the output current, voltage, and power of a PV module, respectively. In order to track the maximum output power, some measures have to be taken to automatically follow the changes in environmental conditions. The MPPT technology has been proposed in response to this issue [3].

Conventional MPPT algorithms include open-circuit voltage method, short-circuit current method, perturbation-observation method, incremental conductance method, and so on [6, 7]. Recently, some intelligent tracking algorithms have appeared, such as fuzzy control algorithms and neural network algorithms. Several widely used MPPT algorithms are induced as follows.

1.5.1 Perturbation and Observation Method

The perturbation and observation method finds the maximum power point by increasing and decreasing the output voltage of PV strings with a small step based

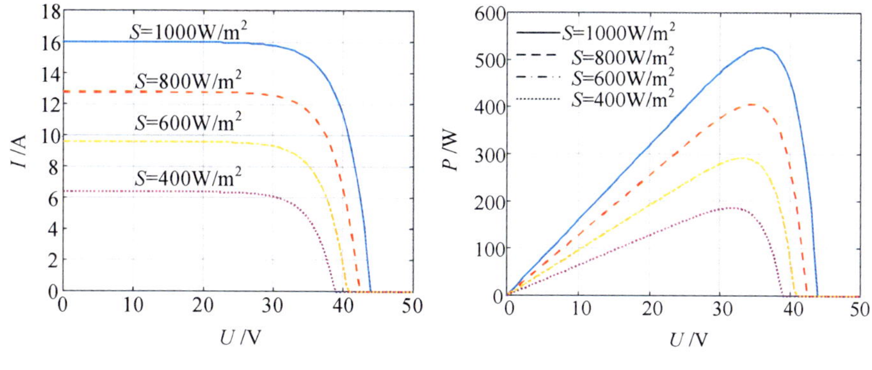

Fig. 1.3 Output characteristic curves of a PV module, **a** I–U curve, **b** P–U curve

on measuring and comparing power variations [8, 9]. Briefly speaking, the algorithm keeps adjusting the output voltage with the same step if the output power variation is positive; otherwise, change the step with opposite direction. Obviously, this algorithm belongs to hill-climbing method, and depends on the *P–U* curves, as shown in Fig. 1.3b. It is the most commonly used MPPT method in practice, but it results in power loss due to small voltage oscillations.

1.5.2 Incremental Conductance Method

The increment conductance method essentially belongs to hill-climbing method too [10]. It is proposed based on that the derivative of the PV cell output power to voltage is zero at the maximum power point. Therefore,

$$\frac{\mathrm{d}P}{\mathrm{d}U} = \frac{\mathrm{d}(UI)}{\mathrm{d}U} = I + U\frac{\mathrm{d}I}{\mathrm{d}U} = 0 \tag{1.1}$$

Then the incremental conductance terms are obtained,

$$\frac{I}{U} + \frac{\mathrm{d}I}{\mathrm{d}U} = G + \Delta G = 0 \tag{1.2}$$

The change of the control signal is decided by detecting instantaneous conductance G and conductance increment ΔG of the PV string output, and comparing these two values with zero. The outstanding advantage of the incremental conductance method is that the amplitude of its voltage oscillation is relatively small. However, the determination of the voltage increment step is much complicated. For instance, the tracking error will be large if the step is too large; on the contrary, the tracking speed will be slower. On the other hand, the incremental conductance method requires high accuracy and response time of the sensors and detection circuits.

1.5.3 Intelligent Algorithms

In practical applications, complex environments such as local shadows often make the *P–U* characteristic curves of PV arrays present multiple power pick points. For conventional direct MPPT control methods based on sampled data, such as above-mentioned perturbation and observation method, and incremental conductance method, they tend to be trapped in local peak points and cause serious power mismatch. As a consequence, the local peak points not only lose energy but also damage PV cells because of hot spot phenomenon. Therefore, how to track the global maximum power point in the case of local shadows is a critical issue needed to be solved.

The main idea is to introduce intelligent algorithms into the conventional MPPT methods to implement a global search task [10, 11]. There are several intelligent MPPT methods listed as follows:

1. Particle swarm optimization algorithm

Particle swarm optimization algorithm is a global optimization method based on multi-extremal function, which resulted from foraging behavior of birds. In the continuous iterative process of particles, the update of positions and velocities is under the influence of its own optimal solution and the group optimal solution, so that the particles move to the position of the group optimal solution to achieve global optimization [12]. Taking MPPT application as an example, the target is to find the best particle (control parameter), so that the objective function (output power) has a global optimal solution, that is, the global maximum power point of the array.

2. Fuzzy algorithm

The essence of fuzzy logic control is a logical reasoning system based on the experience and intuition of the equipment operator, which is suitable for some systems whose mathematical models are difficult to build. For nonlinear output characteristics of PVPG systems affected under factors of temperature and irradiance, a suitable control fuzzy rule table can be formulated based on the PV output characteristics and operating experience. The working voltage can be fast and accurately stabilized at the maximum power point [13–17].

Because traditional fuzzy logic control is not suitable for the global MPPT problem under local shadows, Hopfield neural network can be used to optimize the fuzzy logic controller so that it has a dynamic fuzzy rule table [17]; or using immune algorithm to optimize fuzzy logic control, for instance, combined with the double advantages of fuzzy logic theory and artificial immune theory, the global maximum power point can be achieved under local shadows [16].

3. Neural Network Algorithm

Neural network technology is a kind of control technology that mimics human thinking. It does not depend on the mathematical model of the controlled process, and features strong anti-interference ability and black box learning ability. Therefore, it is quite suitable for the MPPT function of PVPG systems [18, 19]. However, the learning mode of the neural network algorithm requires long-term training, and the relationship between input and output data is difficult to be expressed.

1.6 Islanding Detection

Islanding refers to an electrical phenomenon that the distributed generator continues to supply power for the local loads when the power grid is disconnected from an area covering PVPG and loads [20–22]. The islanding phenomenon is shown in Fig. 1.4,

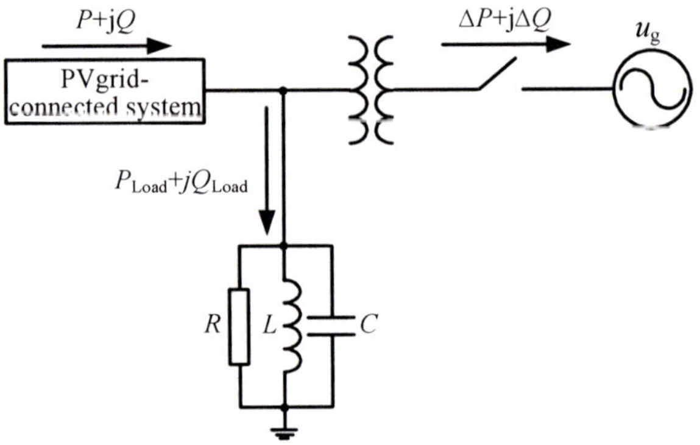

Fig. 1.4 Schematic diagram of isolated islanding powered by PVPG system

where P and Q represent active power and reactive power, respectively, u_g represents the power grid, and R, L, and C are resistor, inductor, and capacitor of local loads, respectively. When the output power of the PVPG matches the load consumption, that is, $\Delta P = \Delta Q = 0$, the voltage and frequency of the local power area will not change basically after the grid-connected circuit breaker is disconnected. As a consequence, the passive protection algorithm is not able to identify the absence of the grid.

When the islanding phenomenon occurs, it will cause serious consequences if the PVPG fails to detect the islanding state timely and quickly, including but not limited:

(1) Users' equipment might be damaged when the voltage and frequency of the islanding area exceed a certain range because of the absence of the grid.
(2) May damage the bus or maintenance people.
(3) Reclosing of circuit breaker may cause rush current and damage the equipment associated with the grid because of mismatch phase angle.

In order to avoid the above-mentioned damages, international standards, such as IEEE Std 929-2000 [23] and UL1741 [24], put forward mandatory requirements for islanding detection methods of distributed generators. The main technical terms are shown in Table 1.1.

According to China's technical standard GB/T 29319-2012 [25], the allowable deviation of the phase voltage is −10% to +7% of the rated voltage, the frequency at the point of common coupling is in the range of 49.5−50.2 Hz, the total harmonic current should be less than 5% of the rated output current, and the DC current component in the grid-connected current cannot exceed 0.5% of its rated output current when the grid is working normally. The response time against the abnormal grid is basically consistent with Table 1.1. In addition, there are some specific limitations as listed in the following:

(1) The PVPG system should have the ability to quickly identify islanding and immediately disconnect from the grid;
(2) The response time of the anti-islanding protection should be shorter than 2 s, and the acting should be compatible with relay protection;
(3) The PVPG should not deliver power in 20–300 s after the grid return to normal.

Table 1.1 IEEE Std 929-2000/UL1741 requirements for maximum response time of islanding detection

Status	Voltage amplitude after power off	Frequency after power off/Hz	Allowed maximum detection time
A	$u < 0.5u_{\text{nom}}$	f_{nom}	6 power frequency cycles
B	$0.5u_{\text{nom}} \leq u \leq 0.88u_{\text{nom}}$	f_{nom}	2 s
C	$0.88u_{\text{nom}} \leq u \leq 1.10u_{\text{nom}}$	f_{nom}	2 s
D	$1.10u_{\text{nom}} \leq u \leq 1.37u_{\text{nom}}$	f_{nom}	2 s
E	$1.37u_{\text{nom}} \leq u$	f_{nom}	2 power frequency cycles
F	u_{nom}	$f < f_{\text{nom}}$–0.7	6 power frequency cycles
G	u_{nom}	$f > f_{\text{nom}} + 0.5$	6 power frequency cycles

u_{nom} refers to the nominal value of the grid voltage amplitude, f_{nom} refers to the nominal value of the grid voltage frequency, the quality factor Q_{f} of RLC loads is not less than 2.5

In order to meet the above-mentioned detection requirements, a large number of detection methods have been proposed and reported [26–30]. According to the implementation styles, they can be classified into two categories: remote islanding detection algorithms and local islanding detection algorithms. The remote islanding detection algorithm uses the communication medium to remotely collect grid-related state data to identify potential islanding. The measures feature high reliability but high cost. Therefore, it is not suitable for distributed PVPG with small and medium power ratings.

The local islanding detection algorithm is used to determine the islanding state by detecting the state of the grid, which is simple to implement and cost-effective. The implementation methods can be classified into two categories, namely passive islanding detection methods and active islanding detection methods. The passive islanding detection method recognizes islanding by monitoring the voltage parameters and their variations, such as over/under voltage, phase abruption, over/under frequency, or abrupt change of frequency. It is simple and cheap, but has a large non-detection zone. The active islanding detection methods inject disturbances into point of common coupling to identify the islanding state through monitoring parameter changes, such as active power disturbances, reactive power disturbances, load injection, and frequency shifting and phase shifting. As a result, the non-detection zone is significantly reduced. However, the injected disturbances would affect the power quality and stability of the power grid.

1.6.1 Summary

This chapter briefly describes the current state of energy around the world, and points out that solar energy, as a renewable energy source, plays an important role in solving energy shortages and environmental pollution. Then, several types of utilization of solar energy are outlined, among which the photovoltaic power generation is the most important one. Finally, a series of key technologies related to grid-connected PVPG system are discussed, such as maximum power point tracking technology and islanding detection technology.

References

1. IEA PVPS, Trends 2018 in photovoltaic applications (2018) [Online]. Available: http://www.iea-pvps.org/fileadmin/dam/public/report/statistics/2018_iea-pvps_report_2018.pdf, IEA Photovoltaic Power Systems Programme, Report IEA PVPS T1–34: 2018. Accessed 15 Jan 2020
2. European Union Science Hub (2004). http://ec.europa.eu/jrc/en. Accessed 15 Jan 2020
3. Zhang X, Cao R (2018) PV grid-connected power generation and its inverter control, 2nd edn. China machine press, Beijing (in Chinese)
4. Solar cells and its application, Hamakawa Yoshihiro, Zhang H, Cui X translating. China Science Press, Beijing (2008) (in Chinese)
5. Luo X, Miu Z, Wei D et al (2016) Photovoltaic power generation technology and design in application, 2016. China Science Press, Beijing (in Chinese)
6. Xiao WD, Elnosh A, Khadkikar V et al (2011) Overview of maximum power point tracking technologies for photovoltaic power systems. Electric Machines & Power Systems 43(12):1329–1351
7. Koutroulis E, Blaabjerg F (2013) Design optimization of transformerless grid-connected PV inverters including reliability. IEEE Trans Power Electron 28(1):325–33
8. Hegazy R, Ali ME (2015) A comprehensive comparison of different MPPT techniques for photovoltaic systems. Sol Energy 112:1–11
9. Ioan VB, Marcel I, Comparative analysis of the perturb-andobserve and incremental conductance MPPT methods In 2013 International Symposium on Advanced Topics in Electrical Engineering, Bucharest
10. Liu FR, Duan SX, Liu F et al (2008) A variable step size INC MPPT method for PV systems. IEEE Trans Industr Electron 55(7):2622–2628
11. Jin XL, Wen SS, Ni HZ, Yang YX et al (2019) Review of maximum power point tracking of photovoltaic system. Chinese Journal of Power Sources 43(03):532–535
12. Nie XH, Lai JJ (2014) A survey on tracking and control approaches for global maximum power point of photovoltaic arrays in partially shaded environment. Power System Technology 38(12):3279–3285
13. Ram PJ, Rajasekar N (2017) A new robust, mutated and fast tracking LPSO method for solar PV maximum power point tracking under partial shaded conditions. Appl Energy 201:45–59
14. Veerachary M, Senjyu T, Uezato K (2003) Neural-network-based maximum-power-point tracking of coupled-inductor interleaved-boost converter- supplied PV system using fuzzy controller. IEEE Trans Industr Electron 50(4):749–758
15. Bendib B, Krim F, Belmili H et al (2014) Advanced fuzzy MPPT controller for a stand-alone PV system. Energy Procedia 50:383–392
16. Liu LQ, Wang ZX, Zhang HHQ (2010) Fuzzy-immune MPPT control of PV generation system under partial shade condition. Electric Power Autom Equip 30(7):96–99

17. Subiyanto MA, Shareef H (2012) Hopfield neural network optimized fuzzy logic controller for maximum power point tracking in a photovoltaic system. Int J Photoenergy 2012:1–13
18. Khaehintung N, Pramotung K, Tuvirat B et al, RISC microcontroller built-in fuzzy logic controller of maximum power point tracking for solar-powered light-flasher applications In 2004 Annual Conference IEEE Industrial Electronic Society, Busan
19. Hiyama T, Kouzuma S, Imakubo T (1995) Identification of optimal operating point of PV modules using neural network for real time maximum power tracking control. IEEE Trans Energy Conv 10(2), 360–36
20. Sedaghati F, Nahavandi A, Badamchizadeh MA et al (2012) PV maximum power-point tracking by using artificial neural network. Mathematical Problems Eng 1:95–102
21. Mango FD, Liserre M, Dell'Aquila A, Overview of anti-islanding algorithms for PV systems. Part II: Active methods In 2006 International Power Electronics and Motion Control Conference, Portoroz
22. Hernandez-Gonzalez G, Iravani R (2006) Current injection for active islanding detection of electronically-interfaced distributed resources. IEEE Trans Power Deliv 21(3), 1698–170
23. IEEE, IEEE recommended practice for utility interface of photovoltaic (PV) systems: (2000) IEEE Std 929–2000
24. Underwriters Laboratories Inc (2010) Standard for inverters, converters, and controllers for use in independent power systems: UL 1741
25. Technical requirements for connecting photovoltaic power station to power system (2012) Standardization Administration of the P.R.C, GB/T 19964–2012
26. Jia K, Zhu ZX, Yang Z et al (2019) Intelligent islanding detection method for grid-connected photovoltaic power system based on improved adaboost algorithm. Power System Technol 43(04):1227–1238
27. Gao SP, Li WH (2016) New islanding detection method for grid-connected pv inverter. Acta Energiae Solaris Sinica 37(10):2526–2532
28. Shen HZ, Wen F, Wang HB et al (2017) Islanding detection for grid-connected inverter based on reactive current control. Trans China Electrotechn Soc 32(16):294–300
29. Khodaparastan M, Vahedi H, Khazaeli F et al (2017) A novel hybrid islanding detection method for inverter-based DGs using SFS and ROCOF. IEEE Trans Power Deliv 32(5):2162–2170
30. Chen X, Li Y (2016) An islanding detection method for inverter-based distributed generators based on the reactive power disturbance. IEEE Trans Power Electron 31(5):3559–3574

Chapter 2
Transformerless Photovoltaic Grid-Connected Inverters and Leakage Current Issue

Abstract As the interface between PV strings and the grid, grid-connected inverters perform functions of converting power generated by PV modules into the grid. Generally, some indexes are used to evaluate its performance, such as conversion efficiency, volume, cost, and grid-in current quality. This chapter mainly focuses on topologies of distributed PV grid-connected inverters, including isolated type and non-isolated type (also called as transformerless type). Especially, the leakage current issue of transformerless grid-connected inverters is deeply discussed. Further, a common-mode voltage model at switching frequency scale has been built, and restriction rules of leakage current have been concluded. Finally, the proposed restriction rules are discussed and used in full-bridge, half-bridge, and common-ground type topologies.

Keywords PV grid-connected inverte · Transformerless · Leakage current · Combination · Matching

2.1 Isolated PV Grid-Connected Inverters

The topology structure of grid-connected inverters is closely related to the efficiency, cost, security, reliability, and grid-in current quality of PVPG system. Generally speaking, grid-connected inverters have single-phase and three-phase structures. According to whether transformer components are included or not, they can be classified into line-frequency isolated, high-frequency isolated, and non-isolated structures [1–3].

2.1.1 Line-Frequency Isolated Structure

A line-frequency transformer is inserted at the AC output side of the inverter to make galvanic isolation between PV modules and the grid, which is named as the line-frequency isolated PVPG system, as shown in Fig. 2.1. This structure ensures personal safety, and is beneficial to match the output voltage and suppress the DC

H. Xiao and X. Wang, *Transformerless Photovoltaic Grid-Connected Inverters*, CPSS Power Electronics Series, https://doi.org/10.1007/978-981-15-8525-8_2

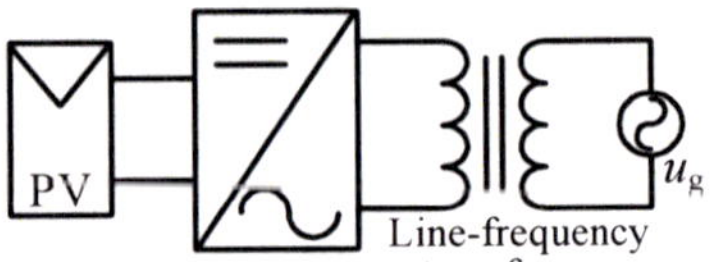

Fig. 2.1 Line-frequency isolated structure

component going into the grid. However, the line-transformer increases the size, weight, and cost of the PVPG system, and reduces the conversion efficiency of it [4, 5]. Grid-connected inverters with line-frequency transformers are applied typically in high-power three-phase and few single-phase PVPG systems; commonly, the conversion efficiency range of the line-frequency PVPG system is from 94 to 96%.

2.1.2 High-Frequency Isolated Structures

The line-frequency isolated structure is excluded in low-power and medium-power PVPG applications because of big size, high weight, and cost. Inserting a high-frequency transformer into the DC/AC conversion link is an alternative method to realize the galvanic isolation and voltage matching. At the same time, the high-frequency transformer significantly reduces the size, weight, and cost.

According to forms of the DC bus in the power conversion link, there are three kinds of high-frequency isolated structures, that is, DC-link form [6–8], pseudo DC-link form [9–16], and without DC-link form [17, 18], as shown in Fig. 2.2a, b, and c, respectively.

The topology circuits of low-power PVPG are significantly enriched because of inducing the high-frequency transformer, and are popularly installed in residential applications at the very beginning. However, the inserted high-frequency transformer makes the conversion circuits sophisticated, and results in low conversion efficiency of 90–95%.

2.2 Transformerless PV Grid-Connected Inverters

An inverter structure with neither line-frequency nor high-frequency transformer is named as transformerless grid-connected inverter (TLI), which brings the advantages of higher efficiency, simple circuit, and reduced weight and cost. In general, TLIs have single-stage and two-stage topologies [19], as shown in Fig. 2.3a, b, respectively.

Two-stage TLIs fit wide input voltage range and simple system design since the pre-stage DC to DC converter and the post-stage inverter can be controlled and optimized separately. Single-stage TLIs require higher input voltage from PV strings, which should be higher than the peak voltage of the grid but offer higher efficiency

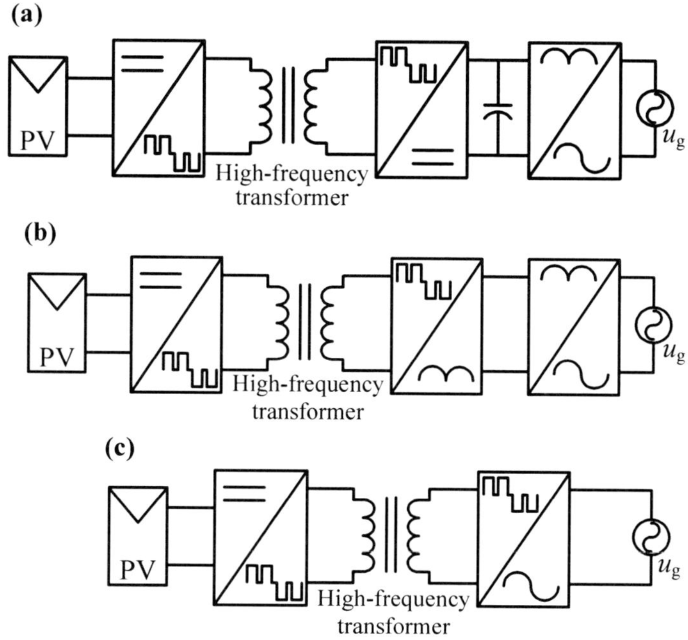

Fig. 2.2 High-frequency isolated structures, **a** structure with DC/DC stage, **b** structure with pseudo DC/DC stage, **c** Structure without DC/DC stage

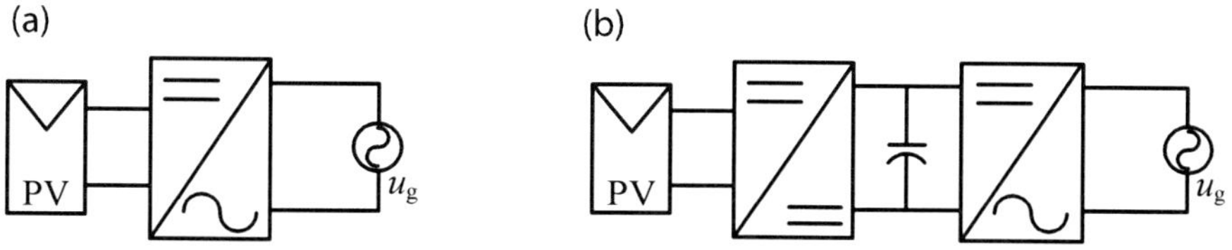

Fig. 2.3 Transformerless grid-connected inverters, **a** single-stage structure, **b** two-stage structure

than two-stage TLIs. With the trend that the output voltage of PV strings rises toward 1500 V (medium voltage), the single-stage TLIs will gain more market share in the future.

Due to the advantages of TLIs, almost all photovoltaic inverter manufacturers have launched their TLI product lines; the companies include Sunways, SMA, Sungrow Power, Growatt, and so on.

2.2.1 Leakage Current Issue

TLIs feature higher efficiency, smaller volume, and lower cost compared with isolated counterparts. However, PVPG without transformer results in a galvanic connection between the PV arrays and the grid [20–22]. Owing to the emergence of parasitic capacitors between the PV arrays and the earth, as shown in Fig. 2.4, high-frequency potential differences induced by switching actions may stimulate leakage current (LC), also called as common-mode current or ground current. The high-frequency LC results in severe conduction and radiation, electromagnetic interference, grid-in current distortion, and energy loss, and it can even jeopardize equipment and individual safety.

In order to guarantee the safety of individual and equipment, the LC of TLIs has to comply with the mandatory standards. For example, the German standard VDE0126-1-1 requires that leakage currents greater than 300 mA must trigger a break within 0.3 s [23]. Besides, irrespective of the rated power of the inverter, any sudden leakage currents should trigger the break at certain time. Table 2.1 shows the sudden change amounts and related break times.

Chinese standard NB/T 32004-2013 also states that PVPG must be quit within 0.3 s and alarms if LC exceeds 300 mA for rated PVPG lower than 30 kVA, and 10 mA/kVA for rated PVPG higher than 30 kVA [24]. Meanwhile, the protection procedure and limitations of LC changes are in accordance with Table 2.1.

Leakage current issue is of great importance because it is directly related to the quality certification and marketing authorization of TLI products. That is the reason why North American, Japan, and Australia forbade TLIs in PVPG system at the beginning. In order to popularize the TLIs, the mechanism of LC must be figured out, so a common-mode analytical model for TLIs is crucial.

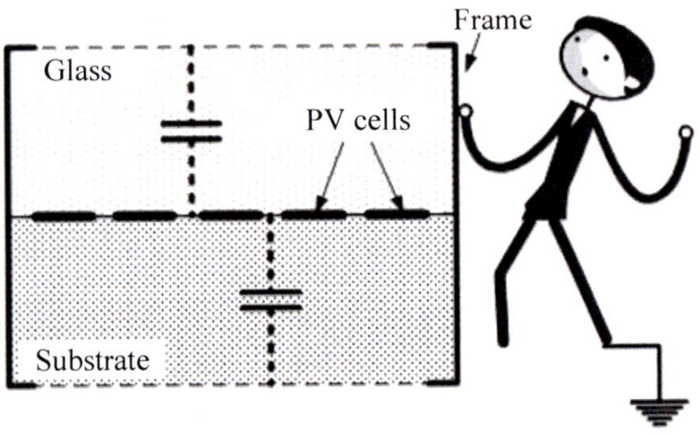

Fig. 2.4 Parasitic capacitors between PV cells and the frame

Table 2.1 LC change amount and break time

RMS value of LC change (mA)	Break time (s)
30	0.3
60	0.15
150	0.04

2.2.2 Elimination Rules of LC

Taking popular distributed PVPG system rated between 2 and 10 kW as an example, a typical system structure is shown in Fig. 2.5, where i_{CM} and i_{DM} represent the common-mode current and differential-mode current, respectively. From the perspective of spectral components, i_{CM} includes a fundamental component at the grid frequency, such as 50 or 60 Hz; a medium-frequency component band around the switching frequency, whose scope is from 10–100 kHz roughly; and a high-frequency component band from 150 kHz to 30 MHz.

Generally speaking, there are different measures for different LC frequency bands. The fundamental frequency of common-mode current can be restricted and lower than limitations of standards without any special design and hardware improvements in distributed PVPG systems due to the large impedance in the common-mode loop and the low grid voltage amplitude. High-frequency common-mode current components belong to electromagnetic interference (EMI) noise, so the EMI filter is still the main measure to suppress them. Medium-frequency common-mode current is the most concerning component, and it may be much higher than the limitations of standards when the switching modulation strategies are not suitable. This section mainly discusses the inducing mechanism and suppression rules of the medium-frequency common-mode current.

1. Common-mode equivalent models of TLIs

A generic TLI model is illustrated in Fig. 2.6, where C_{dc} indicates that the TLI belongs to voltage source inverter type. In Fig. 2.6, the generic TLI is composed of some basic cells (BCs). In detail, BC may represent one or multiple power device combinations with semiconductor switches and diodes while setting BC = 3; or BC = 2 when only capacitors are covered in a BC; or BC = 1 while it is shorted; or BC = 0 when it is an open circuit state. In addition, P and N are input terminals of DC source, respectively. C_{pv1} and C_{pv2} are PV parasitic capacitors against ground; they depend on the material and area of PV panels, soil properties, air humidity, and installation styles [20]. Points 1 and 2 are output terminals of differential-mode voltage; C_1 and C_2 are parasitic capacitors of output terminals, respectively. L_1 and L_2 are two filter inductors; Y

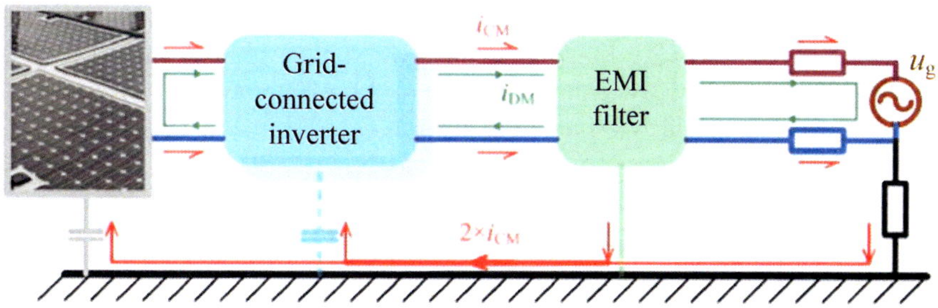

Fig. 2.5 Schematic diagram of a PVPG system without transformer

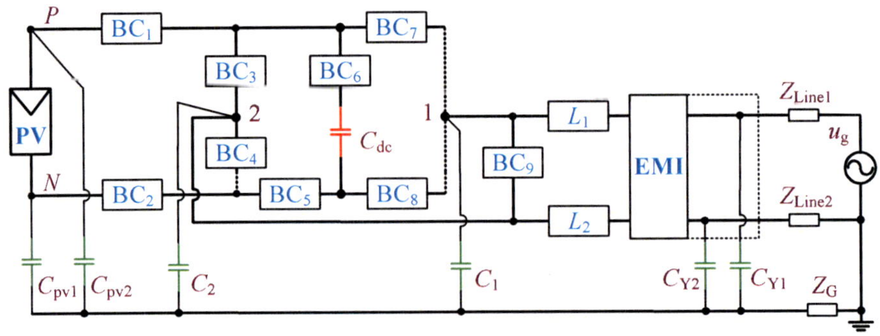

Fig. 2.6 Genetic TLI model with parasitic capacitors

capacitors C_{Y1} and C_{Y2} are part of EMI filter. Z_{Line1} and Z_{Line2} are impedance between point of common coupling (PCC) and the grid; Z_G represents the impedance between grounding point of TLI and neutral line of the grid [21]; and u_g is the AC grid voltage.

Taking terminal N as the reference, the common-mode voltage (CMV) v_{CM} and differential-mode voltage (DMV) v_{DM} are defined as

$$u_{CM} = \frac{u_{1N} + u_{2N}}{2} \tag{2.1}$$

$$u_{DM} = u_{1N} - u_{2N} \tag{2.2}$$

It can be seen that v_{CM} and v_{DM} are related to v_{1N} and v_{2N}. By combining (2.1) and (2.2), v_{1N} and v_{2N} can be expressed as follows:

$$u_{1N} = u_{CM} + \frac{u_{DM}}{2} \tag{2.3}$$

$$u_{2N} = u_{CM} - \frac{u_{DM}}{2} \tag{2.4}$$

In order to obtain the CMV model at switching frequency, BCs of Fig. 2.6 can be substituted with (2.3) and (2.4). Therefore, the DMV branches and elements can be removed based on superposition principle; as a result, only the CMV branches and elements are retained. Meanwhile, the grid voltage source can be shorted for switching frequency CMV equivalent circuit. As an intermediate result, the first CMV equivalent circuit is gained, as shown in Fig. 2.7, where L_{CM} represents the common-mode inductance of EMI filter.

Furthermore, Fig. 2.7 can be torn down to part A and part B, as shown in Fig. 2.8. Then, two DMV branches can be simplified based on Thevenin's theorem. It is worth noting that the Delta–Wye transform needs to be used twice for simplifying branch A, which is illustrated in Fig. 2.9. Finally, the simplified CMV equivalent circuit is obtained, as shown in Fig. 2.10, where u_A and u_B represent the equivalent voltage

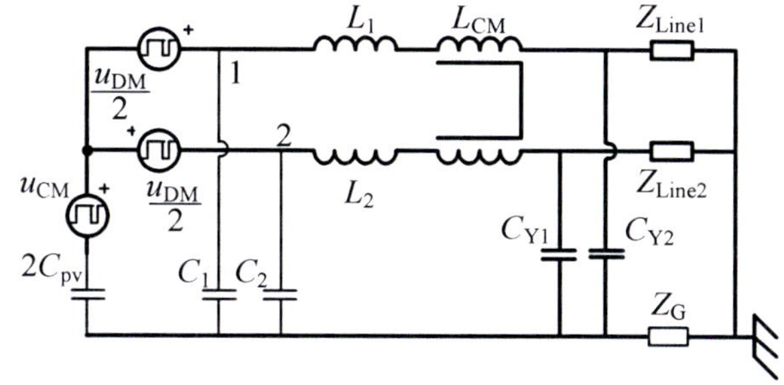

Fig. 2.7 CMV equivalent model 1 for TLI system

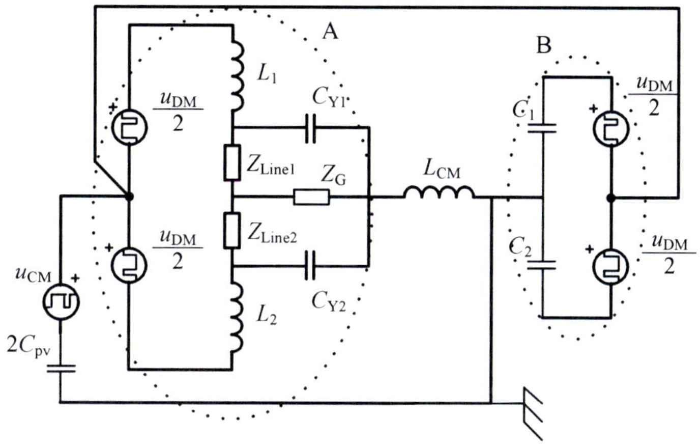

Fig. 2.8 CMV equivalent model 2 for TLI system

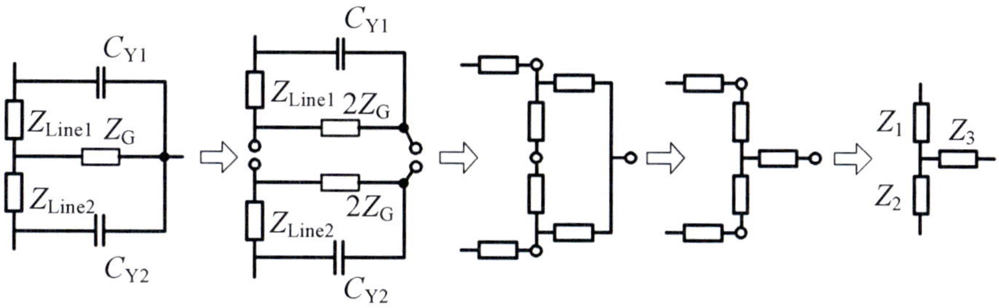

Fig. 2.9 Delta–Wye transform for branch A

sources of branches A and B, respectively.

The impedances Z_1, Z_2, and Z_3 in Fig. 2.9 are expressed as follows:

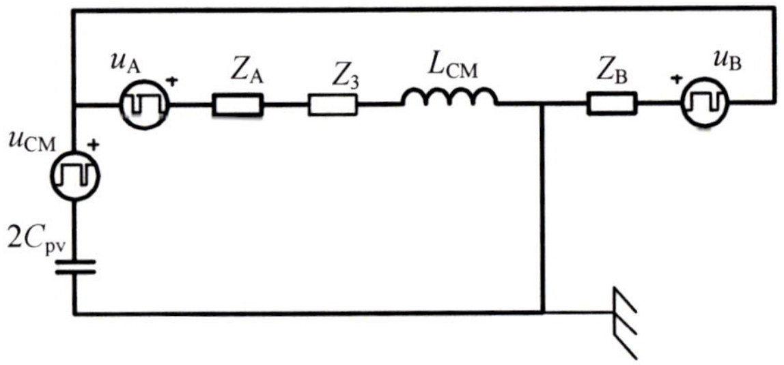

Fig. 2.10 CMV equivalent model 3 for TLI system

$$Z_2 = \frac{Z_{\text{Line2}} Z_{\text{CY2}}}{Z_{\text{Line2}} + 2Z_{\text{G}} + Z_{\text{CY2}}} + \frac{2Z_{\text{Linea}} Z_{\text{G}} Z_{\text{CY2}} + 2Z_{\text{Line2}} Z_{\text{G}} Z_{\text{CY2}} (Z_{\text{Line1}} + 2Z_{\text{G}} + Z_{\text{CY1}}) / (Z_{\text{Line2}} + 2Z_{\text{G}} Z_{\text{CY2}})}{(Z_{\text{Line1}} + Z_{\text{CY1}})(Z_{\text{Line2}} + 2Z_{\text{G}} + Z_{\text{CY2}})(Z_{\text{Line2}} + Z_{\text{CY2}})(Z_{\text{Line1}} + 2Z_{\text{G}} + Z_{\text{CY1}})} \tag{2.5}$$

$$Z_1 = \frac{Z_{\text{Line1}} Z_{\text{CY1}}}{Z_{\text{Line1}} + 2Z_{\text{G}} + Z_{\text{CY1}}} + \frac{2Z_{\text{Line2}} Z_{\text{G}} Z_{\text{CY1}} + 2Z_{\text{Line1}} Z_{\text{G}} Z_{\text{CY1}} (Z_{\text{Line2}} + 2Z_{\text{G}} + Z_{\text{CY2}}) / (Z_{\text{Line1}} + 2Z_{\text{G}} Z_{\text{CY1}})}{(Z_{\text{Line1}} + Z_{\text{CY1}})(Z_{\text{Line2}} + 2Z_{\text{G}} + Z_{\text{CY2}})(Z_{\text{Line2}} + Z_{\text{CY2}})(Z_{\text{Line1}} + 2Z_{\text{G}} + Z_{\text{CY1}})} \tag{2.6}$$

$$Z_3 = \frac{2Z_{\text{CY1}} Z_{\text{CY2}} Z_{\text{G}}}{(Z_{\text{Line1}} + Z_{\text{CY1}})(Z_{\text{Line2}} + 2Z_{\text{G}} + Z_{\text{CY2}}) + (Z_{\text{Line2}} + Z_{\text{CY2}})(Z_{\text{Line1}} + 2Z_{\text{G}} + Z_{\text{CY1}})} \tag{2.7}$$

In Fig. 2.10, the equivalent voltage sources and series impedances of branches A and B are expressed as follows:

$$u_{\text{A}} = \frac{u_{\text{DM}}}{2} \frac{(Z_{\text{L2}} + Z_2) - (Z_{\text{L1}} + Z_1)}{Z_{\text{L1}} + Z_{\text{L2}} + Z_1 + Z_2} \tag{2.8}$$

$$Z_{\text{A}} = \frac{(Z_{\text{L1}} + Z_1)(Z_{\text{L2}} + Z_2)}{Z_{\text{L1}} + Z_{\text{L2}} + Z_1 + Z_2} \tag{2.9}$$

$$u_{\text{B}} = \frac{u_{\text{DM}}}{2} \frac{Z_{\text{C2}} - Z_{\text{C1}}}{Z_{\text{C1}} + Z_{\text{C2}}} \tag{2.10}$$

$$Z_{\text{B}} = \frac{Z_{\text{C1}} Z_{\text{C2}}}{Z_{\text{C1}} + Z_{\text{C2}}} \tag{2.11}$$

2. Eliminating rules of switching frequency LC

The simplest CMV analysis model can be derived by using Thevenin's theorem based on Fig. 2.10, and the result is shown in Fig. 2.11, where $u_{\text{CM\_tot}}$ and Z represent the total equivalent CMV source and impedance of whole CMV path, respectively.

$$u_{\text{CM}-\text{DM}} = \frac{u_{\text{A}} Z_{\text{B}} + u_{\text{B}} (Z_{\text{A}} + Z_3 + Z_{\text{LCM}})}{Z_{\text{A}} + Z_3 + Z_{\text{LCM}} + Z_{\text{B}}} \tag{2.12}$$

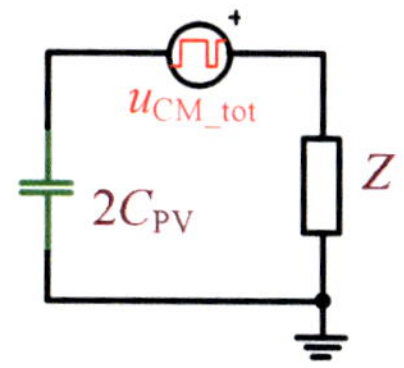

Fig. 2.11 The simplest CMV equivalent model for TLI system

$$Z = \frac{Z_B(Z_A + Z_3 + Z_{LCM})}{Z_A + Z_3 + Z_{LCM} + Z_B} \quad (2.13)$$

$$u_{CM\_tot} = u_{CM} + u_{CM-DM} = u_{CM} + \frac{u_{DM}}{2} \cdot \frac{L_2 - L_1}{L_2 + L_1} \quad (2.14)$$

An important conclusion can be drawn from Fig. 2.11 that u_{CM_tot} must be maintained constant in order to eliminate switching frequency LC in the CMV loop. Furthermore, the generic LC elimination rules can be summarized from three kinds of TLI topologies, respectively.

Rule 1 if $BC_5 = 1, BC_6 = 0, BC_1 = BC_2 = BC_3 = BC_4 = 3, BC_7 = BC_8 = BC_9 = 3$ and $L_1 = L_2 \neq 0$, Fig. 2.6 can be redrawn as Fig. 2.12, which is a generic full-bridge TLI topology. Since $L_1 = L_2$, the second term of (2.14) equals zero, so u_{CM_tot} can be constant as long as u_{CM} maintains constant in (2.1). Therefore, switching frequency LC components can be eliminated if full-bridge TLIs can maintain CMV constant.

Rule 2 if $BC_1 = BC_2 = BC_5 = 1, BC_7 = BC_8 = BC_9 = 3, BC_3 = BC_4 = 2, L_1 \neq 0$ and $L_2 = 0$, Fig. 2.6 can be redrawn as Fig. 2.13, which is a generic half-bridge TLI topology. Substituting $L_2 = 0$, (2.1), and (2.2) into (2.14), we can obtain that u_{CM_tot} is

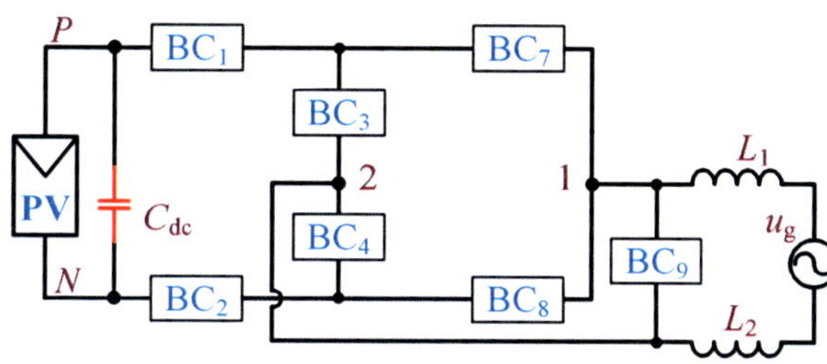

Fig. 2.12 Full-bridge TLI structure

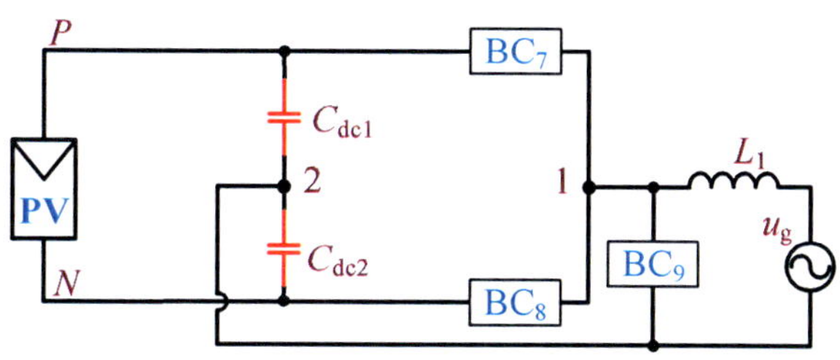

Fig. 2.13 Half-bridge TLI structure

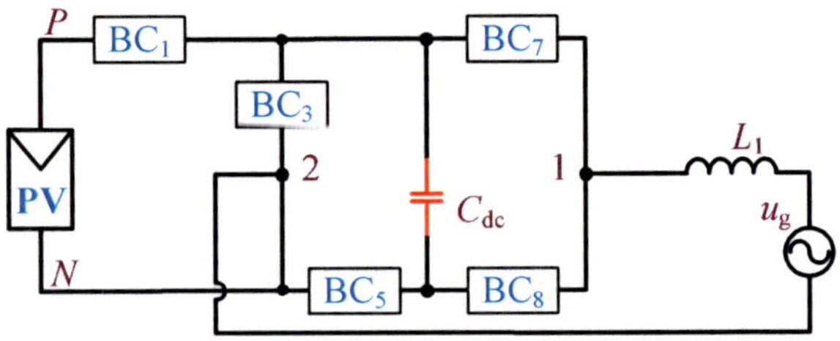

Fig. 2.14 Common-ground TLI structure

equal to u_{2N}. Since u_{2N} equals half of the DC voltage, thanks to the voltage-dividing capacitors, half-bridge TLIs can naturally maintain total CMV u_{CM_tot} constant.

Rule 3 if $BC_9 = 0$, $BC_1 = BC_3 = BC_5 = BC_7 = BC_8 = 3$, $BC_2 = BC_4 = BC_6 = 1$, $L_1 \neq 0$ and $L_2 = 0$, Fig. 2.6 can be redrawn as Fig. 2.14, which is a generic common-ground TLI topology. Especially, the neutral line of AC grid is directly connected to DC input terminal *N*, so u_{2N} equals zero. Then, the expression $u_{DM} = 2u_{CM}$ can be obtained by combining (2.1) and (2.2). Therefore, the value of the total equivalent CMV u_{CM_tot} can be obtained according to (2.14) since $L_2 = 0$, and naturally maintains zero.

Besides, grid voltage may cause line-frequency CMV across parasitic capacitors. The CMV waveforms of full-bridge TLIs, half-bridge TLIs, and common-ground TLIs are illustrated in Fig. 2.15, where u_P and u_N represent the voltages across C_{pv2} and C_{pv1}, respectively [25].

It can be seen that full-bridge TLIs have line-frequency CMV components, which can induce line-frequency LC in the CMV loop. Fortunately, the impedance of parasitic capacitors is high for line-frequency CMV. Meanwhile, TLIs are commonly installed in low-voltage distributed PVPG systems, so the grid voltage amplitude is limited. Therefore, the line-frequency LC will be restricted and lower than the limitations of standards without any special design and hardware.

It is also worth mentioning that the line-frequency CMV is constant in both of half-bridge and common-ground TLIs. As a result, there is no line-frequency LC according to the simplest CMV model in Fig. 2.11 due to the DC current blocking property of capacitors.

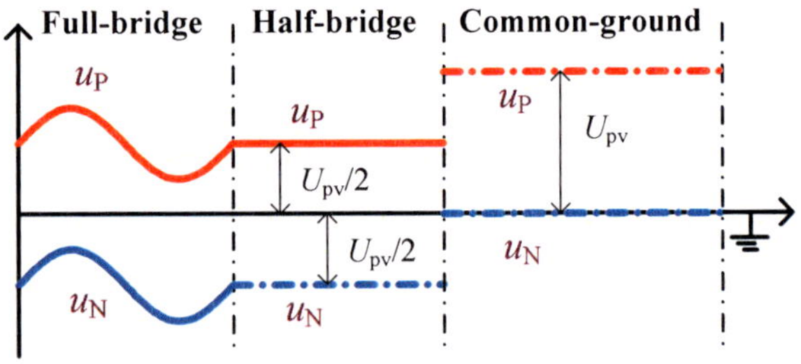

Fig. 2.15 Line-frequency CMV waveforms on parasitic capacitors in three kinds of typical TLI topologies

2.3 LC Suppression Techniques for Full-Bridge TLIs

Full-bridge topologies feature high DC voltage utilization ratio, and they are the preferred structures for residential roof PVPG systems. A classical full-bridge circuit and product appearance are shown in Fig. 2.16. Full-bridge TLIs can work with several kinds of modulation strategies, such as bipolar sinusoidal pulse width modulation (SPWM), unipolar SPWM, and double-frequency unipolar SPWM. The CMV characteristics of full-bridge structure will be discussed in this section based on the derived CMV analysis model.

2.3.1 Combinations of Topology Structures and SPWM Strategies

It can be deduced from *Rule 1* and (2.1) that the sum of u_{1N} and u_{2N} must be constant in order to maintain u_{CM} at a constant value, which can be implemented by using suitable modulation strategies. Table 2.2 lists the pulse amplitudes and frequencies

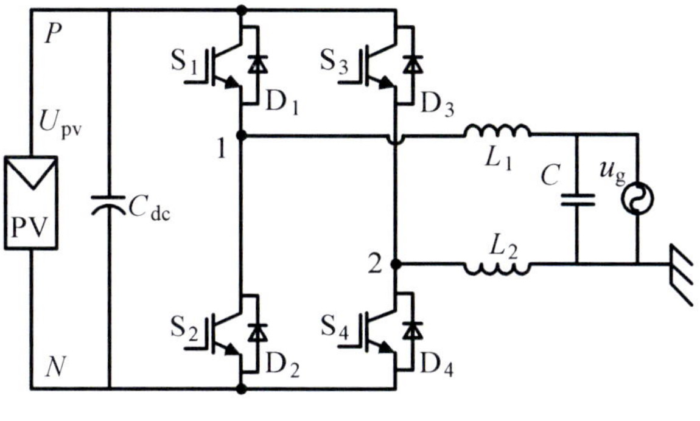

Fig. 2.16 Single-phase full-bridge TLI

Table 2.2 Levels and frequencies of u_{CM} and u_{DM} under different SPWM styles in full-bridge structures

SPWM style	u_{CM} ($\frac{u_{1N}+u_{2N}}{2}$)		u_{DM} ($u_{1N} - u_{2N}$)	
	Levels	Frequency	Levels	Frequency
Bipolar SPWM	$U_{pv}/2$	0	$U_{pv}, -U_{pv}$	f_{SW}
Unipolar SPWM	$U_{pv}/2$, 0	f_{SW}	U_{pv}, 0, $-U_{pv}$	f_{SW}
Double-frequency unipolar SPWM	U_{pv}, $U_{pv}/2$, 0	f_{SW}	U_{pv}, 0, $-U_{pv}$	$2f_{SW}$
Unipolar SPWM with AC bypass [26]	$U_{pv}/2$, uncertain	f_{SW}	U_{pv}, 0, $-U_{pv}$	f_{SW}
Unipolar SPWM with DC bypass [27, 28]	$U_{pv}/2$, uncertain	f_{SW}	U_{pv}, 0, $-U_{pv}$	f_{SW}

of u_{CM} and u_{DM} of full-bridge TLIs working under several typical SPWM styles and optimized topologies, where f_{SW} represents the switching frequency of TLIs.

It can be seen from Table 2.2 that bipolar SPWM has excellent CMV characteristic, but unipolar SPWM and double-frequency unipolar SPWM cannot be directly used in full-bridge TLIs due to the poor CMV characteristics. A reverse conclusion can be found from the perspective of DMV that unipolar SPWM strategies feature higher conversion efficiency and grid-in current quality. Therefore, unipolar SPWM full-bridge topologies with AC bypass and DC bypass have been proposed, under which the PV array can be separated from grid in the freewheeling period by the new freewheeling paths. However, the CMV of topologies proposed in the literatures [26, 27] may still fluctuate at switching frequency scale for the non-negligible stray capacitor of power devices. Based on *Rule 1*, the potential of the freewheeling path must be clamped to half of the input voltage in the freewheeling period in order to eliminate switching frequency LC.

2.3.2 Matching Circuit Parameters

A conclusion that maintaining the sum of u_{CM} and $u_{CM\text{-}DM}$ at a constant value can eliminate switching frequency LC is summarized in the aforementioned CMV model and (2.14). The operation levels of full-bridge TLIs under unipolar and double-frequency unipolar SPWM in one switching period are listed in Table 2.3, and the expressions of *a*, *b*, *c*, *d* derived from (2.8), (2.10), and (2.12) are shown as follows:

$$\begin{cases} a = \dfrac{(Z_{L2} + Z_2) - (Z_{L1} + Z_1)}{Z_{L1} + Z_{L2} + Z_1 + Z_2} \\ b = \dfrac{Z_{C2} - Z_{C1}}{Z_{C1} + Z_{C2}} \\ c = \dfrac{Z_B}{Z_A + Z_3 + Z_{LCM} + Z_B} \\ d = \dfrac{Z_A + Z_3 + Z_{LCM}}{Z_A + Z_3 + Z_{LCM} + Z_B} \end{cases} \tag{2.15}$$

Table 2.3 Operation levels of full-bridge TLIs under unipolar SPWM

u_{1N}	u_{2N}	u_{DM}	u_{CM}	$u_{CM\text{-}DM}$
U_{pv}	U_{pv}	0	U_{pv}	$0 + U_{pv}$
U_{pv}	0	U_{pv}	$U_{pv}/2$	$(1 + ac + bd)U_{pv}/2$
0	U_{pv}	$-U_{pv}$	$U_{pv}/2$	$(1 - ac - bd)U_{pv}/2$
0	0	0	0	$0 + 0$

It can be concluded from Table 2.3 that it is impossible to eliminate switching frequency LC for full-bridge TLIs under unipolar SPWM by matching component parameters, such as L_1, L_2 and/or C_1, C_2. The reason is that the sum of u_{CM} and $u_{CM\text{-}DM}$ in rows 1 and 4 is unequal.

2.4 LC Suppression Techniques for Half-Bridge TLIs

Half-bridge structure has also been extensively applied in TLIs, including two-level and multi-level topologies. For instance, the neutral point clamped (NPC) three-level inverters have already been commercialized in single-phase and three-phase TLI systems, which were developed by Danfoss of Denmark, as shown in Fig. 2.17 [29]. The NPC structure has advantages in dealing with LC and DC current injection (DCCI) issues of TLIs [30], and also has merits of excellent DMV characteristic and low-voltage stresses of power devices.

2.4.1 *Combinations of Topology Structures and SPWM Strategies*

The feasibility of *Rule* 1 in half-bridge structure is chiefly analyzed by a case study of NPC topology. When $L_1 = L_2 \neq 0$, $u_{CM\text{-}DM}$ maintains zero according to expression (2.14), and it is possible that switching frequency LC may be eliminated by choosing proper SPWM style to hold u_{CM} constant. The conventional three-level NPC inverter shown in Fig. 2.17 can output positive and zero levels in the positive half cycle of grid voltage while negative and zero levels in the negative half cycle. This modulation strategy is called unipolar SPWM. However, if it outputs both positive and negative levels in whole cycle of the grid voltage, this modulation strategy is named as bipolar

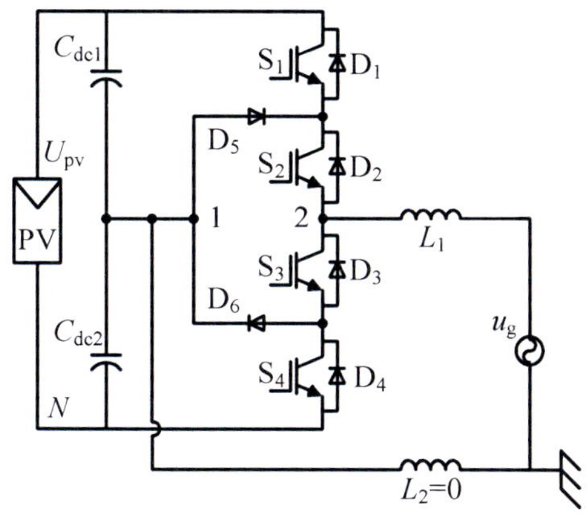

Fig. 2.17 Neutral point clamped three-level inverter and product appearance

Table 2.4 Levels and frequencies of u_{CM} and u_{DM} under different SPWM styles with NPC structures

SPWM style	u_{CM} ($\frac{u_{1N}+u_{2N}}{2}$)		u_{DM} ($u_{1N}-u_{2N}$)	
	Levels	Frequency	Levels	Frequency
Unipolar SPWM	$3U_{pv}/4$, $U_{pv}/2$, $U_{pv}/4$	f_{SW}	$U_{pv}/2$, 0, $-U_{pv}/2$	f_{SW}
Bipolar SPWM	$3U_{pv}/4$, $U_{pv}/4$	f_{SW}	$U_{pv}/2$, $-U_{pv}/2$	f_{SW}

SPWM. The levels and frequencies of u_{CM} and u_{DM} of NPC inverter under the aforementioned two SPWM strategies are listed in Table 2.4.

It can be seen from Table 2.4 that the level of u_{CM} changes at switching frequency scale whether unipolar or bipolar SPWM is applied. The reason is that the potential of point 2, that is, the midpoint of capacitor leg, is constant while the potential of output point 1 varies in an SPWM style; as a consequence, the sum of u_{1N} and u_{2N} cannot maintain constant. Therefore, *Rule* 1 is infeasible for half-bridge inverter structures.

2.4.2 *Matching Circuit Parameters*

According to *Rule* 2, properly matching component parameters of NPC topologies may be an effective way to suppress LC, and will be discussed in detail in this section. Although operation levels under two conventional SPWM styles are given in Table 2.4, only unipolar SPWM is acceptable for three-level NPC grid-connected inverters because bipolar SPWM leads to larger current ripple in the filter inductor and lower conversion efficiency.

The operation levels of three-level NPC grid-connected inverter proposed in [29] are listed in Table 2.5. It can be seen that the expression $u_{CM} + u_{CM\text{-}DM} = U_{pv}/2$ is true under the assumption that parasitic capacitors C_1, C_2 and impedances Z_{Line1}, Z_{Line2}, Z_G are zero. However, capacitors C_1 and C_2 are commonly different; usually $C_1 > C_2$ in NPC structure due to different formation mechanisms. Therefore, u_B is not equal zero and turns out to be a positive value, which brings a negative impact on the expectation that $u_{CM\text{-}DM} = -u_{CM}$. Moreover, the prerequisite of $Z_A + Z_3 + Z_{LCM} << Z_{LB}$ is no longer completely reliable as switching frequency increases even

Table 2.5 Operation levels of three-level NPC inverter with $C_1 = C_2 = 0$, $L_1 = L$, and $L_2 = 0$

u_{1N}	u_{2N}	u_{DM}	u_{CM}	$u_{CM\text{-}DM}$	$u_{CM} + u_{CM\text{-}DM}$
$U_{pv}/2$	U_{pv}	$U_{pv}/2$	$3U_{pv}/4$	$-U_{pv}/4$	$U_{pv}/2$
$U_{pv}/2$	$U_{pv}/2$	0	$U_{pv}/2$	0	$U_{pv}/2$
$U_{pv}/2$	0	$-U_{pv}/2$	$U_{pv}/4$	$U_{pv}/4$	$U_{pv}/2$

$u_{CM\text{-}DM} = -u_{DM/2}$ can also be derived by substituting $Z_{CY1} = Z_{CY2} = \infty$, $Z_{Line1} = Z_{Line2} = 0$, $Z_{L1} = Z_L$, $Z_{L2} = 0$, $Z_{C1} = Z_{C2} = \infty$ into (2.8)–(2.12)

Table 2.6 Possible schemes for LC suppression in NPC topologies

Methods	Theoretic expression	Prerequisite	Feasibility
Full-offset solution	$u_A = -u_{DM}/2$, $u_B = -u_{DM}/2$	$L_1 = L$, $L_2 = 0$, $C_1 << C_2$	Fine
Filter branch offset solution	$u_A = -u_{DM}/2$, $u_B = 0$ $Z_B/(Z_A + Z_3 + Z_{LCM} + Z_B) = 1$	$L_1 = L$, $L_2 = 0$, $C_1 = C_2$ $Z_A + Z_3 + Z_{LCM} << Z_B$	Difficulty of matching $C_1 = C_2$; Performance is affected by switching frequency
Parasitic branch offset solution	$u_A = 0$, $u_B = -u_{DM}/2$ $(Z_A + Z_3 + Z_{LCM})/(Z_A + Z_3 + Z_{LCM} + Z_B) = 1$	$L_1 = L_2 = L/2$ $C_1 << C_2$ $Z_A + Z_3 + Z_{LCM} \gg Z_B$	Need for huge C_2; Performance is affected by branch A

though C_1 and C_2 are exactly the same, so the $u_{CM\text{-}DM} = -u_{CM}$ approximately makes sense according to (2.12).

According to the analysis in [29], the equivalent voltage source u_A shown in Fig. 2.10 can counteract the high-frequency pulse of u_{CM}. However, the compensation effect is not perfect because the structure of three-level NPC inverter is asymmetric, that is, capacitor leg and switch leg have different parasitic capacitors against the earth [30]. Several possible schemes derived from the CMV analytical model are listed in Table 2.6. However, only the full-offset solution can be realistically achieved. "Filter branch offset solution" falls inferior because of its difficulty in determining the value of C_1, and this eliminating solution is an approximate method.

Although matching $C_2 = C_1$ can be approximately realized by adding some extra capacitors, for instance, at least 10 times greater than the parasitic capacitor, in doing so, high-frequency current across C_1 will be rapidly increased due to the high-frequency pulse of point 1, and the realizability of expression $Z_A + Z_3 + Z_{LCM} << Z_{LB}$ is weakened as well. "Parasitic branch offset solution" is also difficult to achieve since a huge C_2 is required to satisfy the prerequisite $Z_A + Z_3 + Z_{LCM} \gg Z_{LB}$.

2.4.3 LC Suppression Techniques for Common-Ground TLIs

The neutral line of AC grid is directly connected to DC input terminal N in common-ground TLIs, and the best CMV characteristic is achieved according to *Rule 3* [31–33].

2.4.4 Combinations of Topology Structures and SPWM Strategies

It can be seen from Fig. 2.14 that the sum of u_{1N} and u_{2N} changes at switching frequency scale in the common-ground structure since u_{2N} maintains zero while u_{1N} varies in an SPWM style, which then leads to an inconstant CMV. Therefore, the elimination methods applicable in full-bridge structure such as constructing DC or AC bypass do not work in common-ground structure.

2.4.5 Matching Circuit Parameters

According to the aforementioned *Rule 3*, the constant total equivalent CMV u_{CM_tot} for switching frequency LC elimination can be achieved when $L_2 = 0$ and $u_{2N} = 0$. Based on this conclusion, the virtual DC bus inverter (VDCBI) [31] shown in Fig. 2.18 can work under unipolar SPWM and has a three-level output voltage. The voltage of the capacitor C_{dc2} is supported by DC link voltage and charged in the freewheeling period of negative half cycle through S_1 and S_3.

The operation levels of VDCBI are listed in Table 2.7. It can be seen that the sum of u_{CM} and $u_{CM\text{-}DM}$ remains zero in the whole cycle. Considering the practical circuit structure, it is still necessary to apply the full-offset solution presented in Table 2.6 to achieve the best performance of LC suppression.

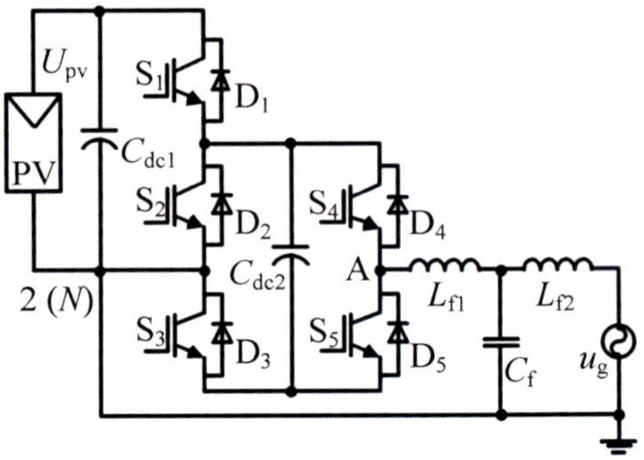

Fig. 2.18 Single-phase virtual DC bus common-ground TLI

Table 2.7 Operation levels of VDCBI with $C_1 = C_2 = 0$, $L_1 = L$ and $L_2 = 0$

u_{1N}	u_{2N}	u_{DM}	u_{CM}	$u_{CM\text{-}DM}$	$u_{CM} + u_{CM\text{-}DM}$
U_{pv}	0	U_{pv}	$U_{pv}/2$	$-U_{pv}/2$	0
0	0	0	0	0	0
$-U_{pv}$	0	$-U_{pv}$	$-U_{pv}/2$	$U_{pv}/2$	0

2.5 Summary

A switching-frequency CMV equivalent model for TLI system has been developed in this chapter, and the parasitic parameters are taken into full consideration during the derivation. Moreover, a total equivalent CMV source is calculated and then three rules of LC elimination are summarized from the model. Finally, the applications of these rules in full-bridge, half-bridge, and common-ground topologies are discussed in detail, respectively.

References

1. Kjaer SB, Pedersen JK, Blaadjerg F (2005) A review of single-phase grid-connected inverters for photovoltaic modules. IEEE Trans Ind Appl 41(5):1292–1306
2. Li Q, Peter W (2008) A review of the single phase photovoltaic module integrated converter topologies with three different DC link configurations. IEEE Trans Power Electron 23(3):1320–1333
3. Carrasco JM, Franquelo LG, Bialasiewicz JT et al (2006) Power-electronic systems for the grid integration of renewable energy sources: a survey. IEEE Trans Industr Electron 53(4):1002–1016
4. Zhao D (2009) Research on long-term changes of solar energy in china and its calculation methods. Nanjing University of Information Science and Technology, Nanjing
5. Zhang YM (2007) Status and prospects of China's photovoltaic generation industry. New Energy New Materi 1:1–6
6. Andersen M, Alvsten B. 200 W low cost module integrated utility interface for modular photovoltaic energy systems In 1995 IEEE International conference on industrial electronics, control and instrumentation, Orlando, pp 572–577
7. Martins DC, Demonti R, Photovoltaic energy processing for utility connected system. In: 2001 IEEE international conference on industrial electronics, control and instrumentation, Denver, pp 1292–1296
8. Martins DC, Demonti R, Grid connected PV system using two energy processing stages In 2002 IEEE Photovoltaic specialists conference, New Orleans, pp 1649–1652
9. Herrmann U, Langer H G, Heinz VDB, Low cost DC to AC converter for photovoltaic power conversion in residential applications. In: 1993 IEEE power electronics specialist conference, Seattle, pp 588–594
10. Martins DC, Demonti R, Interconnection of a photovoltaic panels array to a single-phase utility line from a static conversion system. In: 2000 IEEE power electronics specialist conference, Galway, pp 1207–1211
11. Mekhilef S, Rahim NA, Omar AM, A new solar energy conversion scheme implemented using grid-tied single phase inverter. In 2000 IEEE intelligent systems and technologies for the New Millennium, Kuala Lumpur, pp 524–527
12. Shimizu T, Wade K, Nakamura N (2006) Flyback-type single-phase utility interactive inverter with power pulsation decoupling on the dc input for an ac photovoltaic modele system. IEEE Trans Power Electron 21(5):1264–1272
13. Kjar SB, Blaabjerg F, Design optimization of a single phase inverter for photovoltaic applications. In: 2003 IEEE annual conference on power electronics specialist, Acapulco, pp 1183–1190
14. Kasa N, Iida T, Chen L (2005) Flyback inverter controlled by sensorless current MPPT for photovoltaic power system. IEEE Trans Industr Electron 52(4):1145–1152

15. Kasa N, Iida T, Bhat A KS, Zero-voltage transition flyback inverter for small scale photovoltaic power system. In: 2005 IEEE power electronics specialists conference, Recife, pp 2098–2103
16 Fernandez A, Sebastian J, Hernando MM et al, Single stage inverter for a direct ac connection of a photovoltaic cell module. In: 2006 IEEE power electronics specialists conference, Jeju, pp 93–98
17. Yatsuki S, Wade K, Shimizu T, A novel AC photovoltaic module system based on the impedance—admittance conversion theory. In: 2001 IEEE power electronics specialists conference, Vancouver, pp 2191–2196
18. Souza KCA de, Castro MR de, Antunes F, A DC/AC converter for single-phase grid-connected photovoltaic systems. In: 2002 IEEE annual conference of the industrial electronics society, Sevilla, pp 3268–3273
19. Bae HS, Park JH, Cho B H et al, New control strategy for 2-stage utility-connected photovoltaic power conditioning system with a low cost digital processor. In: 2005 IEEE power electronics specialists conference, Recife, pp 2925–2929
20. Myrzik JMA, Calais M, String and module integrated inverters for single-phase grid connected photovoltaic systems: a review. In: 2003 IEEE Bologna Power Tech Conference, Bologna, pp 23–26
21. Xiao HF, Xie SJ (2010) Leakage current analytical model and application in single-phase transformerless photovoltaic grid-connected inverter. IEEE Trans Electromagn Compat 52(4):902–913
22. Calais M, Agelidis VG, Multilevel converters for single-phase grid connected photovoltaic systems—an overview. In: 1998 IEEE international symposium on industrial electronics, Pretoria, pp 224–229
23. VDE 0126 1 1, Automatic disconnection device between a generator and the public low voltage grid. DIN VDE Normo (2008)
24. China Electrical Equipment Industry Association (2013) Technical specifications for photovoltaic grid-connected inverters: NB/T 32004-2013. China Electric Power Press, Beijing
25. Barater D, Lorenzani E, Concari C et al (2016) Recent advances in single-phase transformerless photovoltaic inverters. IET Renew Power Gener 10(2):260–273
26. Heribert S, Christoph S, Jurgen K (2003) Inverter for transforming a DC voltage into an AC current or an AC voltage. European, EP1369985 B1
27. Yang B, Li WH, Gu YJ et al (2012) Improved transformerless inverter with common-mode leakage current elimination for a photovoltaic grid-connected power system. IEEE Trans Power Electron 27(2):752–762
28. Victor M, Greizer F, Bremicker et al (2008) Method of converting a direct current voltage from a source of direct current voltage, more specifically from a photovoltaic source of direct current voltage, into a alternating current voltage. America, US 7411802 B2
29. Nabae A, Takahashi I, Akagi H (1981) A new neutral-point-clamped PWM inverter. IEEE Trans Ind Appl 17(5):518–523
30. Roberto G, Eugenio G, Jesus L et al (2008) Transformerless single-phase multilevel-based photovoltaic inverter. IEEE Trans on Industrial Electron 55(7):2694–2702
31. Gu YJ, Li WH, Zhao Y et al (2013) Transformerless inverter with virtual DC bus concept for cost-effective grid-connected PV power systems. IEEE Trans Power Electron 28(2):793–805
32. Siwakoti YP, Blaabjerg F (2018) Common-ground-type transformerless inverters for single-phase solar photovoltaic systems. IEEE Trans Industr Electron 65(3):2100–2111
33. Ardashir JF, Sabahi M, Hosseini SH et al (2016) Transformerless inverter with charge pump circuit concept for PV application. IEEE J Emerg Sel Topics Power Electron, Early Access

Chapter 3
Full-Bridge Transformerless PV Grid-Connected Inverters

Abstract The CMV analysis model and three rules of LC elimination are discussed in Chap. 2. One of the conclusions is that *Rule* 1 can be used in full-bridge TLIs. This chapter discusses the application of *Rule* 1 in full-bridge TLIs, including LC elimination by only decoupling the PV arrays with the grid and clamping the potential of the freewheeling path, respectively. In general, decoupling circuits may disconnect the PV arrays with the grid at AC side or DC side, while the clamping circuits may be unidirectional or bidirectional. The number of added switches in clamping circuits includes one, two, or more.

Keywords Transformerless PV grid-connected inverter · Full-bridge · Common-mode voltage · Decoupling · Clamping

3.1 Conventional Full-Bridge TLI

Conventional four-switch full-bridge TLI is the most mature and popular topology in industrial applications, and the circuit is shown in Fig. 3.1a. C_{dc} is the DC side filter capacitor, S_1–S_4 are power switches, L_1, L_2, and C are AC side filter inductors and capacitor, respectively. It features simple and symmetrical circuit structure, high DC voltage utilization, abundant modulation strategies, and so on.

Figure 3.1b, c shows two common modulation strategies, named as bipolar SPWM and unipolar SPWM, respectively. v_m is the reference, v_{tri} is the carrier, and u_{12} is the output voltage.

3.1.1 Bipolar SPWM Modulation

When the bipolar SPWM full-bridge TLI works with unity power factor, that is, the grid-in current and the grid voltage are in phase all the time, four operating modes can be gained, as shown in Fig. 3.2. Mode 1 and mode 3 are energy delivery stages,

H. Xiao and X. Wang, *Transformerless Photovoltaic Grid-Connected Inverters*, CPSS Power Electronics Series, https://doi.org/10.1007/978-981-15-8525-8_3

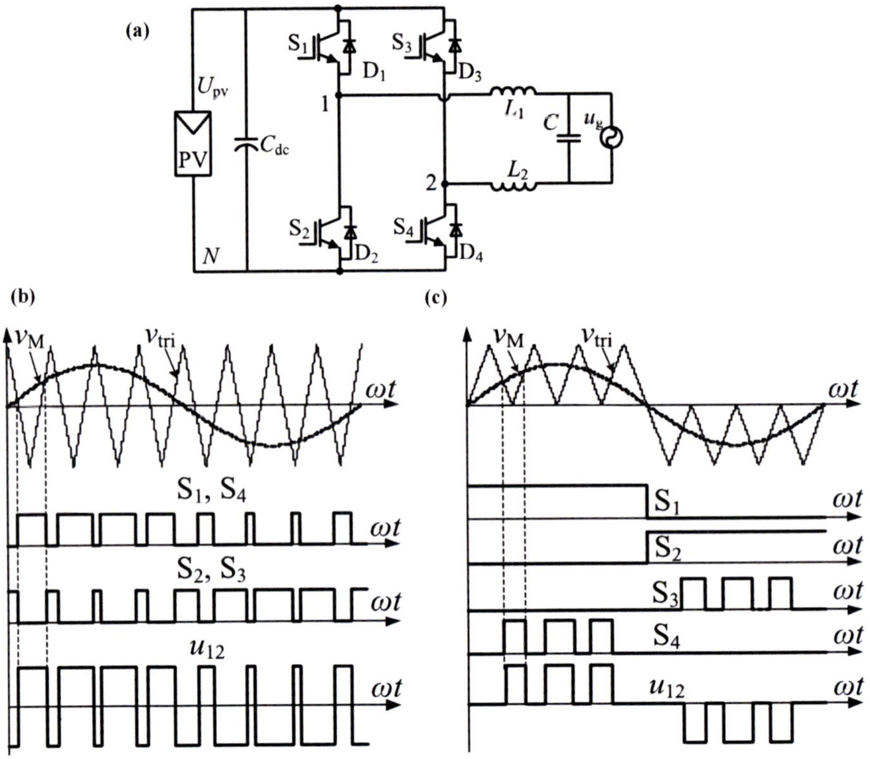

Fig. 3.1 Conventional full-bridge TLI and its modulation strategies, **a** conventional full-bridge TLI, **b** bipolar SPWM, **c** unipolar SPWM

so energy is fed forward into the grid from DC side. Mode 2 and mode 4 are energy return stages, and the energy is fed back into DC side from AC filter.

According to the definitions of DMV and CMV, u_{DM} and u_{CM} were expressed as (2.1) and (2.2) in Chap. 2. For convenience of reading, they are duplicated here as (3.1) and (3.2), respectively.

$$u_{DM} = u_{1N} - u_{2N} \tag{3.1}$$

$$u_{CM} = \frac{u_{1N} + u_{2N}}{2} \tag{3.2}$$

Based on the aforementioned operating modes, the potential of points 1 and 2 against N as well as u_{DM} and u_{CM} is shown in Table 3.1. It can be seen that DMV is a two-level voltage, and CMV is a constant value.

Figures 3.3 and 3.4 are the simulation waveforms of DMV and CMV in time domain and frequency domain, respectively. It can be seen that CMV is a constant value, and the root mean square (RMS) of DMV is zero at switching frequency, that

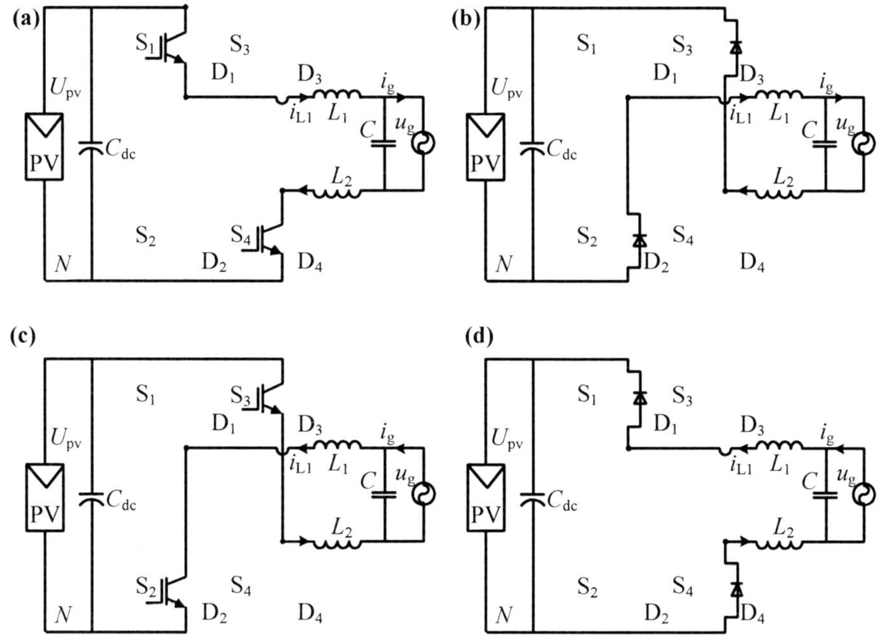

Fig. 3.2 Equivalent circuits of operating modes under bipolar SPWM modulation, **a** mode 1, **b** mode 2, **c** mode 3, **d** mode 4

Table 3.1 DMV and CMV of different working stages with bipolar SPWM

Modes	u_{1N}	u_{2N}	u_{DM}	u_{CM}
(1)	U_{pv}	0	U_{pv}	$U_{pv}/2$
(2)	0	U_{pv}	$-U_{pv}$	$U_{pv}/2$
(3)	0	U_{pv}	$-U_{pv}$	$U_{pv}/2$
(4)	U_{pv}	0	U_{pv}	$U_{pv}/2$

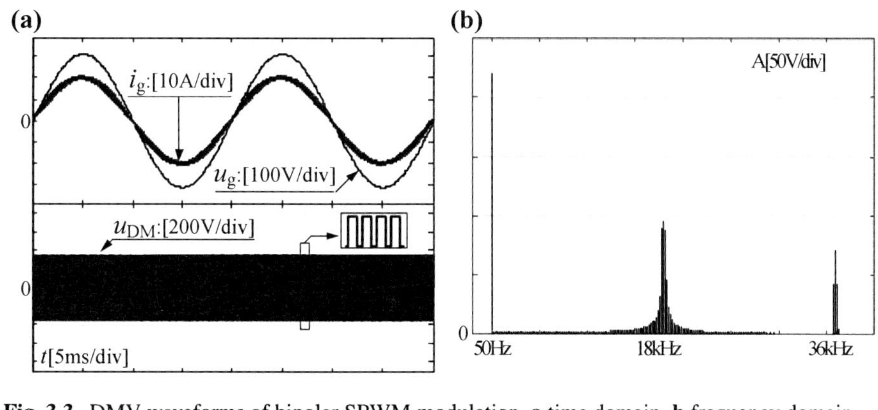

Fig. 3.3 DMV waveforms of bipolar SPWM modulation, **a** time domain, **b** frequency domain

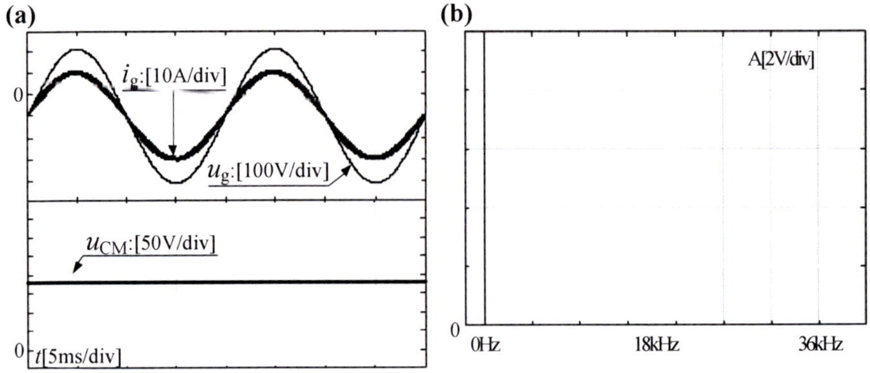

Fig. 3.4 CMV waveforms of bipolar SPWM modulation, **a** time domain, **b** frequency domain

is, the high-frequency LC will be zero. However, DMV is a two-level waveform at switching frequency scale. The RMS of DMV is about 100 V at switching frequency, and DMV characteristic is poor.

In short, CMV can maintain constant during the whole line-frequency period by using bipolar SPWM modulation. As a result, it features the best LC performance. However, poor DMV characteristic results in too many harmonic components in the output voltage, and energy return stages lead to too much reactive power swapping. Obviously, bipolar SPWM modulation causes low conversion efficiency and big filter volume.

3.1.2 Unipolar SPWM Modulation

It also has four operating modes when full-bridge TLI works with unity power factor adopting unipolar SPWM, as shown in Fig. 3.5. Mode 1 and mode 3 are energy delivery stages, and the energy is fed forward into the grid from DC side. Mode 2 and mode 4 are freewheeling stages, and the energy is also released into the grid from AC filter.

Based on these modes, the potential of points 1 and 2 against N as well as u_{DM} and u_{CM} is shown in Table 3.2. It can be seen that DMV is a three-level voltage, but CMV contains switching frequency components.

Figures 3.6 and 3.7 are the simulation waveforms of DMV and CMV in time domain and frequency domain, respectively. It is worth noting that the RMS of DMV is about 75 V at switching frequency, which is reduced compared with Fig. 3.3b, so the DMV characteristic is getting better. However, the RMS of CMV is about 50 V at switching frequency, which is significantly enlarged compared to Fig. 3.4. Thus, the CMV performance of unipolar SPWM modulation is not acceptable.

In summary, for bipolar SPWM modulation, the CMV performance is the best but the DMV performance is worse. On the contrary, the DMV performance is better but

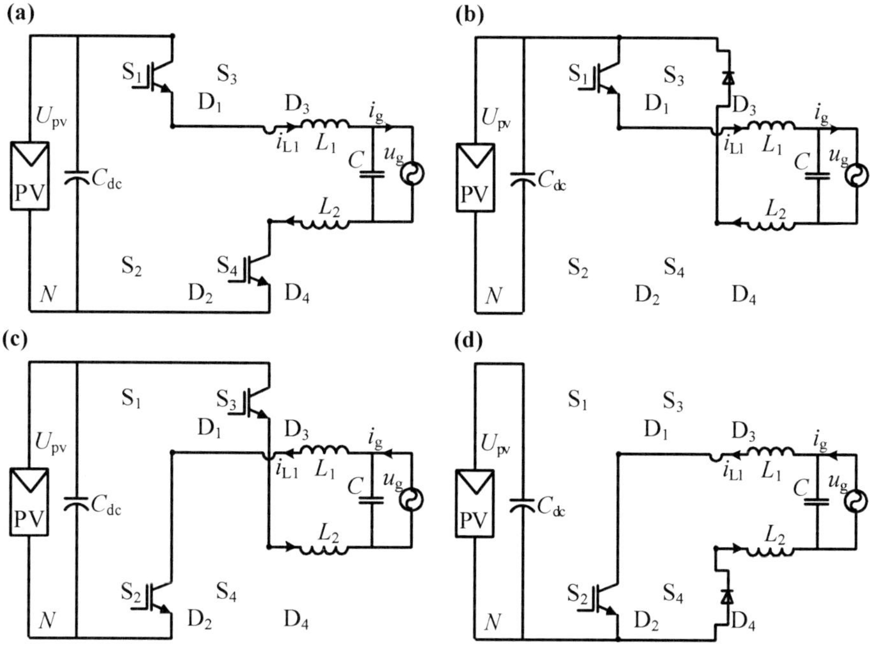

Fig. 3.5 Equivalent circuits of operating modes under unipolar SPWM modulation, **a** mode 1, **b** mode 2, **c** mode 3, **d** mode 4

Table 3.2 DMV and CMV of different working stages with unipolar SPWM

Modes	u_{1N}	u_{2N}	u_{DM}	u_{CM}
(1)	U_{pv}	0	U_{pv}	$U_{pv}/2$
(2)	U_{pv}	U_{pv}	0	U_{pv}
(3)	0	U_{pv}	$-U_{pv}$	$U_{pv}/2$
(4)	0	0	0	0

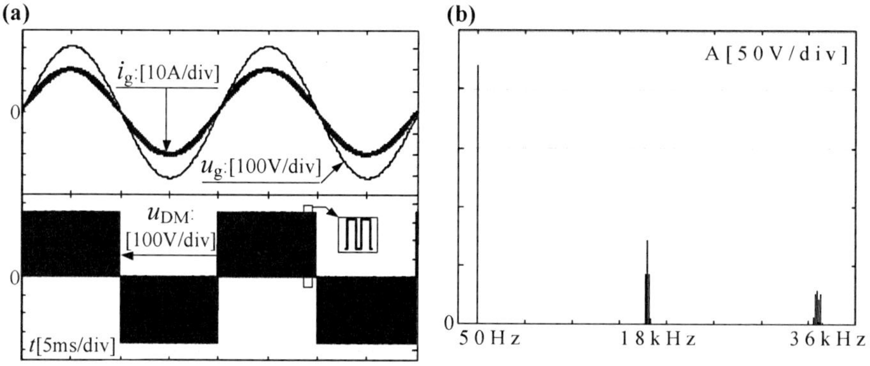

Fig. 3.6 DMV waveforms of unipolar SPWM modulation, **a** time domain, **b** frequency domain

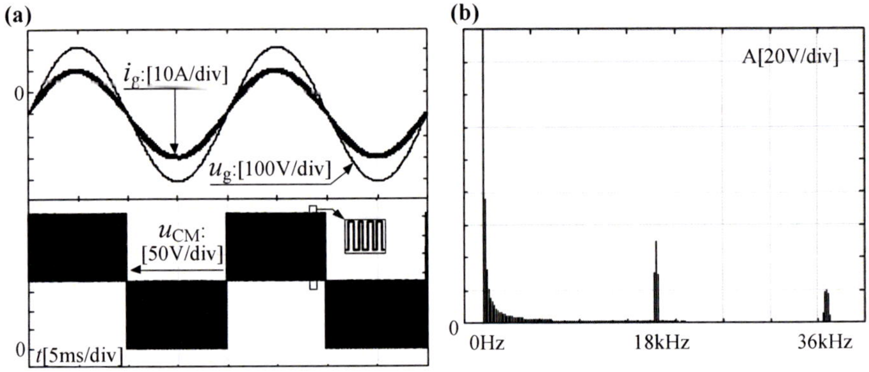

Fig. 3.7 CMV waveforms of unipolar SPWM modulation, **a** time domain, **b** frequency domain

CMV is unacceptable when unipolar SPWM modulation is adopted in conventional four-switch full-bridge TLI. In order to figure out this contradiction, a lot of improved full-bridge TLIs are proposed by combining the advantages of the two modulation strategies, including decoupling type and clamping type full-bridge TLIs.

3.2 Full-Bridge TLIs with Decoupling Branches

High efficiency and reliable inverter conception (HERIC) was first proposed by Fraunhofer Institute of Solar Energy Systems (Fraunhofer ISE), which was presented to use in distributed PVPG systems [1]. Later on, it was patented by Sunways Inc. and commercially used in single-phase PVPG, including NT 3000, NT 4200, and NT 5000, as well as three-phase PVPG, including PT 30k, PT 33k, and so on. For detailed information refer to www.sunways.de.

3.2.1 HERIC Inverter

1. Operating principle

The main circuit of HERIC is illustrated in Fig. 3.8a, and it can be seen that a decoupling branch made of S_5/D_5 and S_6/D_6 is added into conventional four-switch full-bridge TLI at AC side. In this circuit, S_5/D_5 and S_6/D_6 are connected in anti-series between points 1 and 2, also called as "AC decoupling type". The modulation strategy for six switches is shown in Fig. 3.8b. It is worth noting that S_1–S_4 work with high frequency, and the added S_5 and S_6 work with line-frequency. As a result, a three-level output voltage is obtained, and high conversion efficiency is maintained.

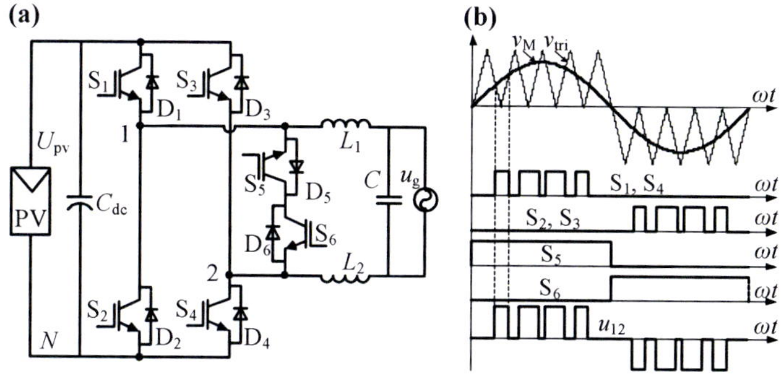

Fig. 3.8 High efficiency and reliable inverter conception, **a** main circuit, **b** modulation strategy

When HERIC works with unity power factor, there are four operating modes based on the directions of grid-in current and grid voltage, as shown in Fig. 3.9. Among them, Fig. 3.9a, b belongs to the positive half grid cycle, while Fig. 3.9c, d belongs to the negative half grid cycle.

Mode 1: Refer to Fig. 3.9a; at this stage, the energy is delivering from the PV side into the grid through S_1, S_4, and AC filter. The operation mode is the same as conventional four-switch full-bridge TLI.

Mode 2: Refer to Fig. 3.9b; S_1–S_4 are all off while S_5 is on and S_6 is off according to Fig. 3.8b. At this stage, the freewheeling current flows through D_6, S_5, and the filter, and keeps feeding the grid. The different thing from Fig. 3.5b is that the freewheeling path is disconnected from PV side; as a result, the LC loop is cut off.

Mode 3: Refer to Fig. 3.9c; at this stage, the energy is delivering from the PV side to the grid through S_2, S_3, and the filter. It is a symmetric mode with Fig. 3.9a.

Mode 4: Refer to Fig. 3.9d; S_1–S_4 are all off while S_6 is on and S_5 is off according to Fig. 3.8b. At this stage, the freewheeling current flows through D_5, S_6, and the filter, and keeps feeding the grid. The same function of disconnecting from PV side as Mode 2 does.

Based on the modes analyzed, the potential of points 1 and 2 against N as well as u_{DM} and u_{CM} is shown in Table 3.3. It can be seen that DMV maintains unipolar SPWM performance, and CMV has two uncertain states because the freewheeling path is just disconnected from PV side in Mode 2 and Mode 4. In other words, the CMV is not a perfect one in HERIC.

In summary, HERIC disconnects PV side and the grid side during freewheeling periods by introducing a decoupling branch. LC will be significantly decreased due to CMV source is weakened. According to the press reports, HERIC topology has been commercialized by Sunways Inc., its European efficiency was 95%, and the peak efficiency was up to 95.6%.

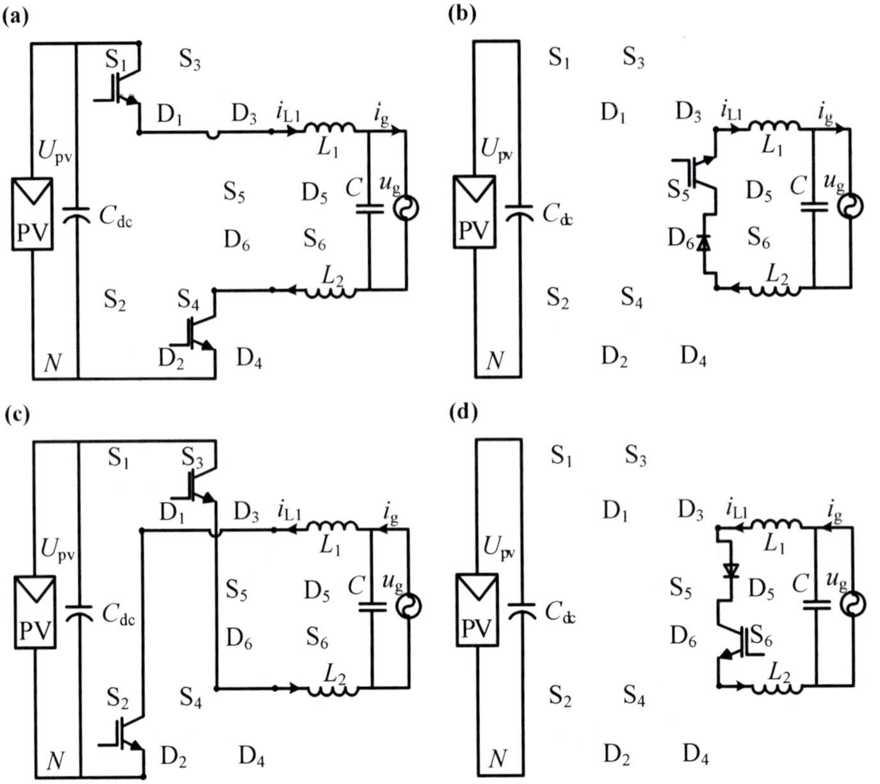

Fig. 3.9 Equivalent circuits of operating modes of HERIC, **a** mode 1, **b** mode 2, **c** mode 3, **d** mode 4

Table 3.3 DMV and CMV of different operating modes for HERIC

Modes	u_{1N}	u_{2N}	u_{DM}	u_{CM}
(1)	U_{pv}	0	U_{pv}	$U_{pv}/2$
(2)	$U_{pv}/2$	$U_{pv}/2$	0	Uncertain
(3)	0	U_{pv}	$-U_{pv}$	$U_{pv}/2$
(4)	$U_{pv}/2$	$U_{pv}/2$	0	Uncertain

2. Simulation results

A simulation model is built by using Matlab/Simulink to verify the operating principle and the performance analysis aforementioned, and the key parameters of HERIC circuit are shown in Table 3.4.

The simulation waveforms of grid voltage u_g, grid-in current reference i_{ref}, and grid-in current i_g are shown in Fig. 3.10. The simulation waveforms of DMV u_{DM} in time domain and frequency domain are shown in Fig. 3.11a, b, respectively. It can

Table 3.4 Key parameters of HERIC circuit

Parameters	Value	Parameters	Value
Input voltage U_{pv}/V	360	Switching frequency f_s/kHz	18
Grid voltage u_g/V	220	Filter inductors L_1, L_2/mH	0.86
Grid frequency/Hz	50	Filter capacitor C/μF	4.7
Rated power P/W	3000	PV parasitic capacitor C_{pv}/nF	300

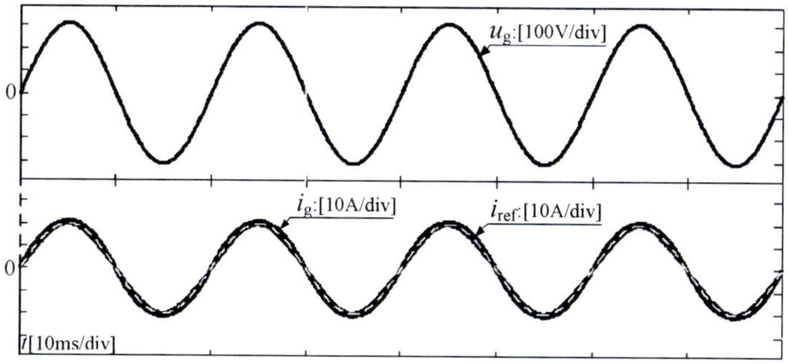

Fig. 3.10 Simulation waveforms of u_g, i_{ref}, and i_g of HERIC

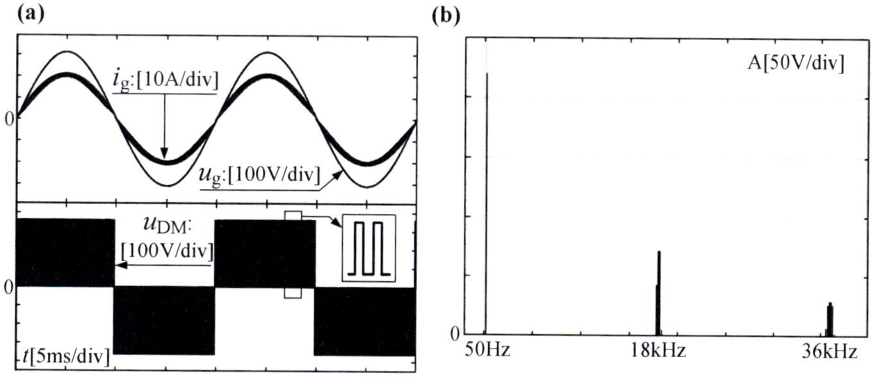

Fig. 3.11 DMV waveforms of HERIC, **a** time domain, **b** frequency domain

be seen that the RMS of DMV is about 75 V at switching frequency, which is equal to Fig. 3.6b with typical unipolar SPWM performance.

The simulation waveforms of CMV u_{CM} in time domain and frequency domain are shown in Fig. 3.12a, b, respectively. It can be seen that the RMS of CMV is about 2.5 V at switching frequency, which is quite close to Fig. 3.4b with typical bipolar SPWM performance.

The simulation waveforms of LC i_{CM} in time domain and frequency domain are shown in Fig. 3.13a, b, respectively. It can be seen that the RMS of LC is about

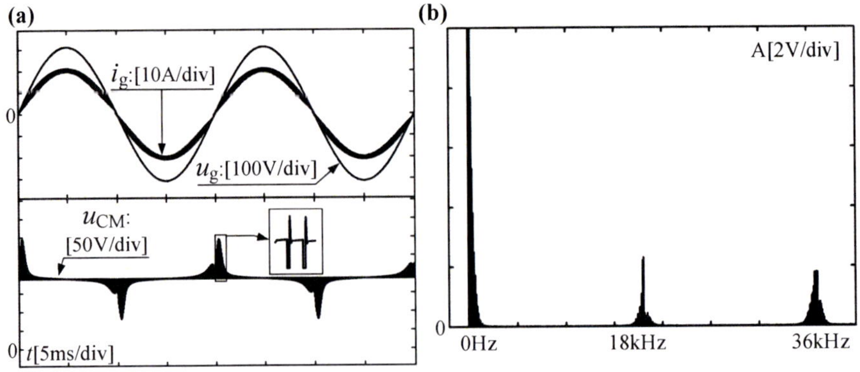

Fig. 3.12 CMV waveforms of HERIC, **a** time domain, **b** frequency domain

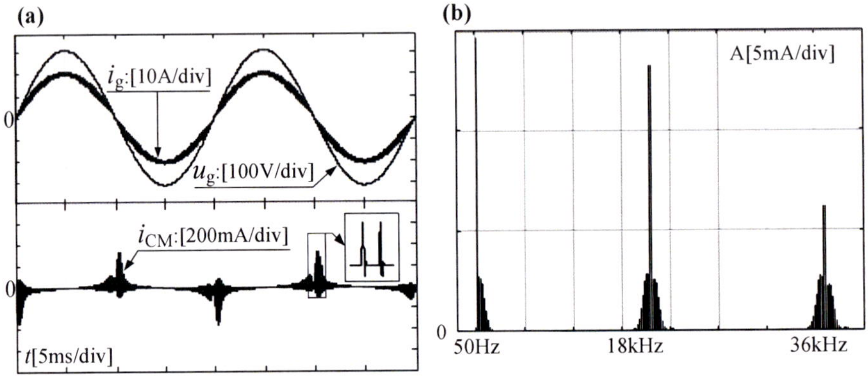

Fig. 3.13 LC waveforms of HERIC, **a** time domain, **b** frequency domain

13 mA at switching frequency, which meets the limitation of DIN VDE 0126-1-1 [2] and NB/T 32004-2013 [3] of 30 mA. It is worth noting that the line-frequency component of LC is obvious, which is induced by the fundamental component of the grid voltage according to the analysis of Fig. 2.15 in Chap. 2.

3. Experimental results

In order to verify the theoretical analysis and simulation waveforms, a prototype of HERIC having the same parameters with Table 3.4 is built in laboratory, as shown in Fig. 3.14.

The experimental waveforms of DMV u_{DM} in time domain and frequency domain are shown in Fig. 3.15a, b, respectively. It can be seen that the RMS of DMV is about 70 V at switching frequency, which is in accordance with the simulation result of Fig. 3.11b.

The experimental waveforms of CMV u_{CM} in time domain and frequency domain are shown in Fig. 3.16a, b, respectively. It can be seen that there exist some CMV

Fig. 3.14 Experimental platform of 3 kW HERIC prototype

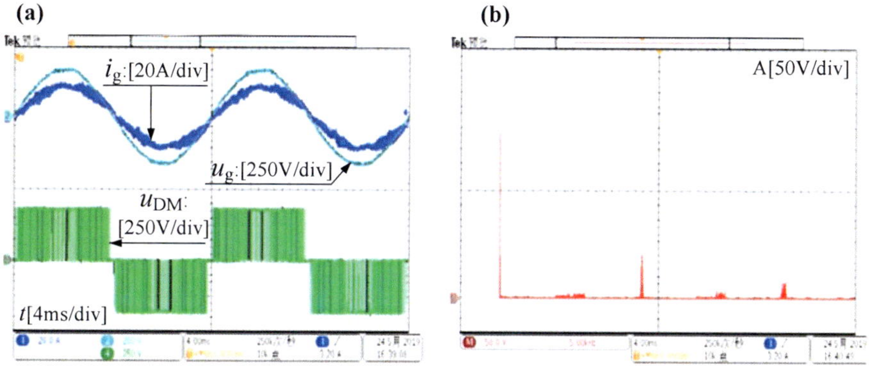

Fig. 3.15 DMV waveforms of HERIC, **a** time domain, **b** frequency domain

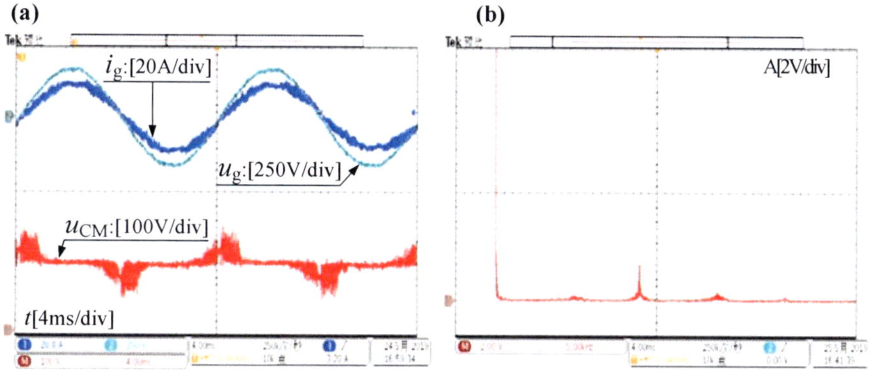

Fig. 3.16 CMV waveforms of HERIC, **a** time domain, **b** frequency domain

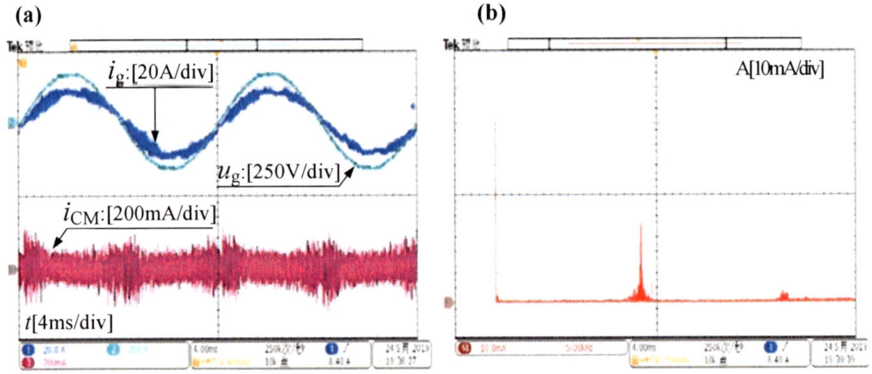

Fig. 3.17 LC waveforms of HERIC, **a** time domain, **b** frequency domain

components at switching frequency, and this is in accordance with Fig. 3.12b; the reason is from the uncertain modes during the freewheeling periods as Table 3.3 lists. Furthermore, the LC i_{CM} in time domain and frequency domain are shown in Fig. 3.17a, b, respectively. It can be seen that the LC is about 22 mA at switching frequency, and it is a little higher than the simulation result but still meets the standard limitation.

With the popularization of distributed PVPG systems in Germany, patented HERIC has been commercialized by Sunways Inc., and made a great success in PVPG applications. The great market demand also attracted the attention of another PV inverter manufacturer, named as SMA. In order to hold the key techniques of TLIs, a patented topology named as H5 has been invented [4].

3.2.2 H5 Inverter

1. Operating principle

The main circuit of H5 is shown in Fig. 3.18a. A new power switch S_5 is inserted at the positive DC bus, which is called DC decoupling type TLI [4]. The modulation strategy is shown in Fig. 3.18b. Apparently, S_1–S_4 work in unipolar modulation strategy and added S_5 works in high frequency. The three-level output voltage remains.

When H5 works in unity power factor, there are four operating modes according to the grid voltage and energy transmission direction, as shown in Fig. 3.19. Among them, Fig. 3.19a, b belongs to the positive grid cycle, while Fig. 3.19c, d belongs to the negative grid cycle.

Mode 1: Refer to Fig. 3.19a; in this stage, energy is delivering from PV side into the grid through S_1, S_4, S_5, and the filter. The operating principle is the same as it is in conventional four-switch full-bridge TLI.

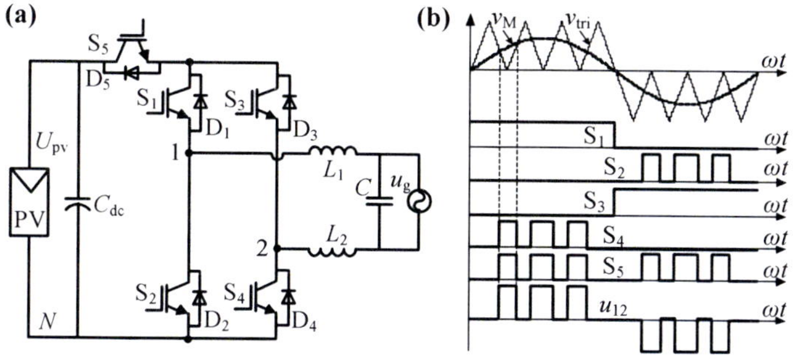

Fig. 3.18 H5 inverter, **a** main circuit, **b** modulation strategy

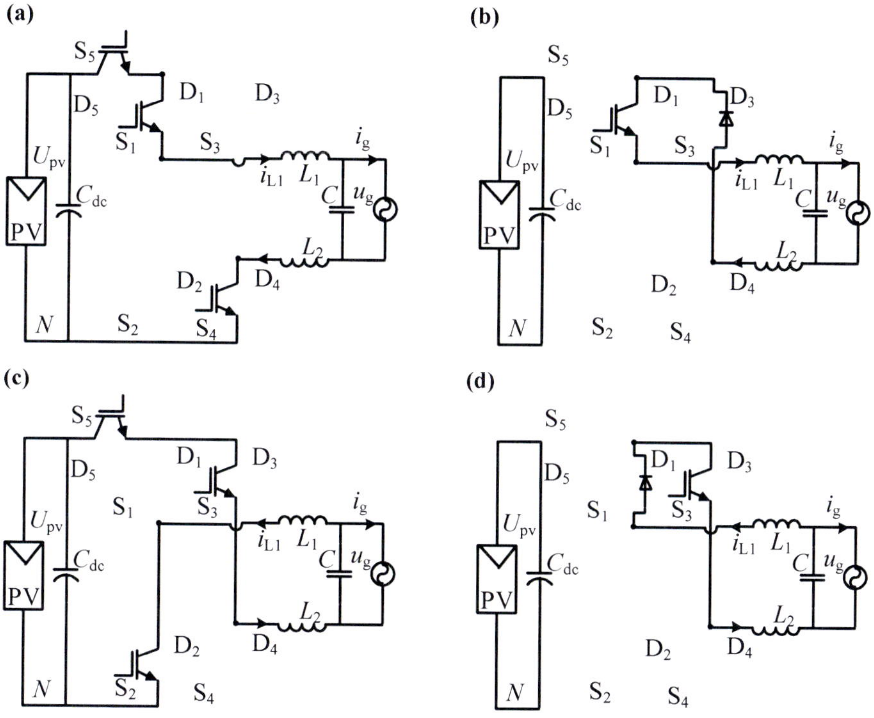

Fig. 3.19 Equivalent circuits of operating modes of H5, **a** mode 1, **b** mode 2, **c** mode 3, **d** mode 4

Mode 2: Refer to Fig. 3.19b; S_2–S_5 are all closed while S_1 is conducting from Fig. 3.18b. At this stage, the freewheeling current flows through D_3, S_1, and the filter, and keeps feeding grid. The freewheeling circuit is disconnected from PV side completely and the LC loop is cut off compared to conventional four-switch full-bridge TLI.

Table 3.5 DMV and CMV levels of different operating modes in H5

Modes	u_{1N}	u_{2N}	u_{DM}	u_{CM}
(1)	U_{pv}	0	U_{pv}	$U_{pv}/2$
(2)	$U_{pv}/2$	$U_{pv}/2$	0	Uncertain
(3)	0	U_{pv}	$-U_{pv}$	$U_{pv}/2$
(4)	$U_{pv}/2$	$U_{pv}/2$	0	Uncertain

Mode 3: Refer to Fig. 3.19c; in this stage, the energy is delivering from the PV side to the grid through S_2, S_3, S_5, and the filter. It is a symmetric mode with Fig. 3.19a.

Mode 4: Refer to Fig. 3.19d; only S_3 is conducting from Fig. 3.18b. At this stage, the freewheeling current flows through D_1, S_3, and the filter and still feeding the grid. The same purpose is realized as the Mode 2.

Based on these modes, the potential of points 1 and 2 against N as well as u_{DM} and u_{CM} is shown in Table 3.5. From this table, it is clear that DMV maintains unipolar SPWM performance, and CMV has two uncertain states because the freewheeling path is just disconnected from PV side in Mode 2 and Mode 4. The CMV is not a perfect one in H5.

In conclusion, a decoupling branch is introduced in the DC side to disconnect the PV with grid during freewheeling stages in H5. LC will decrease due to CMV source is weakened. The current flows through three power switches in power transmission stages, and two power switches conduct current in freewheeling stages. The power loss improved compared to HERIC. As the press reported, H5 topology has been commercialized in SunnyBoy 4000/5000 series by SMA Inc. Its European efficiency is 97.7% and the peak efficiency was up to 98%.

2. Simulation results

A simulation model is built by Matlab/Simulink to verify the performance analysis aforementioned, and the key parameters of H5 are shown in Table 3.4.

The simulation waveforms of u_g, grid-in current reference i_{ref}, and grid-in current i_g are shown in Fig. 3.20. The simulation waveforms of DMV u_{DM} in time domain and frequency domain are shown in Fig. 3.21a, b, respectively. It can be seen that the RMS of DMV is about 75 V at switching frequency, which is equal to Fig. 3.6b with typical unipolar SPWM performance.

The simulation waveforms of CMV u_{CM} in time domain and frequency domain are shown in Fig. 3.22a, b, respectively. It can be seen that the RMS of CMV is about 2 V at switching frequency, which is quite close to Fig. 3.4b with typical bipolar SPWM performance.

The simulation waveforms of LC i_{CM} in time domain and frequency domain are shown in Fig. 3.23a, b, respectively. It can be seen that the RMS of LC is about 13 mA at switching frequency, which meets the limitation of DIN VDE 0126-1-1 [2] and NB/T 32004-2013 [3] of 30 mA. It is worth noting that the line-frequency component of LC is obvious, which is induced by the fundamental component of the grid voltage according to the analysis of Fig. 2.15 in Chap. 2.

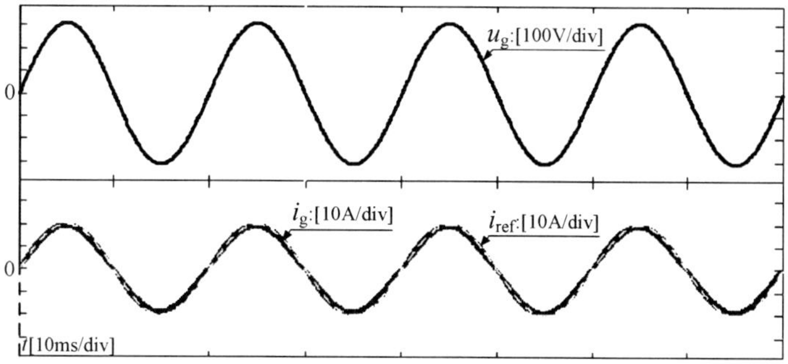

Fig. 3.20 Simulation waveforms of u_g, i_{ref}, and i_g of H5

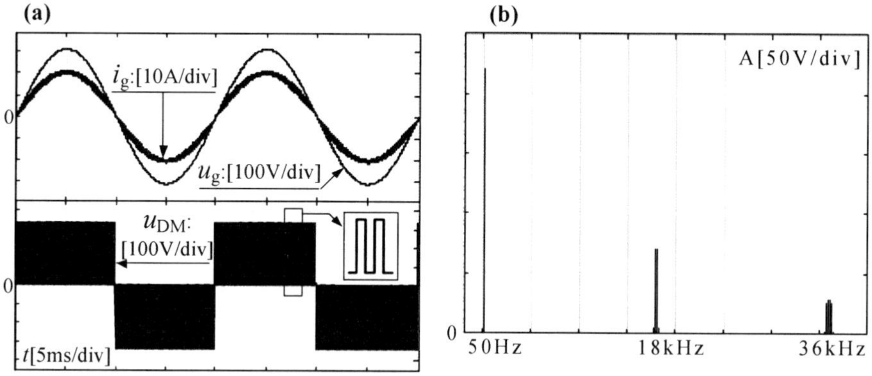

Fig. 3.21 DMV waveforms of H5, **a** time domain, **b** frequency domain

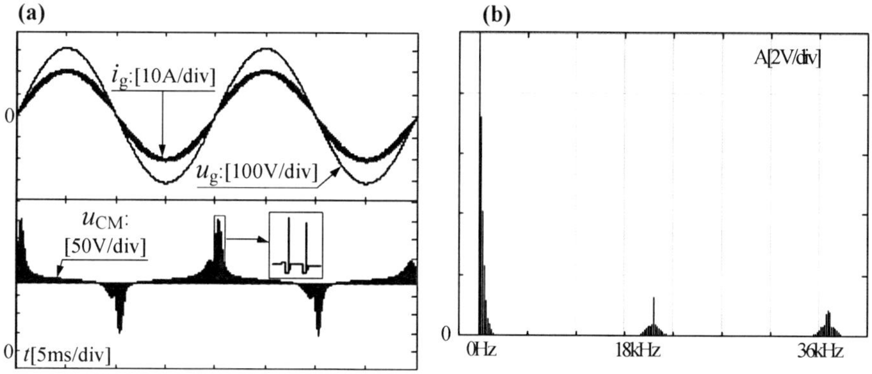

Fig. 3.22 CMV waveforms of H5, **a** time domain, **b** frequency domain

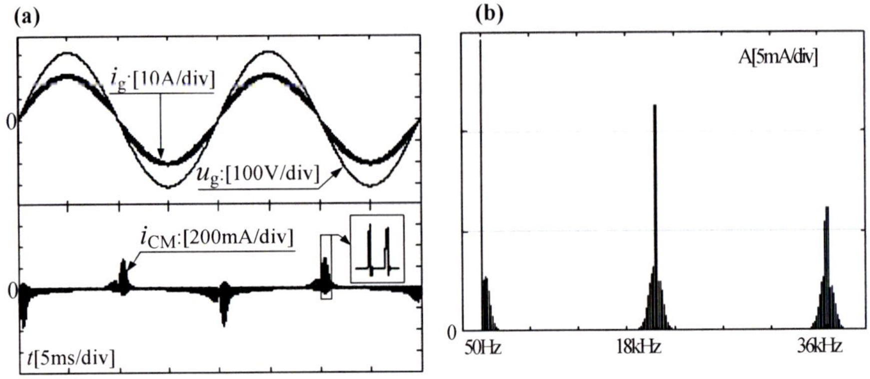

Fig. 3.23 LC waveforms of H5, **a** time domain, **b** frequency domain

3. Experimental results

In order to verify the theoretical analysis and simulation waveforms, a prototype of H5 is built using parameters in Table 3.6.

The experimental waveforms of u_g, i_g, u_{CM}, i_{CM} are shown in Fig. 3.24a, b, respectively. u_{CM} is still changed with switching frequency and the LC is about 2.8 mA at switching frequency and can meet the national standards.

Researchers have invented various improved full-bridge TLIs with the rapid development of PVPG systems in the world, especially in North America and Asia. Following the decoupling ideas, many AC coupling H6 as well as DC decoupling H6 topologies are proposed.

Table 3.6 Key parameters of H5 prototype

Parameters	Value	Parameters	Value
Input voltage U_{pv}/V	340–700	Filter inductors L_1, L_2/mH	4
Grid voltage u_g/V	240	Filter capacitor C/μF	6.6
Grid frequency/Hz	50	Common-mode inductor L_{CM}	2 × 2 W-43615-TC
Rated power P/W	1000	Wire/mm	Wire: 2 mm
Switching frequency f_s/kHz	20	Turns	Turns: 10 + 10
DC side capacitors C_{dc1}, C_{dc2}/μF	470 μF/400 V	Common-mode capacitor C_{Y1}, C_{Y2}/nF	2.2
MOSFET S_1–S_6	IXFN36N100	PV parasitic capacitor C_{pv1}, C_{pv2}/μF	0.1

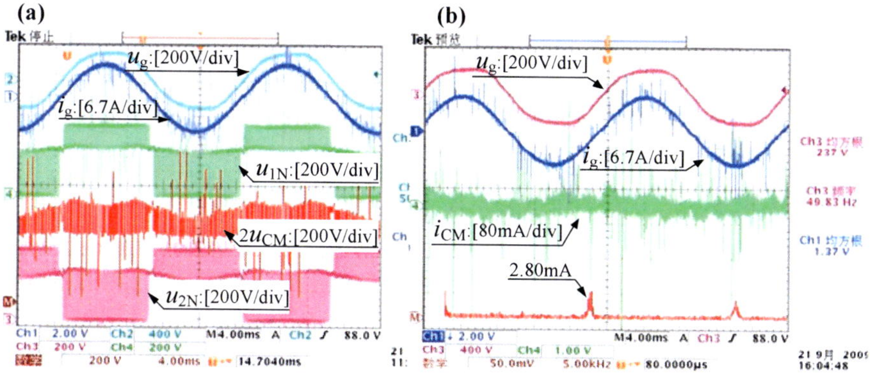

Fig. 3.24 Common mode performances, **a** CMV waveforms, **b** LC waveforms

3.2.3 *H6-I Inverter*

H6-I was proposed by future energy electronics center (FEEC) of America [5].

1. Operating principle

The main structure of H6-I is shown in Fig. 3.25a, where two new switches S_5 and S_6 are added to the inverter legs, respectively. This inverter is also classified in AC decoupling type [6]. The modulation strategy is shown in Fig. 3.25b. Apparently, S_1–S_4 work in high frequency and S_5–S_6 work in line-frequency. The output voltage is also a three-level voltage and maintains high conversion efficiency.

There are four operating modes when H6-I works in unity power factor based on the directions of grid voltage and grid-in current, as shown in Fig. 3.26. Among them, Fig. 3.26a, b belongs to the positive grid cycle, while Fig. 3.26c, d belongs to the negative grid cycle.

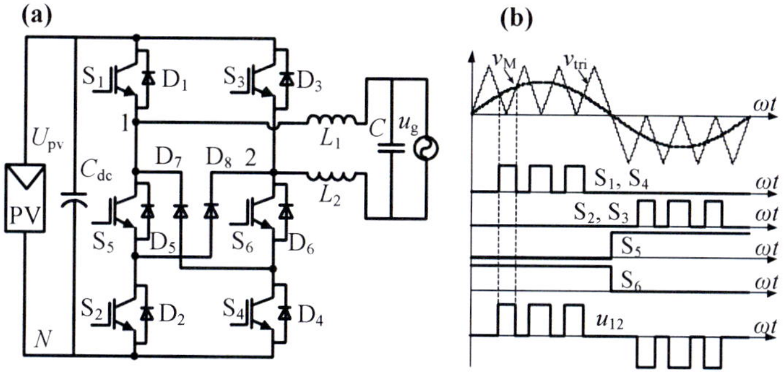

Fig. 3.25 H6-I inverter, **a** main circuit, **b** modulation strategy

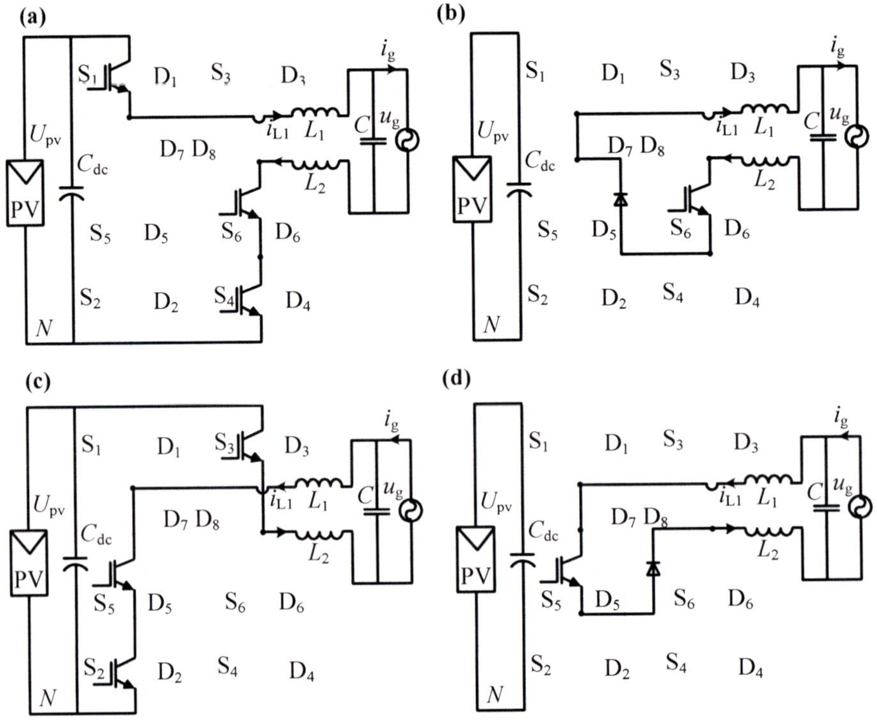

Fig. 3.26 Equivalent circuits of operating modes of H6-I, **a** mode 1, **b** mode 2, **c** mode 3, **d** mode 4

Mode 1: Refer to Fig. 3.26a; in this stage, the energy is delivering from PV side to the grid through S_1, S_4, S_6, and the filter. The operating principle is the same as it is in conventional four-switch full-bridge TLI.

Mode 2: Refer to Fig. 3.26b; only S_6 is conducting from Fig. 3.25b. At this stage, the freewheeling current flows through S_6, D_7, and the filter while keeping feeding grid. Freewheeling circuit is disconnected with PV side completely and the LC loop is cut off compared to conventional full-bridge TLI.

Mode 3: Refer to Fig. 3.26c; at this stage, the energy is delivering from the PV side to the grid through S_2, S_3, S_5, and the filter.

Mode 4: Refer to Fig. 3.26d; only S_5 is conducting from Fig. 3.25b. In this stage, the freewheeling current flows through S_5, D_8, and the filter. The same function is realized with Mode 2.

Based on these modes, the potential of points 1 and 2 against N as well as u_{DM} and u_{CM} is shown in Table 3.7. It can be seen that DMV maintains unipolar SPWM performance, while CMV has two uncertain states because the freewheeling path is just disconnected from PV side in Mode 2 and Mode 4. In other words, the CMV is not a perfect one in H6-I.

Table 3.7 DMV and CMV of different operating modes in H6-I

Modes	u_{1N}	u_{2N}	u_{DM}	u_{CM}
(1)	U_{pv}	0	U_{pv}	$U_{pv}/2$
(2)	$U_{pv}/2$	$U_{pv}/2$	0	Uncertain
(3)	0	U_{pv}	$-U_{pv}$	$U_{pv}/2$
(4)	$U_{pv}/2$	$U_{pv}/2$	0	Uncertain

In conclusion, two power switches are introduced in H6-I compared to conventional four-switch full-bridge TLI. The PV side and grid are disconnected during freewheeling stages in H6-I. LC will decrease due to the weak CMV source. The current flows through three power switches in power transmission stages, while two power switches in freewheeling stages. The power loss improved compared to HERIC.

2. Simulation results

A simulation model is built by Matlab/Simulink to verify the performance analysis aforementioned, and the key parameters of H6-I are shown in Table 3.4.

The simulation waveforms of u_g, grid-in current reference i_{ref}, and grid-in current i_g are shown in Fig. 3.27. The simulation waveforms of DMV u_{DM} in time domain and frequency domain are shown in Fig. 3.28a, b, respectively. It can be seen that DMV is a three-level voltage and the RMS of it is about 75 V at switching frequency.

The simulation waveforms of CMV u_{CM} in time domain and frequency domain are shown in Fig. 3.29a, b, respectively. It can be seen that the RMS of CMV is about 2 V at switching frequency.

The simulation waveforms of LC i_{CM} in time domain and frequency domain are shown in Fig. 3.30a, b, respectively. It can be seen that the RMS of LC is about 12 mA at switching frequency, which meets the limitation of national standards.

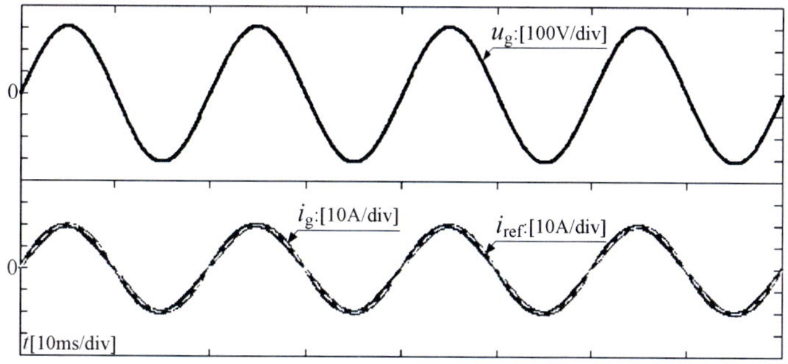

Fig. 3.27 Simulation waveforms of u_g, i_{ref}, and i_g of H6-I

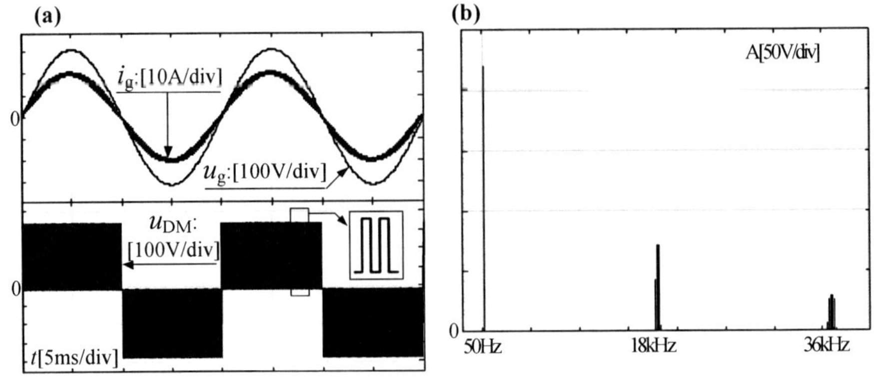

Fig. 3.28 DMV waveforms of H6-I, **a** time domain, **b** frequency domain

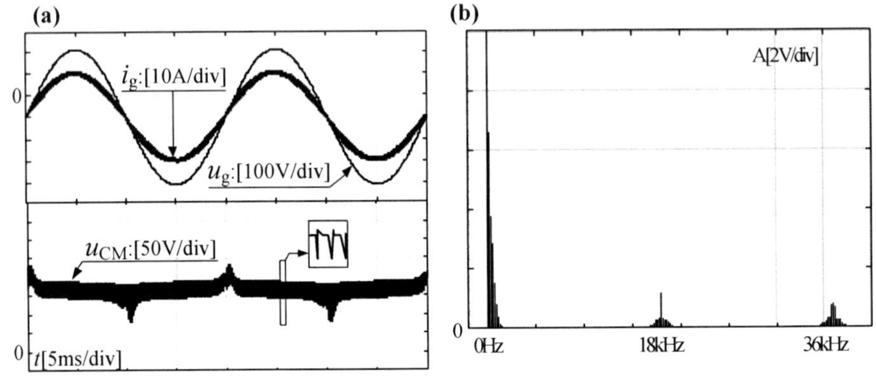

Fig. 3.29 CMV waveforms of H6-I, **a** time domain, **b** frequency domain

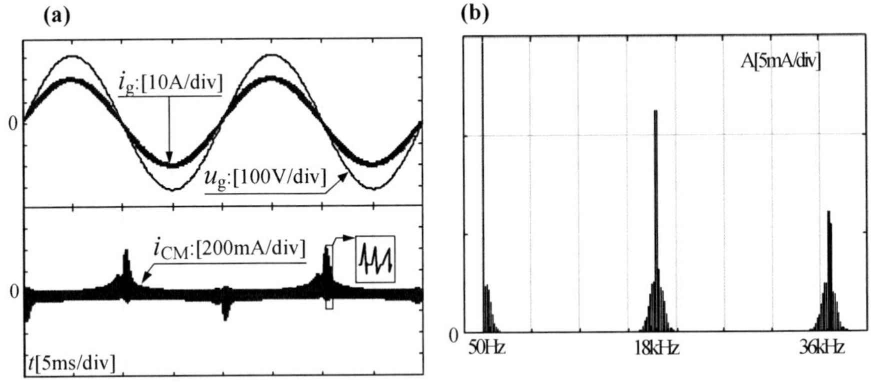

Fig. 3.30 LC waveforms of H6-I, **a** time domain, **b** frequency domain

3.2.4 H6-II Inverter

1. Operating principle

H6-II was proposed in [7], and its main circuit is shown in Fig. 3.31a. The structure of H6-II is similar to H6-I. It reduces two diodes compared to H6-I. This inverter is also AC side decoupling circuit. The modulation strategy is shown in Fig. 3.31b. The output voltage is a three-level voltage.

There are four operating modes based on the directions of grid voltage and grid-in current when H6-II works in unity power factor, as shown in Fig. 3.32. Among them, Fig. 3.32a, b belongs to the positive grid cycle, while Fig. 3.32c, d belongs to the negative grid cycle.

Mode 1: Refer to Fig. 3.32a; in this stage, the energy transfers from the PV side to the grid through S_1, S_4, S_6, and the filter.

Mode 2: Refer to Fig. 3.32b; only S_6 is conducting from Fig. 3.31b. In this stage, the freewheeling current flows through S_6, D_5, and the filter. The freewheeling circuit is disconnected with PV side completely and the LC loop is broken compared to conventional full-bridge TLI.

Mode 3: Refer to Fig. 3.32c; in this stage, the energy transfers from PV side to the grid through S_2, S_3, S_5, and the filter.

Mode 4: Refer to Fig. 3.32d; only S_5 is conducting from Fig. 3.31b. In this stage, the freewheeling current flows through S_5, D_6, and the filter. The freewheeling circuit is disconnected with DC side completely. It has the same function as Mode 2.

Based on these modes, the potential of points 1 and 2 against N as well as u_{DM} and u_{CM} is shown in Table 3.8. It can be seen that DMV maintains unipolar SPWM performance, while CMV has two uncertain states because the freewheeling path is just disconnected from PV side in Mode 2 and Mode 4. In other words, the CMV is not a perfect one in H6-II.

In conclusion, two power switches are introduced in H6-II compared to conventional four-switch full-bridge TLI. The PV side and grid side are disconnected during freewheeling stages in H6-II. LC will decrease due to the weak CMV source. The

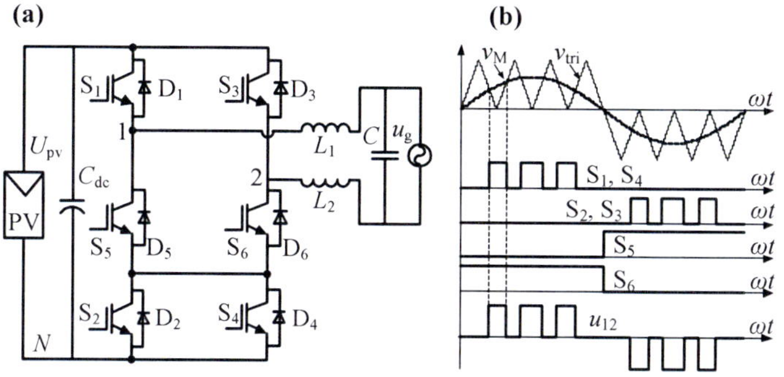

Fig. 3.31 H6-II inverter, **a** main circuit, **b** modulation strategy

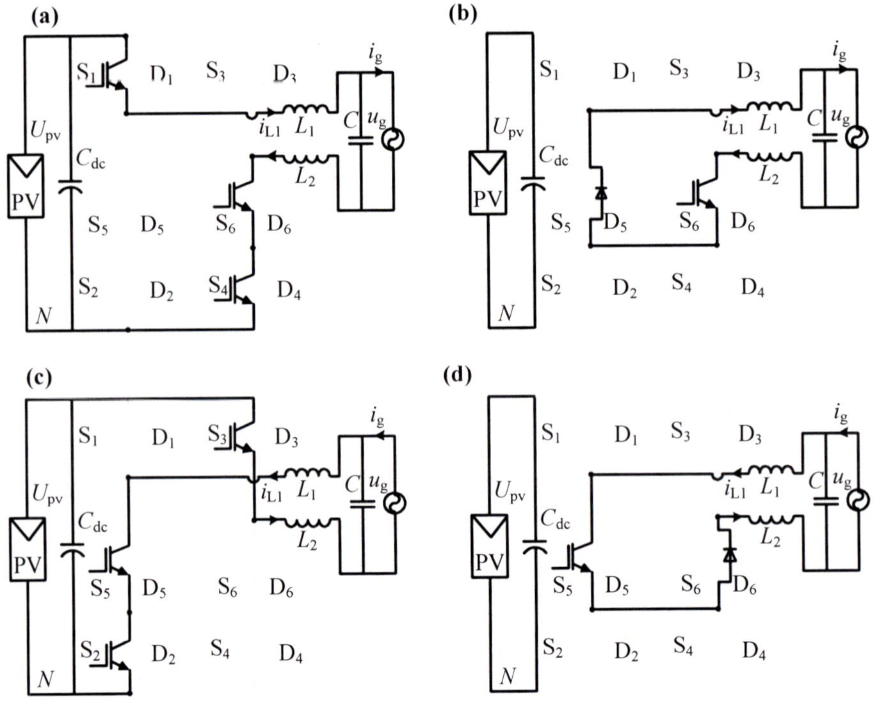

Fig. 3.32 Equivalent circuits of operating modes of H6-II, **a** mode 1, **b** mode 2, **c** mode 3, **d** mode 4

Table 3.8 DMV and CMV of different operating modes in H6-II

Modes	u_{1N}	u_{2N}	u_{DM}	u_{CM}
(1)	U_{pv}	0	U_{pv}	$U_{pv}/2$
(2)	$U_{pv}/2$	$U_{pv}/2$	0	Uncertain
(3)	0	U_{pv}	$-U_{pv}$	$U_{pv}/2$
(4)	$U_{pv}/2$	$U_{pv}/2$	0	Uncertain

current flows through three power switches in power transmission stages, and two power switches conduct in freewheeling stages. Its working principles are similar to H6-I.

2. Simulations results

A simulation model is built by Matlab/Simulink to verify the performance analysis aforementioned, and the key parameters of H6-II are shown in Table 3.4.

The simulation waveforms of u_g, i_{ref}, and i_g are shown in Fig. 3.33. The simulation waveforms of DMV u_{DM} in time domain and frequency domain are shown in Fig. 3.34a, b. u_{DM} is a three-level voltage and the RMS of it is about 75 V at switching

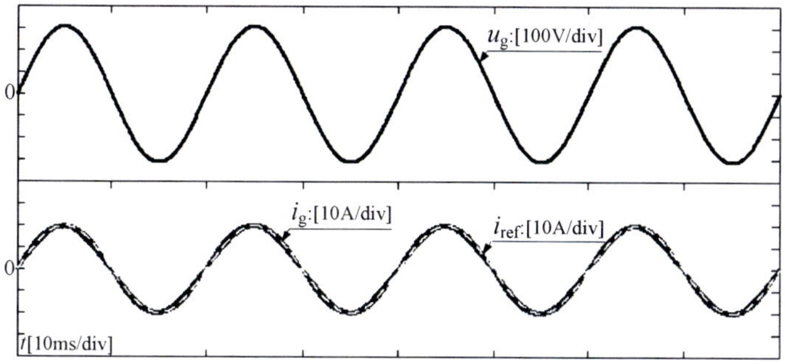

Fig. 3.33 Simulation waveforms of u_g, i_{ref}, and i_g of H6-II

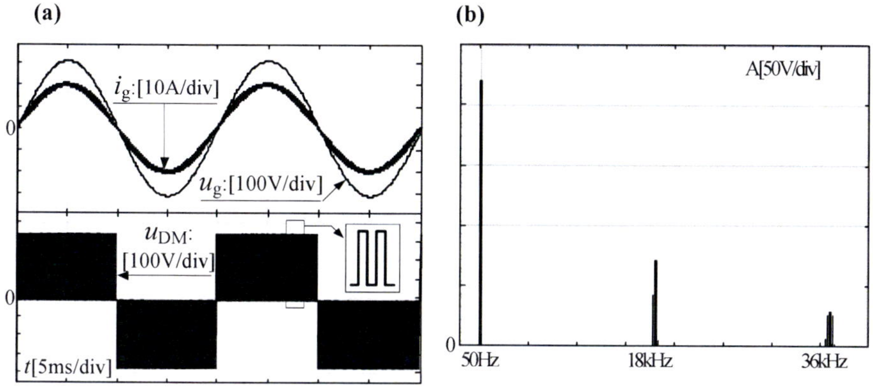

Fig. 3.34 DMV waveforms of H6-II, **a** time domain, **b** frequency domain

frequency.

The simulation waveforms of CMV u_{CM} in time domain and frequency domain are shown in Fig. 3.35a, b, respectively. It can be seen that the RMS of CMV is about 2 V at switching frequency.

The simulation waveforms of LC i_{CM} in time domain and frequency domain are shown in Fig. 3.36a, b. From Fig. 3.36, the LC is about 14 mA at switching frequency. Its common-mode performance is similar to H6-I.

3.2.5 H6-III Inverter

1. Operating principle

H6-III was proposed in [8], and its main circuit is shown in Fig. 3.37a. Two power switches S_5 and S_6 as well as two independent diodes D_6 and D_7 are introduced in H6-III compared to conventional four-switch full-bridge TLI. S_5 and S_6 are inserted

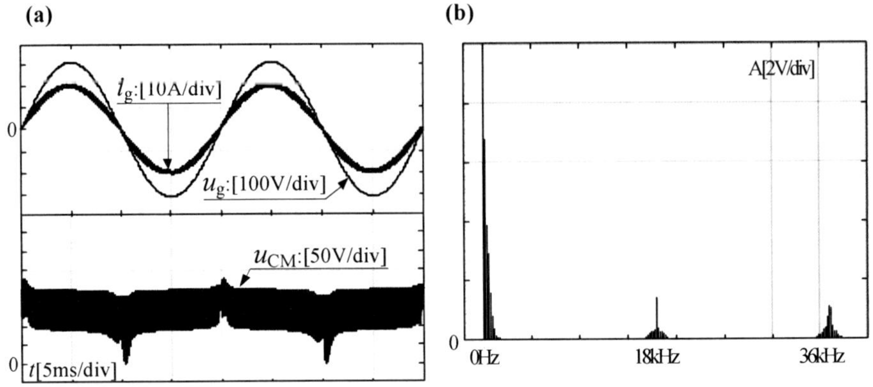

Fig. 3.35 CMV waveforms of H6-II, **a** time domain, **b** frequency domain

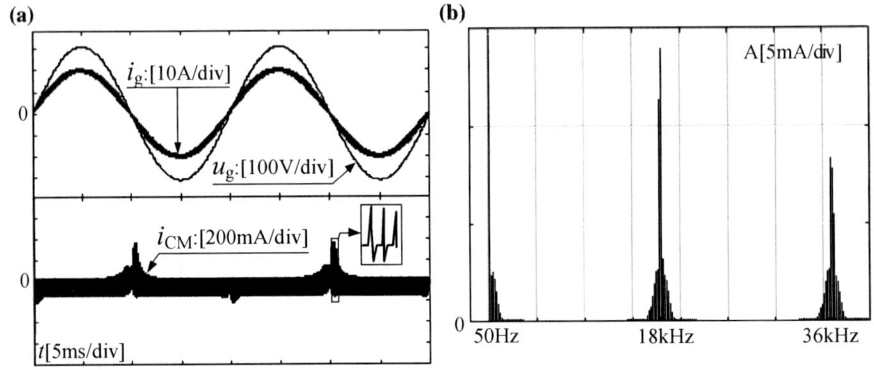

Fig. 3.36 LC waveforms of H6-II, **a** time domain, **b** frequency domain

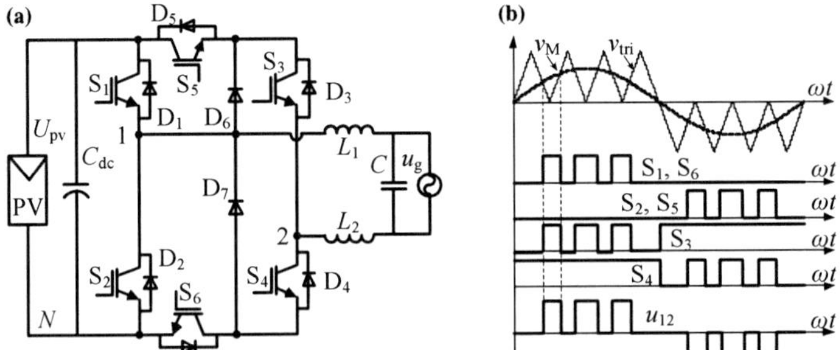

Fig. 3.37 H6-III inverter, **a** main circuit, **b** modulation strategy

into positive and negative DC buses, respectively. Although the structure is different from H6 topologies mentioned above, H6-III is still AC side decoupling circuit [9–11]. The modulation strategy is shown in Fig. 3.37b. The output voltage is also a three-level voltage.

There are four operating modes based on the directions of grid voltage and grid-in current when H6-III works in unity power factor, as shown in Fig. 3.38. Figure 3.38a, b belongs to the positive grid cycle, while Fig. 3.38c, d belongs to the negative grid cycle.

Mode 1: Refer to Fig. 3.38a; in this stage, the energy transfers from the PV side to the grid side through S_1, S_4, S_6, and the filter.

Mode 2: Refer to Fig. 3.38b; only S_4 is conducting from Fig. 3.37b. In this stage, the freewheeling current flows through S_4, D_7, and the filter. The freewheeling circuit is disconnected with PV side completely and the LC loop is broken.

Mode 3: Refer to Fig. 3.38c; in this stage, the energy transfers from the PV side to the grid through S_2, S_3, S_5, and the filter.

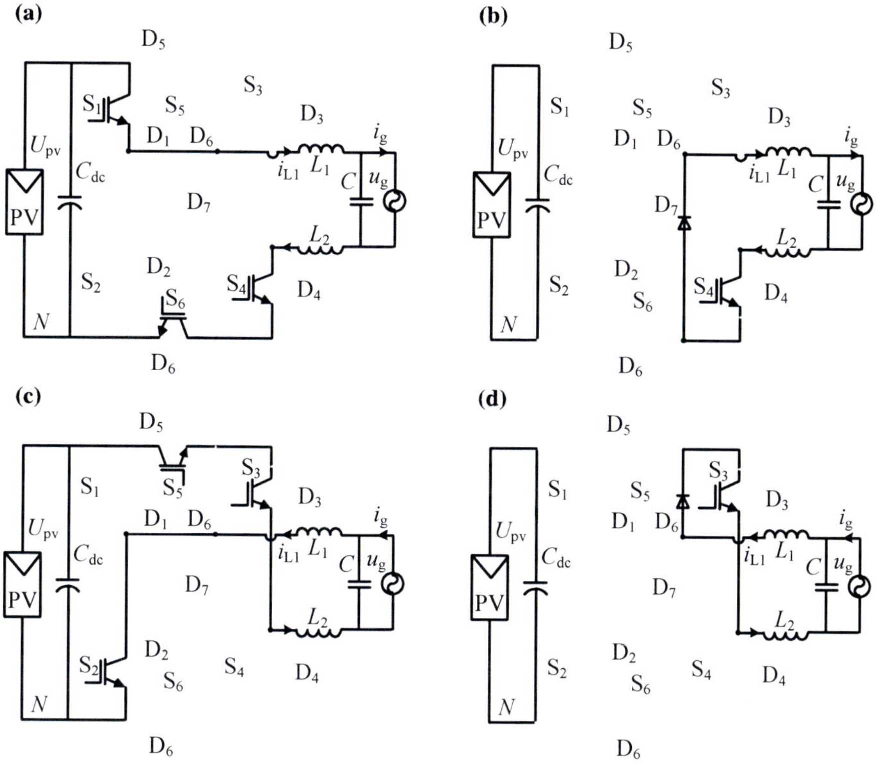

Fig. 3.38 Equivalent circuits of operating modes of H6-III, **a** mode 1, **b** mode 2, **c** mode 3, **d** mode 4

Table 3.9 DMV and CMV of different operating modes in H6-III

Modes	u_{1N}	u_{2N}	u_{DM}	u_{CM}
(1)	U_{pv}	0	U_{pv}	$U_{pv}/2$
(2)	$U_{pv}/2$	$U_{pv}/2$	0	Uncertain
(3)	0	U_{pv}	$-U_{pv}$	$U_{pv}/2$
(4)	$U_{pv}/2$	$U_{pv}/2$	0	Uncertain

Mode 4: Refer to Fig. 3.38d; only S_3 is conducting from Fig. 3.37b. In this stage, the freewheeling current flows through S_3, D_6, and the filter. The LC loop is also broken compared to conventional full-bridge TLI.

Based on these modes, the potential of points 1 and 2 against N as well as u_{DM} and u_{CM} is shown in Table 3.9. From this table, it can be seen that the DMV maintains unipolar SPWM performance, while the CMV has two uncertain states because the freewheeling path is just disconnected from PV side in Mode 2 and Mode 4. In other words, the CMV is not a perfect one in H6-III.

In conclusion, two power switches are inserted into the DC bus to disconnect the PV side and grid side during freewheeling stages in H6-III. LC will decrease due to the disconnected CMV loop and CMV source. The current flows through three power switches in power transmission stages, and two power switches conduct in freewheeling stages. Its working principles and power loss are similar to H6-I or H6-II.

2. Simulations results

A simulation model is built by Matlab/Simulink to verify the performance analysis aforementioned, and the key parameters of H6-III are shown in Table 3.4.

The simulation waveforms of u_g, i_{ref}, and i_g are shown in Fig. 3.39. The simulation waveforms of DMV u_{DM} in time domain and frequency domain are shown

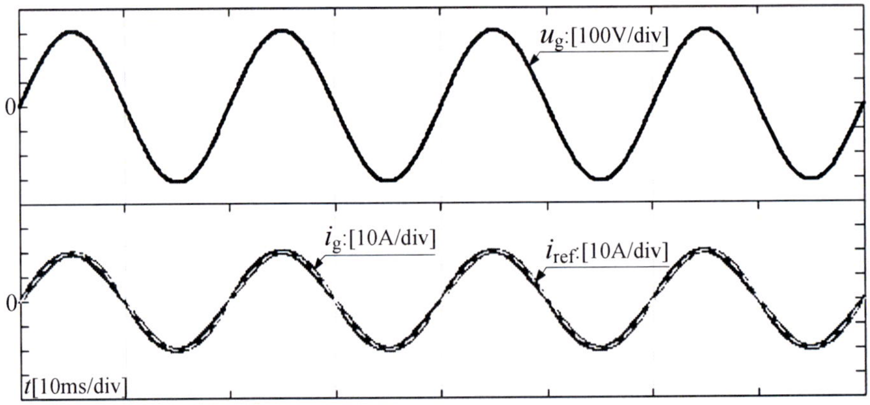

Fig. 3.39 Simulation waveforms of u_g, i_{ref}, and i_g of H6-III

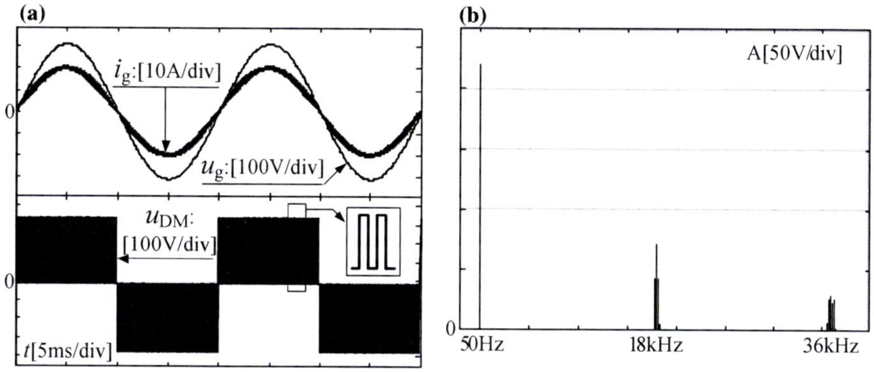

Fig. 3.40 DMV waveforms of H6-III, **a** time domain, **b** frequency domain

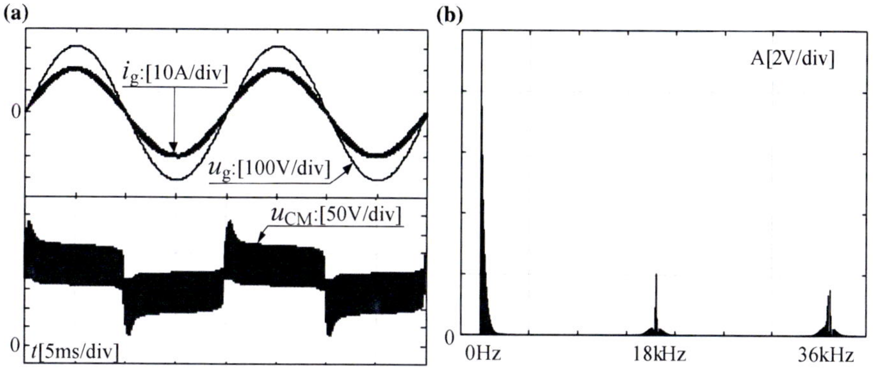

Fig. 3.41 CMV waveforms of H6-III, **a** time domain, **b** frequency domain

in Fig. 3.40a, b, respectively. DMV is a three-level voltage and the RMS of it is about 75 V at switching frequency. The simulation waveforms of CMV u_{CM} in time domain and frequency domain are shown in Fig. 3.41a, b, respectively. It can be seen that the RMS of CMV is about 2 V at switching frequency and still contains certain high-frequency components. The simulation waveforms of LC i_{CM} in time domain and frequency domain are shown in Fig. 3.42a, b, respectively. From Fig. 3.42, the LC is about 20 mA at switching frequency, which is similar to H6-II.

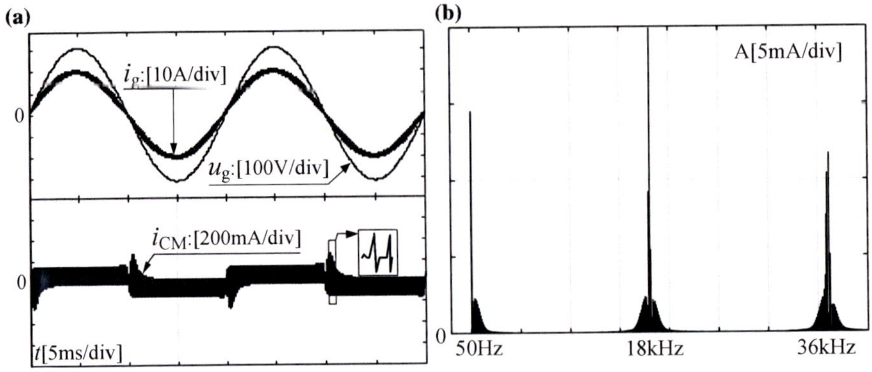

Fig. 3.42 LC waveforms of H6-III, **a** time domain, **b** frequency domain

3.2.6 *H6-IV Inverter*

1. Operating principle

H6-IV was proposed in [12], and its main circuit is shown in Fig. 3.43a. Two power switches S_5 and S_6 are introduced into positive and negative DC buses in H6-IV compared to conventional full-bridge inverter. Different from H6-III, this inverter is DC side decoupling type TLI. The modulation strategy is shown in Fig. 3.43b. S_3–S_6 work in high frequency. The output voltage is a three-level voltage.

There are four operating modes based on the directions of grid voltage and grid-in current when H6-IV works in unity power factor, as shown in Fig. 3.44. Figure 3.44a, b belongs to the positive grid cycle, while Fig. 3.44c, d belongs to the negative grid cycle.

Mode 1: Refer to Fig. 3.44a; in this stage, the energy transfers from the PV side to the grid through S_1, S_4, S_5, S_6, and the filter.

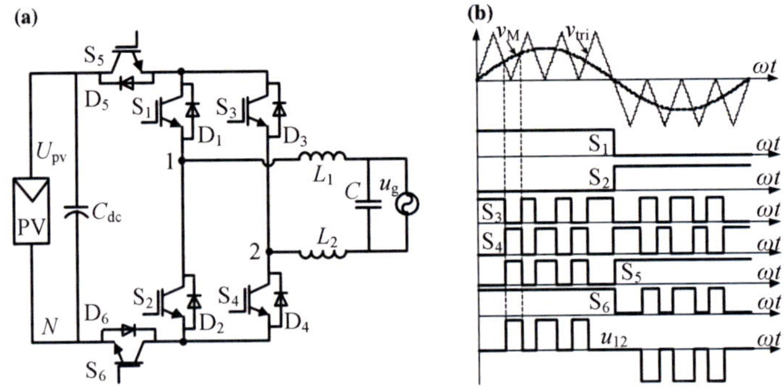

Fig. 3.43 H6-IV inverter, **a** main circuit, **b** modulation strategy

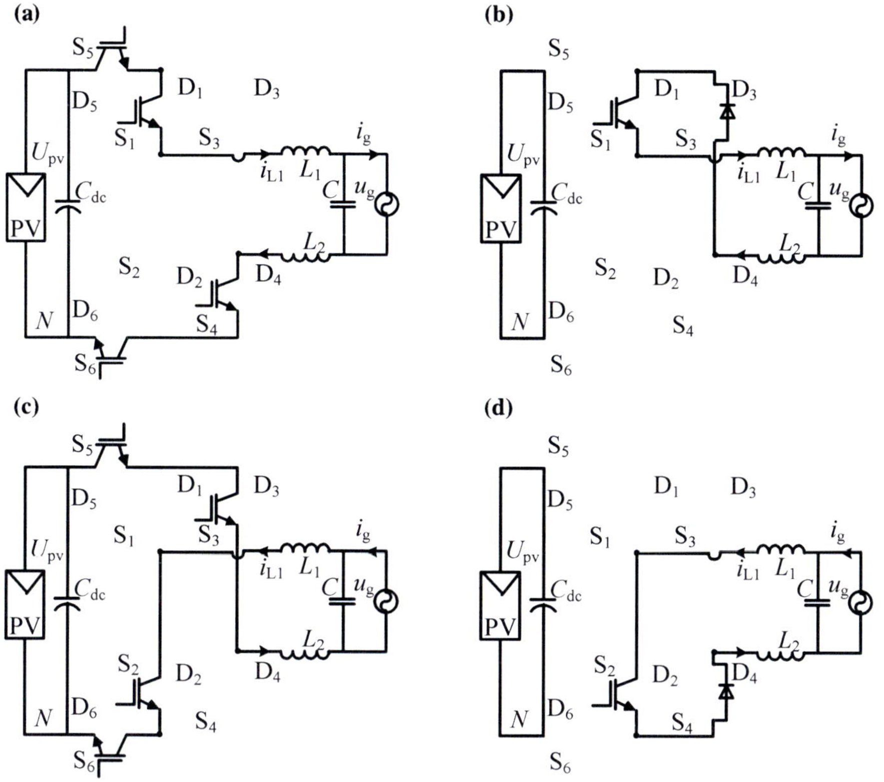

Fig. 3.44 Equivalent circuits of operating modes of H6-IV, **a** mode 1, **b** mode 2, **c** mode 3, **d** mode 4

Mode 2: Refer to Fig. 3.44b; S_1, S_3, and S_6 are conducting from Fig. 3.43b. In this stage, the freewheeling current flows through S_1, D_3, and the filter. The freewheeling circuit is disconnected with PV side completely and the LC loop is cut off.

Mode 3: Refer to Fig. 3.44c; in this stage, the energy transfers from the PV side to the grid through S_2, S_3, S_5, S_6, and the filter.

Mode 4: Refer to Fig. 3.44d; S_2, S_4, and S_5 are conducting from Fig. 3.43b. In this stage, the freewheeling current flows through S_2, D_4, and the filter. The LC loop is also broken compared to conventional full-bridge TLI.

Based on these modes, the potential of points 1 and 2 against N as well as u_{DM} and u_{CM} is shown in Table 3.10. From this table, it can be seen that the DMV maintains unipolar SPWM performance, while the CMV has two uncertain states because the freewheeling path is just disconnected from PV side in Mode 2 and Mode 4. In other words, the CMV is not a perfect one in H6-IV.

Table 3.10 DMV and CMV of different operating modes in H6-IV

Modes	u_{1N}	u_{2N}	u_{DM}	u_{CM}
(1)	U_{pv}	0	U_{pv}	$U_{pv}/2$
(2)	$U_{pv}/2$	$U_{pv}/2$	0	Uncertain
(3)	0	U_{pv}	$-U_{pv}$	$U_{pv}/2$
(4)	$U_{pv}/2$	$U_{pv}/2$	0	Uncertain

In conclusion, the PV and grid are disconnected by the added two power switches during freewheeling stages in H6-IV. LC will decrease due to weak CMV source. The current flows through four power switches in power transmission stages, and two power switches conduct in freewheeling stages. The power loss improved compared to HERIC.

2. Simulations results

A simulation model is built by Matlab/Simulink to verify the performance analysis aforementioned, and the key parameters of H6-II are shown in Table 3.4.

The simulation waveforms of u_g, i_{ref}, and i_g are shown in Fig. 3.45. The simulation waveforms of DMV u_{DM} in time domain and frequency domain are shown in Fig. 3.46a, b, respectively. u_{DM} is a three-level voltage and the RMS of it is about 75 V at switching frequency.

The simulation waveforms of CMV u_{CM} in time domain and frequency domain are shown in Fig. 3.47a, b, respectively. It can be seen that the RMS of CMV is about 2 V at switching frequency and contains little high-frequency components. The simulation waveforms of LC i_{CM} in time domain and frequency domain are shown in Fig. 3.48a, b, respectively. From Fig. 3.48, the LC is about 15 mA at switching frequency. It still meets the standard.

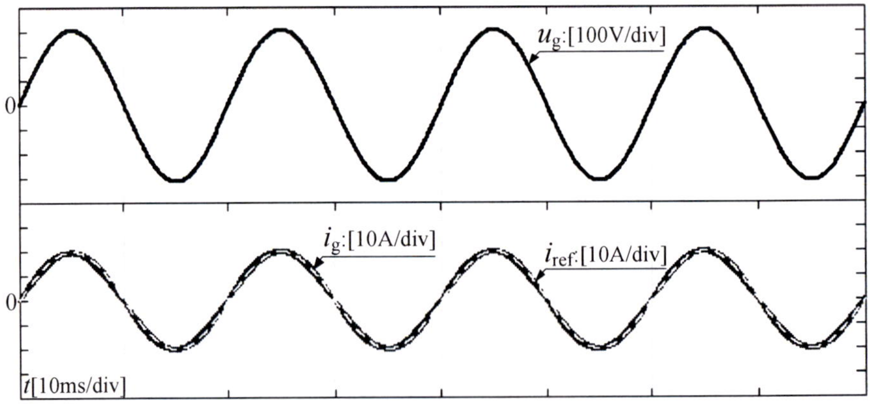

Fig. 3.45 Simulation waveforms of u_g, i_{ref}, and i_g of H6-IV

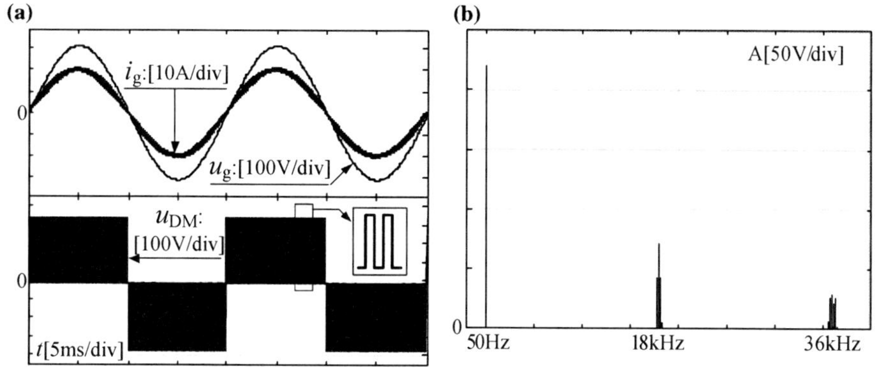

Fig. 3.46 DMV waveforms of H6-IV, **a** time domain, **b** frequency domain

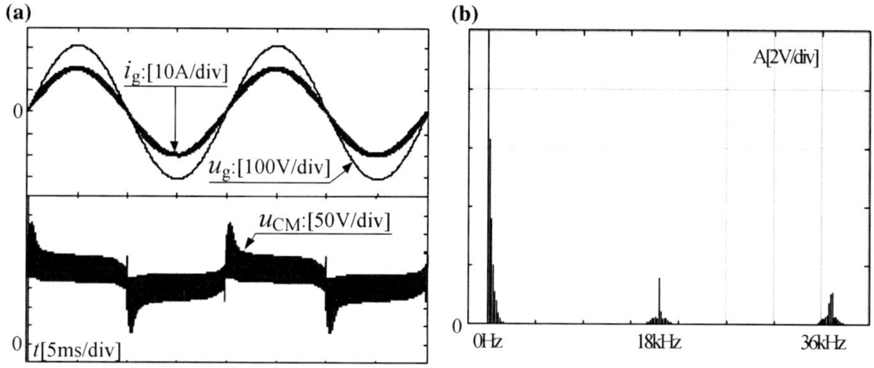

Fig. 3.47 CMV waveforms of H6-IV, **a** time domain, **b** frequency domain

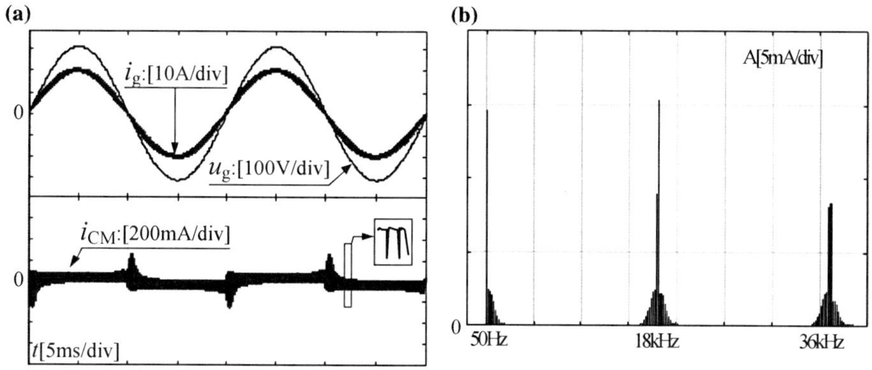

Fig. 3.48 LC waveforms of H6-IV, **a** time domain, **b** frequency domain

3.2.7 H6-V Inverter

1. Operating principle

H6-V was proposed in [13], and its main circuit is shown in Fig. 3.49a. The two power switches S_5 and S_6 are introduced into H6-V. The power switch S_5 is located on the positive DC bus, while the switch S_6 connects the positive DC bus and neutral point 2 of the inverter leg. This inverter is also DC side decoupling circuit. The modulation strategy is shown in Fig. 3.49b. The output voltage is also a three-level voltage.

There are four operating modes based on the directions of grid voltage and grid-in current when H6-V works in unity power factor, as shown in Fig. 3.50. Figure 3.50a, b belongs to the positive grid cycle, while Fig. 3.50c, d belongs to the negative grid cycle.

Mode 1: Refer to Fig. 3.50a; in this stage, the energy transfers from the PV side to the grid through S_1, S_4, S_5, and the filter.

Mode 2: Refer to Fig. 3.50b; S_1 is conducting from Fig. 3.49b. In this stage, the freewheeling current flows through S_1, D_3, and the filter. The freewheeling circuit is disconnected with PV side completely and the LC loop is cut off compared to conventional full-bridge TLI.

Mode 3: Refer to Fig. 3.50c; in this stage, the energy transfers from the PV side to the grid through S_2, S_6, and the filter.

Mode 4: Refer to Fig. 3.50d; S_3 is conducting from Fig. 3.49b. In this stage, the freewheeling current flows through S_3, D_1, and the filter. The LC loop is also open as Mode 2.

Based on these modes, the potential of points 1 and 2 against N as well as u_{DM} and u_{CM} is shown in Table 3.11. From this table, it can be seen that the DMV maintains unipolar SPWM performance, and the CMV has two uncertain states because the freewheeling path is just disconnected from PV side in Mode 2 and Mode 4. In other words, the CMV is not a perfect one in H6-V.

In conclusion, two power switches are introduced at DC side in H6-V. The PV and grid are disconnected during freewheeling stages in H6-V. LC will decrease due

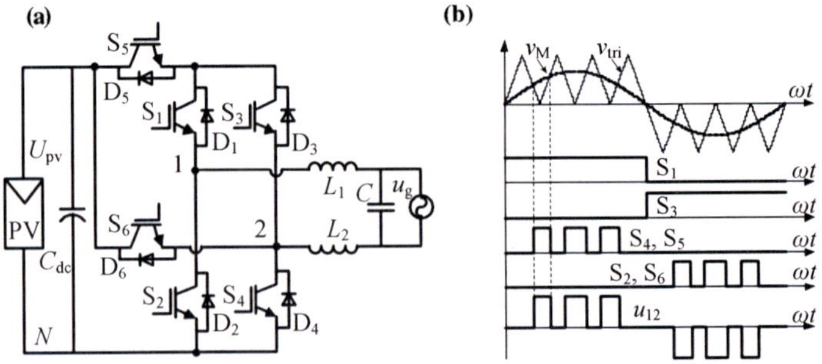

Fig. 3.49 H6-V inverter, **a** main circuit, **b** modulation strategy

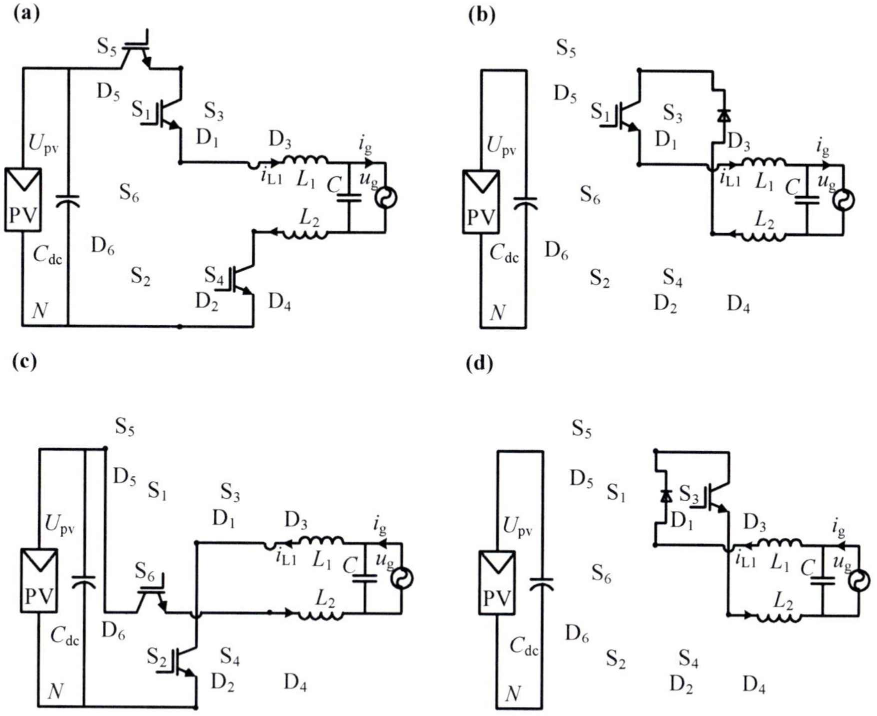

Fig. 3.50 Equivalent circuits of operating modes of H6-V, **a** mode 1, **b** mode 2, **c** mode 3, **d** mode 4

Table 3.11 DMV and CMV of different operating modes in H6-V

Modes	u_{1N}	u_{2N}	u_{DM}	u_{CM}
(1)	U_{pv}	0	U_{pv}	$U_{pv}/2$
(2)	$U_{pv}/2$	$U_{pv}/2$	0	Uncertain
(3)	0	U_{pv}	$-U_{pv}$	$U_{pv}/2$
(4)	$U_{pv}/2$	$U_{pv}/2$	0	Uncertain

to the disconnected CMV loop. Its working principles and power loss are similar to H5.

The current flows through three power switches in energy transmission stage in grid positive cycle. However, the current flows through two power switches in energy transmission stage in grid negative cycle during both power transmission and freewheeling stages. The power loss is reduced compared to H6-I or H6-II.

2. Simulations results

A simulation model is built by Matlab/Simulink to verify the performance analysis aforementioned, and the key parameters of H6-II are shown in Table 3.4.

The simulation waveforms of u_g, grid-in current reference i_{ref}, and grid-in current i_g are shown in Fig. 3.51. The simulation waveforms of DMV u_{DM} in time domain and frequency domain are shown in Fig. 3.52a, b, respectively. u_{DM} is a three-level voltage. The simulation waveforms of CMV u_{CM} in time domain and frequency domain are shown in Fig. 3.53a, b, respectively. It can be seen that the RMS of CMV is about 2 V at switching frequency.

The simulation waveforms of LC i_{CM} in time domain and frequency domain are shown in Fig. 3.54a, b, respectively. From Fig. 3.54, the LC is about 12 mA at switching frequency. It meets the standard and close to it in H6-IV.

Only active power switches are added to the inverter legs in traditional full-bridge TLIs, and thus short-circuit risk exists. To improve the reliability of full-bridge TLIs, researchers introduced the buck structure into full-bridge TLIs, such as double inductors dual buck inverter (DIDBI) [14–16] and high efficiency and reliability transformerless inverter (HERTLI) [17].

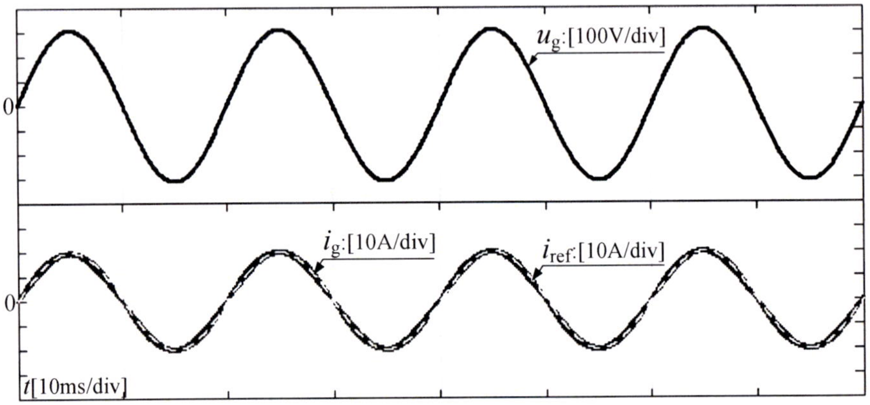

Fig. 3.51 Simulation waveforms of u_g, i_{ref}, and i_g of H6-V

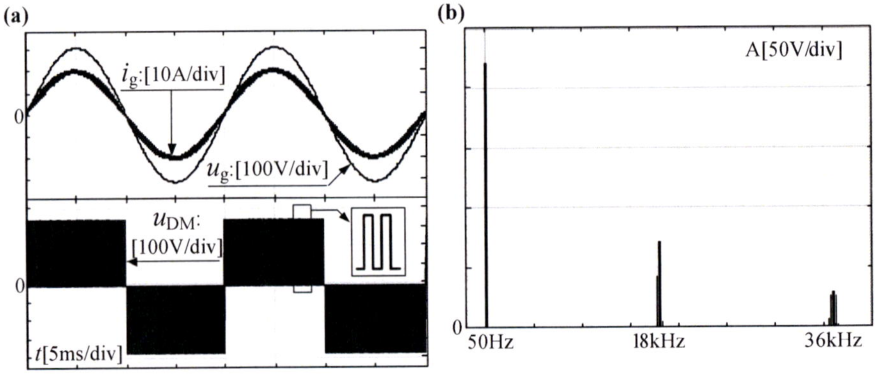

Fig. 3.52 DMV waveforms of H6-V, **a** time domain, **b** frequency domain

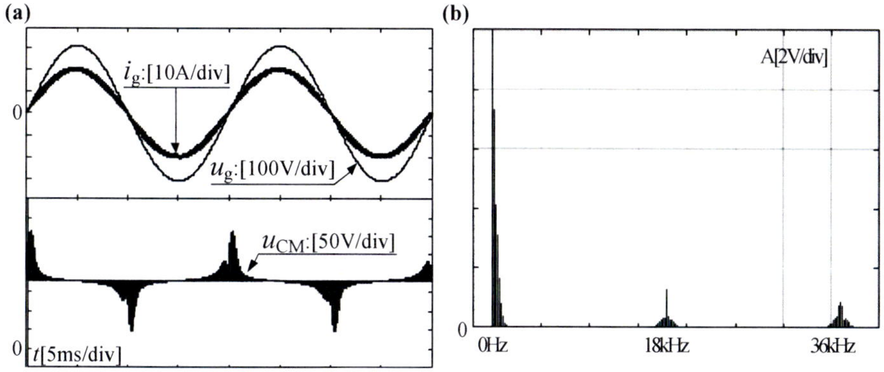

Fig. 3.53 CMV waveforms of H6-V, **a** time domain, **b** frequency domain

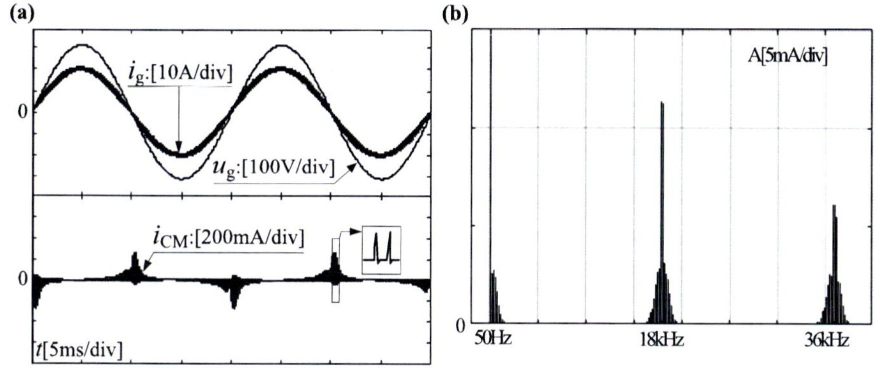

Fig. 3.54 LC waveforms of H6-V, **a** time domain, **b** frequency domain

3.2.8 Double Inductors Dual Buck Inverter

DIDBI was patented by Xantrex in 2005, and the main circuit is shown in Fig. 3.55a. It is also called Xantrex inverter. Four inductors L_1, L_2, L_3, and L_4 are used in DIDBI. These inductors can be discrete components or coupled inductors. Its structure is more complicated than conventional full-bridge TLI, but their operating principles are the same.

1. Operating principle

Two modulation strategies can be used for DIDBI, which are shown in Fig. 3.55b and 3.55c, respectively. In modulation strategy I, as shown in Fig. 3.55b, S_1 and S_4 work in high frequency in positive cycle of grid, and S_2 and S_3 work in high frequency in negative cycle of grid. In this situation, the output voltage is a two-level voltage. In modulation strategy II, S_1 and S_4 work complementary with S_2 and S_3 during the

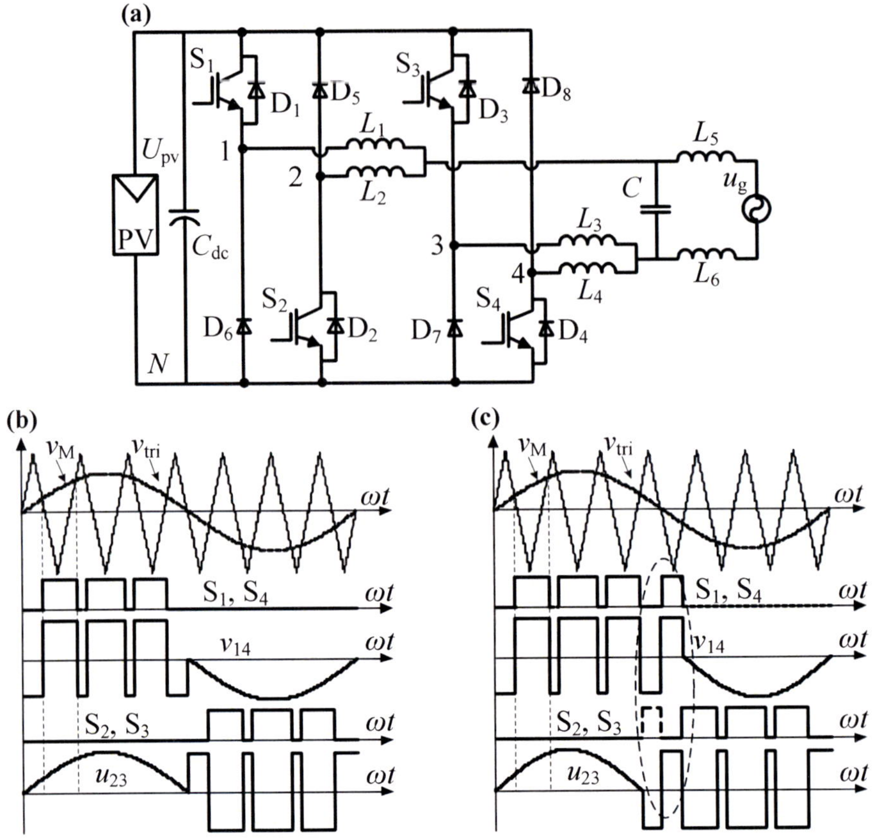

Fig. 3.55 Double inductors dual-buck inverter, **a** main circuit, **b** modulation strategy I, **c** modulation strategy II

grid-in current crossing zero area, and the other modulation strategies are the same with modulation strategy I, as shown in Fig. 3.55c. DIDBI can work in non-unity power factor.

There are four operating modes based on the directions of grid voltage and grid-in current when DIDBI works in unity power factor and using modulation strategy I, as shown in Fig. 3.56. Figure 3.56a, b belongs to the positive grid cycle, while Fig. 3.56c, d belongs to the negative grid cycle.

Mode 1: Refer to Fig. 3.56a; in this stage, the energy transfers from the PV side to the grid through S_1, S_4, L_1, and L_4.

Mode 2: Refer to Fig. 3.56b; all the active power switches are closed from Fig. 3.55b. In this stage, the current flows through D_6, D_8, L_1, and L_4. The energy is transferred back to the DC side. It is the same as with conventional full-bridge TLI using bipolar modulation strategy.

Mode 3: Refer to Fig. 3.56c. In this stage, the energy transfers from the PV side to the grid through S_2, S_3, L_2, and L_3.

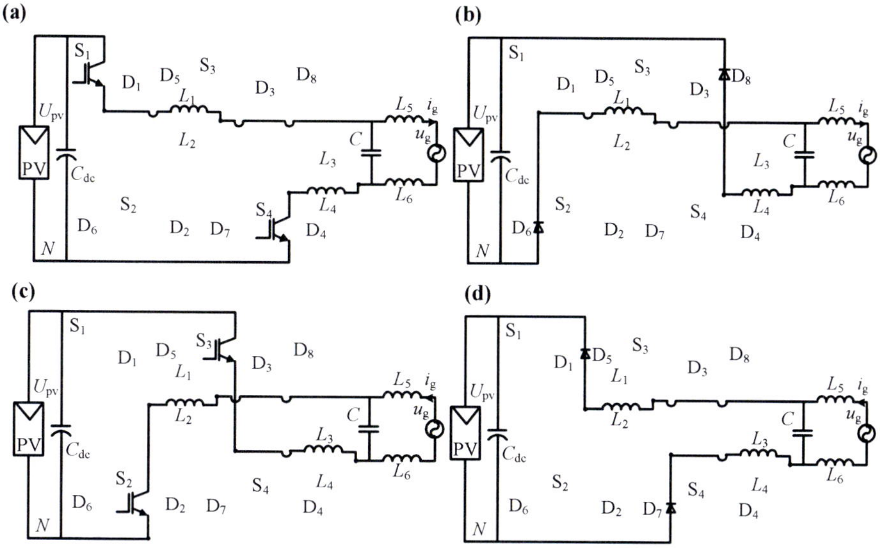

Fig. 3.56 Equivalent circuits of operating modes of DIDBI using modulation strategy I, **a** mode 1, **b** mode 2, **c** mode 3, **d** mode 4

Mode 4: Refer to Fig. 3.56d; all the active power switches are closed from Fig. 3.55b. In this stage, the current flows through D_5, D_7, L_2, and L_3. The energy is feeding back to the DC side as Mode 2.

Based on these modes, the potential of points 1 and 2 against N as well as u_{DM} and u_{CM} is shown in Table 3.12. Because the current flows through different ways in the positive and negative periods, the calculating methods are also different, as shown in Table 3.12. From this table, it is clear that the CMV of DIDBI is constant while DMV is two-level voltage. The CMV is perfect while DMV is poor in DIDBI.

From the operating modes in Fig. 3.56, the output voltage is a two-level voltage and CMV is constant using modulation strategy I. Its operating modes are the same as conventional full-bridge adopting bipolar modulation strategy. There are many harmonics in the output voltage. Four inductors are used in DIDBI, which results in low utilization of components. The inductors should work in DCM and lead to high grid-in current distortion. Therefore, modulation strategy II is proposed to reduce the current distortion. In modulation strategy II, switches S_2 and S_3 are triggered after S_1

Table 3.12 DMV and CMV of different operating modes in DIDBI

Modes	u_{1N}	u_{2N}	u_{3N}	u_{4N}	u_{DM}	u_{CM}
(1)	U_{pv}	–	–	0	U_{pv}	$U_{pv}/2$
(2)	0	–	–	U_{pv}	$-U_{pv}$	$U_{pv}/2$
(3)	–	0	U_{pv}	–	$-U_{pv}$	$U_{pv}/2$
(4)	–	U_{pv}	0	–	U_{pv}	$U_{pv}/2$

and S_4 are closed during the current crossing zero range. Then the inductors current will decrease slowly and flows through S_1 and S_4 to S_2 and S_3 naturally. The current distortion will be reduced in this situation. And another two modes will appear, as shown in Fig. 3.57a, b, respectively. The current will flow through four inductors to commutating smoothly.

2. Simulations results

A simulation model is built by Matlab/Simulink to verify the performance analysis aforementioned, and the key parameters of H6-II are shown in Table 3.13.

The simulation waveforms of u_g, grid-in current reference i_{ref}, and grid-in current i_g are shown in Fig. 3.58 when using modulation strategy I. The simulation waveforms of DMV u_{14} and u_{23} are shown in Fig. 3.59a, b, respectively. From Fig. 3.59, u_{14} and u_{23} are two-level voltages in the positive and negative grid cycle, respectively.

The simulation waveforms of inductors current i_{L1} and i_{L2} are shown in Fig. 3.60a, b. It can be seen that L_1 and L_2 are working in the positive and negative grid cycle, respectively. So do L_3 and L_4. The simulation waveforms of LC i_{CM} in time domain and frequency domain are shown in Fig. 3.61a, b, respectively. From Fig. 3.61, the switching frequency component is suppressed because of two-level output voltage.

The simulation waveforms of i_g are shown in Fig. 3.62a and the simulation waveforms of i_g, i_{L1}, and i_{L2} during the current crossing zero range using modulation

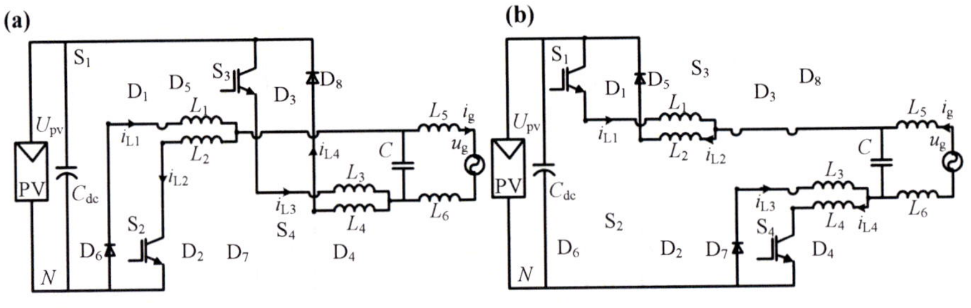

Fig. 3.57 Equivalent circuits of new operating modes of DIDBI using modulation strategy II, **a** mode 1, **b** mode 2

Table 3.13 Simulation parameters of DIDBI

Parameters	Value	Parameters	Value
Input voltage U_{pv}/V	200	Switching frequency f_s/kHz	40
Grid voltage u_g/V	120	Inductors L_1, L_2, L_3, L_4/mH	0.75
Grid frequency/Hz	50	Filter inductors L_5, L_6/mH	0.25
Rated power P/W	2500	PV parasitic capacitor C_{pv}/nF	300

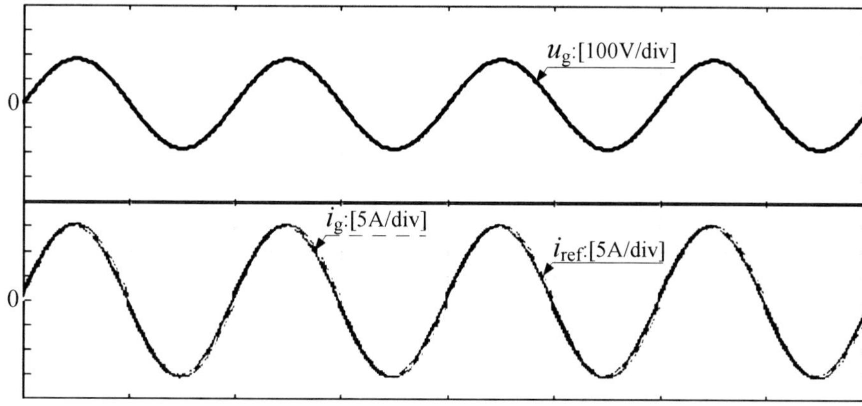

Fig. 3.58 Simulation waveforms of u_g, i_{ref}, and i_g of DIDBI under unity power factor

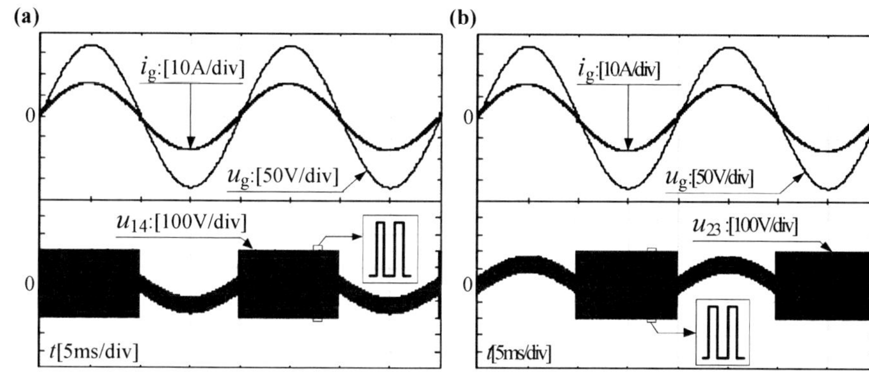

Fig. 3.59 DMV waveforms of DIDBI, **a** u_{14}, **b** u_{23}

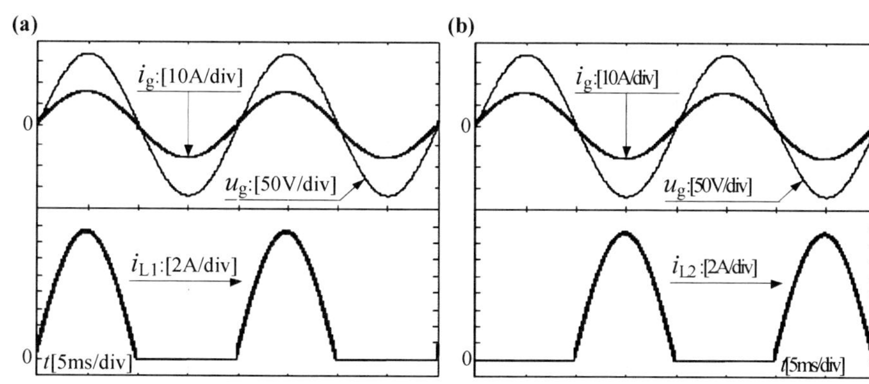

Fig. 3.60 Inductors current waveforms of DIDBI, **a** i_{L1}, **b** i_{L2}

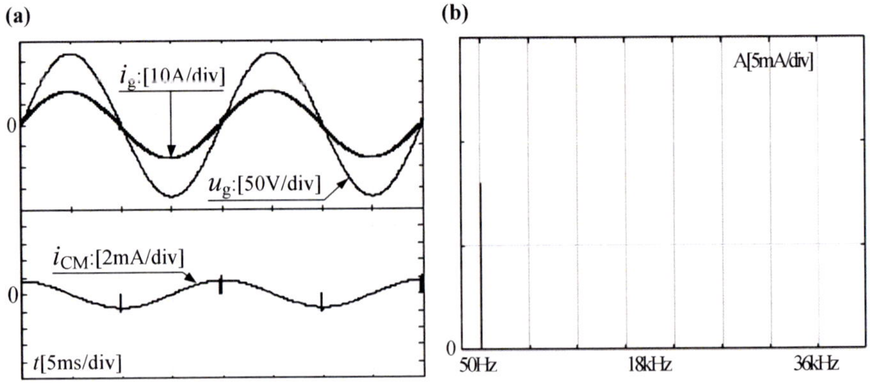

Fig. 3.61 LC waveforms of DIDBI, **a** time domain, **b** frequency domain

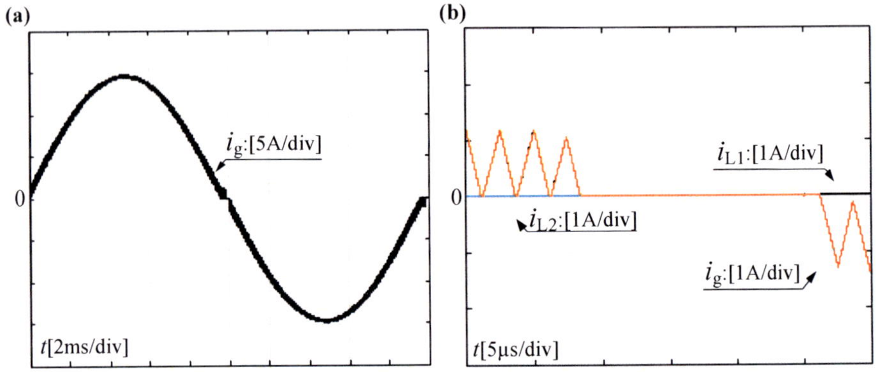

Fig. 3.62 Grid-in current and inductors current waveforms of DIDBI using modulation strategy I, **a** i_g, **b** i_g, i_{L1}, and i_{L2} during current crossing zero range

strategy I are shown in Fig. 3.62b. From Fig. 3.62, it is clear that there is a long time interval between i_{L1} decreasing to zero and i_{L2} increasing.

The simulation waveforms of i_g are shown in Fig. 3.63a, and the simulation waveforms of i_g, i_{L1}, and i_{L2} during the current crossing zero range using modulation strategy II is shown in Fig. 3.63b. As seen, i_{L1} and i_{L2} are not zero during this area. The current distortion is reduced. The current is continuous in this situation and it can work in non-unity power factor, as shown in Fig. 3.64. The simulation waveforms of leading and lagging power factor are shown in Fig. 3.64a, b, respectively. So DIDBI can work in non-unity power factor to support grid with reactive power.

The output voltage of DIDBI is a two-level voltage. The loss is high and needs complicated filter. Then, researchers proposed unipolar high efficiency and reliability transformerless inverter [17–19].

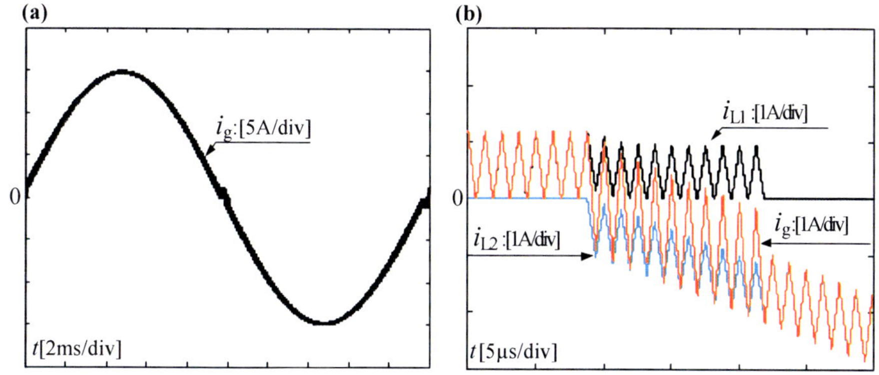

Fig. 3.63 Grid-in current and inductors current waveforms of DIDBI using modulation strategy II, **a** i_g, **b** i_g, i_{L1}, and i_{L2} during current crossing zero range

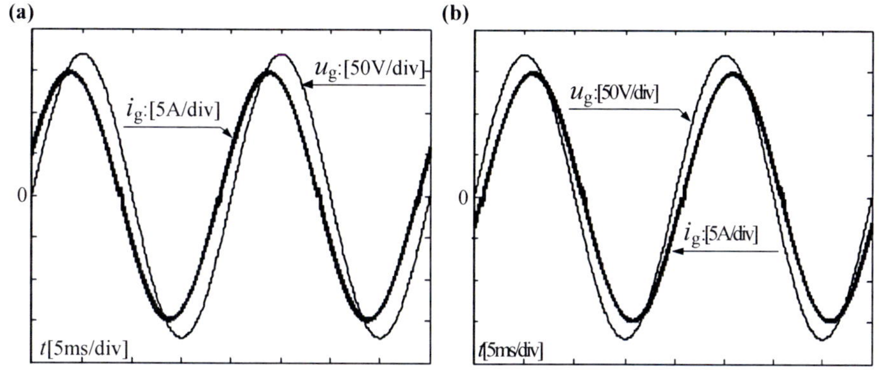

Fig. 3.64 Simulation waveforms of u_g, i_{ref}, and i_g of DIDBI under non-unity power factor, **a** leading power factor, **b** lagging power factor

3.2.9 High Efficiency and Reliability Transformerless Inverter

1. Operating principle

HERTLI [17] is shown in Fig. 3.65a. Two coupled inductors are used in this circuit [18] which can also be replaced by discrete inductors. The modulation strategy is shown in Fig. 3.65b. Apparently, S_1–S_4 work in unipolar modulation strategy and S_5–S_6 work in fundamental frequency. The output voltage is also three-level voltage. This inverter is AC side decoupling type TLI.

There are four operating modes according to the directions of grid voltage and grid-in current when HERTLI works in unity power factor, as shown in Fig. 3.66.

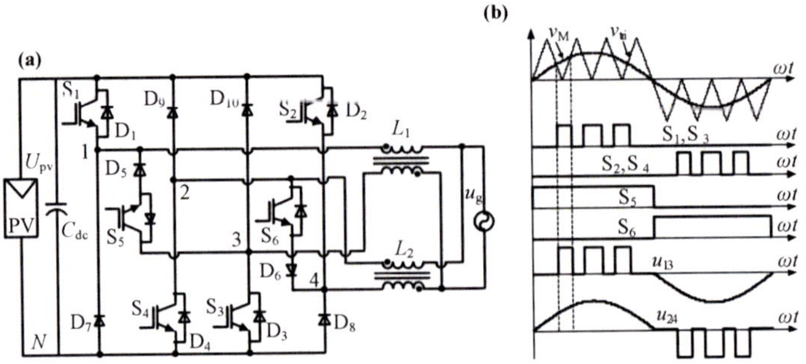

Fig. 3.65 High efficiency and reliability transformerless inverter, **a** main circuit, **b** modulation strategy

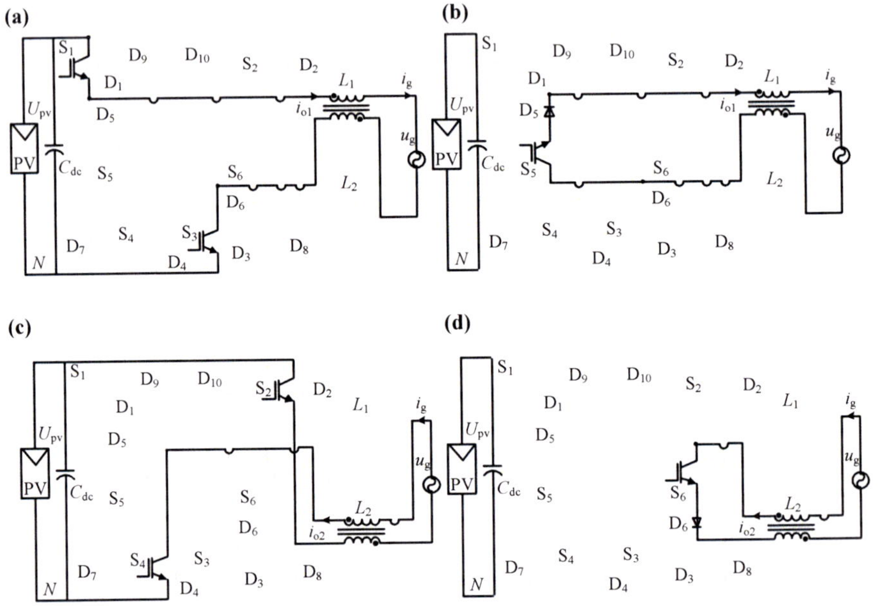

Fig. 3.66 Equivalent circuits of operating modes of HERTLI, **a** mode 1, **b** mode 2, **c** mode 3, **d** mode 4

Figure 3.66a, b belongs to the positive grid cycle, while Fig. 3.66c, d belongs to the negative grid cycle.

Mode 1: Refer to Fig. 3.66a; in this stage, the energy transfers from the PV side to the grid through S_1, S_3, and L_1. In this mode, current in primary winding of L_1 is i_{o1}.

Mode 2: Refer to Fig. 3.66b; only S_5 is conducting from Fig. 3.65b. In this stage, the freewheeling current flows through S_5, D_5, and L_1. The freewheeling circuit

Table 3.14 DMV and CMV levels of different operating modes in HERTLI

Modes	u_{1N}	u_{2N}	u_{3N}	u_{4N}	u_{DM}	u_{CM}
(1)	U_{pv}	–	0	–	U_{pv}	$U_{pv}/2$
(2)	$U_{pv}/2$	–	$U_{pv}/2$	–	0	Uncertain
(3)	–	0	–	U_{pv}	$-U_{pv}$	$U_{pv}/2$
(4)	–	$U_{pv}/2$	–	$U_{pv}/2$	0	Uncertain

is disconnected with PV side completely and the LC loop is cut off compared to conventional full-bridge TLI.

Mode 3: Refer to Fig. 3.66c; in this stage, the energy transfers from the PV side to the grid through S_2, S_4, and L_2. In this mode, current in primary winding of L_2 is i_{o2}.

Mode 4: Refer to Fig. 3.66d; only S_6 is conducting and S_2 and S_4 are closed from Fig. 3.65b. In this stage, the freewheeling current flows through S_6, D_6, and L_2. The LC loop is also cut off.

Based on these modes, the potential of points 1, 2, 3, and 4 against N as well as u_{DM} and u_{CM} is shown in Table 3.14. Because the current flows through different ways in positive and negative period, the calculating methods are also different, as shown in Table 3.14. From this table, CMV of HERTLI is uncertain because the freewheeling path is just disconnected from PV side in freewheeling modes. LC will appear in this situation based on the first principle in Chap. 2.

Table 3.15 shows the number of components of H5, H6-I, and HERTLI in power transmission and freewheeling stages. From this table, the number of conducting components is the least in HERTLI. In [17], the authors give the total power loss of H5, H6-I, and HERTLI at 5 kW and 20 kHz, as shown in Table 3.16. From this table, the total power loss of HERTLI is the least in different output power.

In conclusion, shooting through is avoided by power switches S_1–S_6 series with power diodes in HERTLI. The operating reliability is improved. Two coupled inductors are used in HERTLI. Reverse recovery problems of power diodes are alleviated. The dead time for current crossing zero is avoided and distortion is reduced. But, the design of coupled inductors is complicated and thus cost is improved. CMV of HERTLI is still uncertain. Besides, the power loss is almost equal to it of HERIC, but the utilization of the components of HERIC is higher. From this point, HERIC is more popular in industry.

Table 3.15 Device numbers of conducting durations in different TLIs

Operating stages	Switching components	H5	H6-I	HERTLI
Power transmission stage	Power switches	3	3	2
	Power diodes	0	0	0
Freewheeling stage	Power switches	1	1	1
	Power diodes	1	1	1

Table 3.16 Total power losses in different TLIs

$P_o(\%)$	H5(W)	H6-I(W)	HERTLI(W)
100	76.99	68.65	53.56
75	41.24	43.04	34.55
50	30.67	23.89	20.12
30	17.92	13.23	11.87
20	12.79	9.45	8.85
10	8.48	7.80	6.56

Table 3.17 Simulation parameters of HERTLI

Parameters	Value	Parameters	Value
Input voltage U_{pv}/V	380	Switching frequency f_s/kHz	20
Grid voltage u_g/V	240	Filter inductors L_1, L_2/mH	0.95
Grid frequency/Hz	50	Filter capacitor C/μF	2.2
Rated power P/W	3500	PV parasitic capacitor C_{pv}/nF	300

2. Simulations results

A simulation model is built by Matlab/Simulink to verify the performance analysis aforementioned, and the key parameters of HERTLI are shown in Table 3.17.

The simulation waveforms of u_g, grid-in current reference i_{ref}, and grid-in current i_g are shown in Fig. 3.67. The simulation waveforms of DMV u_{13} and u_{24} are shown in Fig. 3.68a, b, respectively. And u_{13} and u_{24} are three-level voltage in the positive and negative grid cycle, respectively. The simulation waveforms of inductors current i_{o1} and i_{o2} are shown in Fig. 3.69a, b, respectively. It is clear that L_1 and L_2 are working

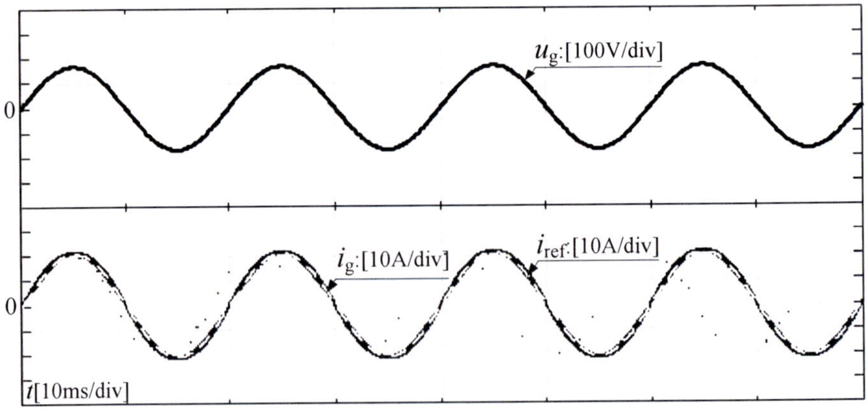

Fig. 3.67 Simulation waveforms of u_g, i_{ref}, and i_g of HERTLI

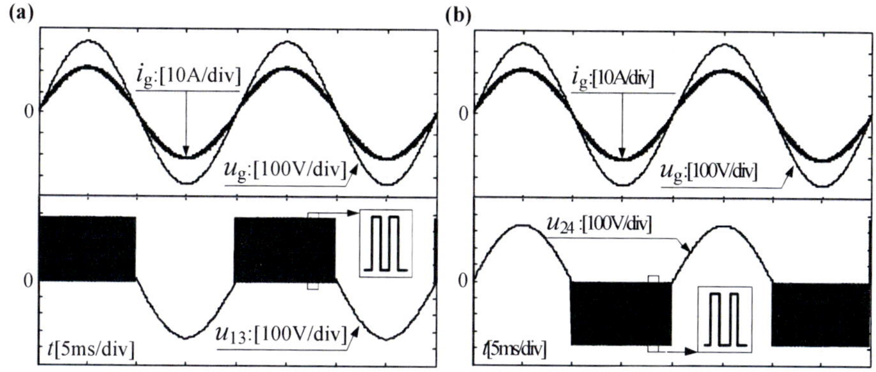

Fig. 3.68 DMV waveforms of HERTLI, **a** u_{13}, **b** u_{24}

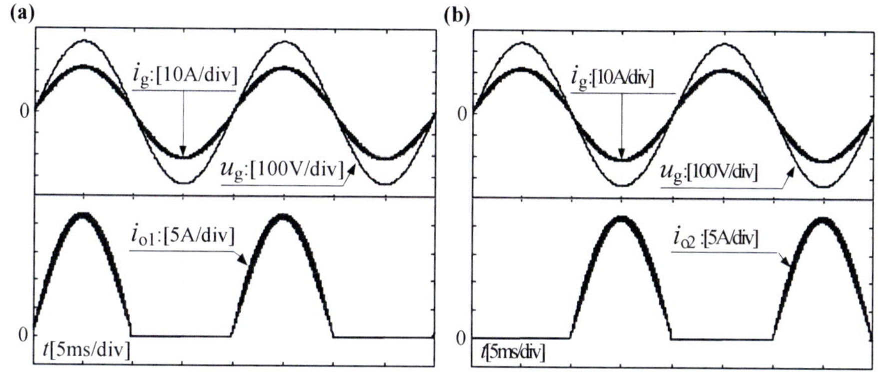

Fig. 3.69 Inductors current waveforms of HERTLI, **a** primary current of L_1 i_{o1}, **b** primary current of L_2 i_{o2}

in the positive and negative grid cycle, respectively. The simulation waveforms of LC i_{CM} in time domain and frequency domain are shown in Fig. 3.70a, b, respectively. From Fig. 3.70, the LC is about 11 mA at switching frequency. It meets the standard and similar to HERIC.

Some of the TLIs above have been used in distributed PV generation. But, LC still exists in decoupling type TLIs from the experimental result in the field. The reason is that the potential of freewheeling circuit is oscillating because of PV parasitic capacitors. CMV is uncertain during freewheeling stages. From the first principle of Sect. 2.2.2, CMV should be clamped at half of the DC input to keep it constant [20, 21]. In [21], the CMV clamping conception is first proposed. New clamping cells are presented. And novel TLI based on H5 is also presented with its modulation strategy. From then on, researchers have done much research on clamping circuits and utilize it in the TLIs introduced above [21–30].

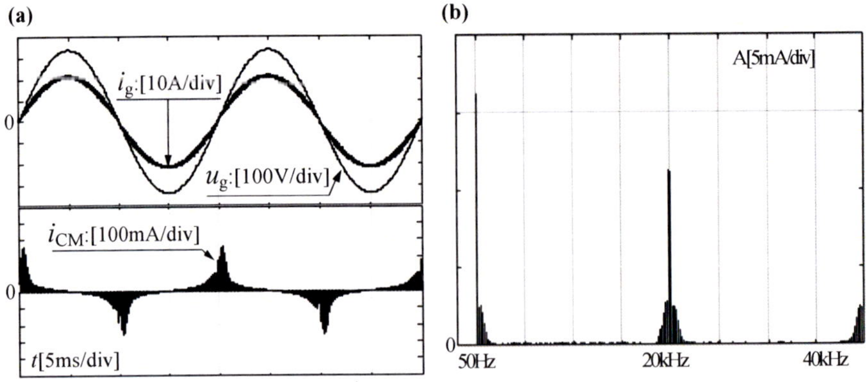

Fig. 3.70 LC waveforms of HERTLI, **a** time domain, **b** frequency domain

3.3 Full-Bridge TLIs with Clamping Branches

3.3.1 Clamping Circuits

The freewheeling circuit is the key structure in decoupling type full-bridge TLIs. The AC decoupling structures are shown in Fig. 3.71 [20]. Two power switches can be in series or parallel together for flowing current bidirectional in Fig. 3.71a–c. The structure can also be replaced by one power switches combined with a rectifier bridge in Fig. 3.71d.

Decoupling circuits should be combined with clamping structure to keep CMV constant. The clamping structure can be classified into unidirectional and bidirectional clamping with power switches or diodes. Unidirectional clamping can only clamp CMV at constant when CMV is decreasing or increasing, while bidirectional clamping can make CMV constant whenever the potential decreases or increases. Only one diode is included in a unidirectional clamping structure because power switches are always parallel with a diode. The clamping structure with only one diode is shown in Fig. 3.72 based on Fig. 3.71.

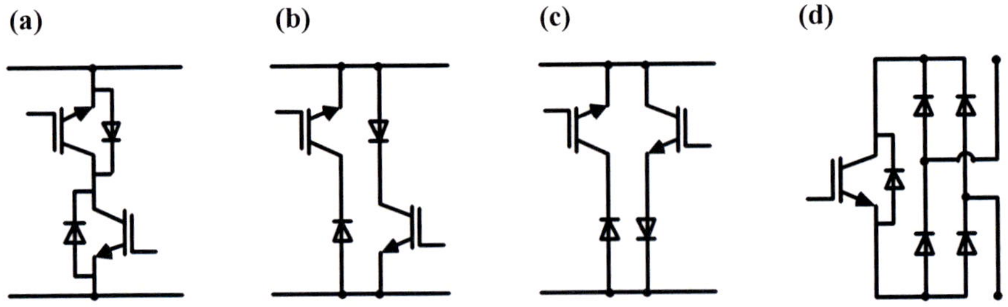

Fig. 3.71 Freewheeling circuits, **a** structure 1, **b** structure 2, **c** structure 3, **d** structure 4

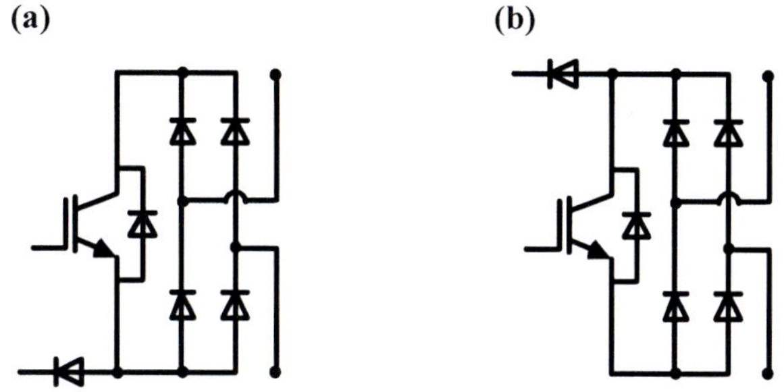

Fig. 3.72 Single direction clamping circuits with single power switch, **a** structure 1, **b** structure 2

Furthermore, bidirectional clamping with one power switch structure is got by replacing the diode with a power switch in Fig. 3.72, as shown in Fig. 3.73c–e. Bidirectional clamping with two power switches structure is shown in Fig. 3.74a. The two power switches can also be connected back to back in Fig. 3.74b to form a bidirectional clamping structure. Those components can be rearranged and replaced by four power switches, as shown in Fig. 3.74c, d.

The clamping structures above can also be used in DC side decoupling type TLIs. Clamping type full-bridge TLIs will be discussed in next section.

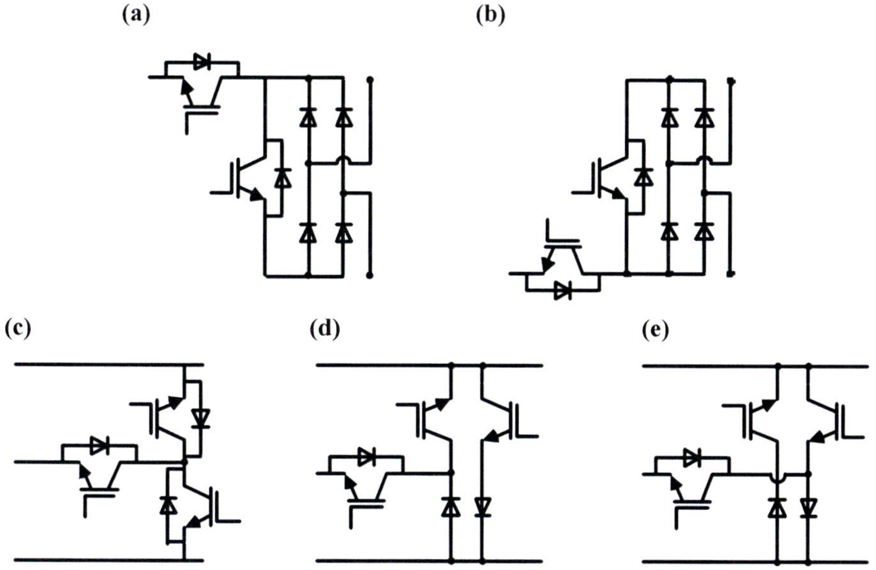

Fig. 3.73 Bidirectional clamping circuits with single power switch, **a** structure 1, **b** structure 2, **c** structure 3, **d** structure 4, **e** structure 5

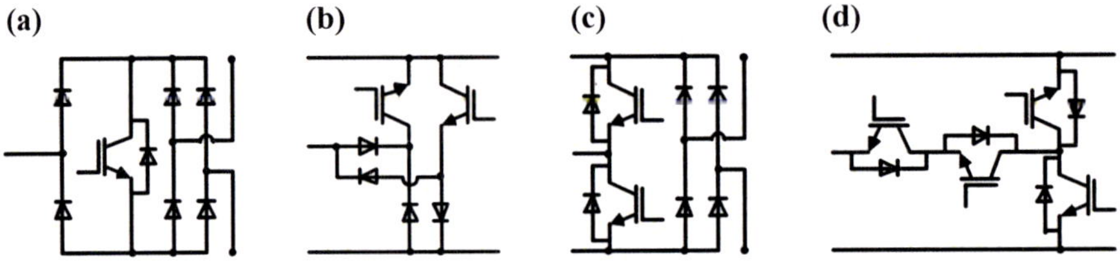

Fig. 3.74 Bidirectional clamping circuits with two power switches, **a** structure 1, **b** structure 2, **c** structure 3, **d** structure 4

3.3.2 Optimized H5 Transformerless Inverter

1. Operating principle

The optimized H5 transformerless inverter (oH5) was proposed in [21], the main circuit of it is shown in Fig. 3.75a. A new power switch S_6 is added to connect the DC bus and the neutral point of DC input based on H5. The oH5 is clamping type TLI but also a DC side decoupling circuit. The modulation strategy is shown in Fig. 3.75b. S_5 and S_6 work in high frequency. S_1–S_4 work in fundamental or high frequency depending on the grid voltage. The output voltage is a three-level voltage and maintains high efficiency as H5. In unity power factor, there are four operating modes based on the directions of grid voltage and grid-in current. The modes are similar to H5, as shown in Fig. 3.19. Dead time should be considered in the driving logic between S_5 and S_6 for avoiding short circuit.

The operating principles of the clamping circuit are shown in Fig. 3.76. The clamping mode when freewheeling circuit potential rises is shown in Fig. 3.76a, while the clamping mode when freewheeling circuit potential decreases is shown in Fig. 3.76b. In freewheeling stages, LC will flow into DC side through S_6 if its potential rises and it will flow from DC side through D_6 if its potential decreases. Thus, CMV is clamped at half of DC input voltage. It is pointed out that during dead

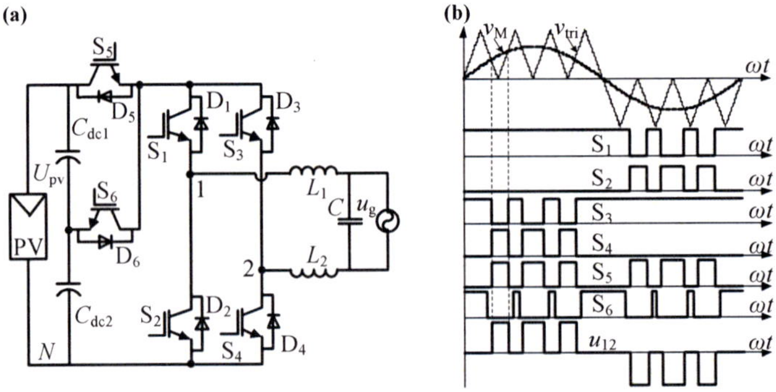

Fig. 3.75 Optimized H5 transformerless inverter, **a** main circuit, **b** modulation strategy

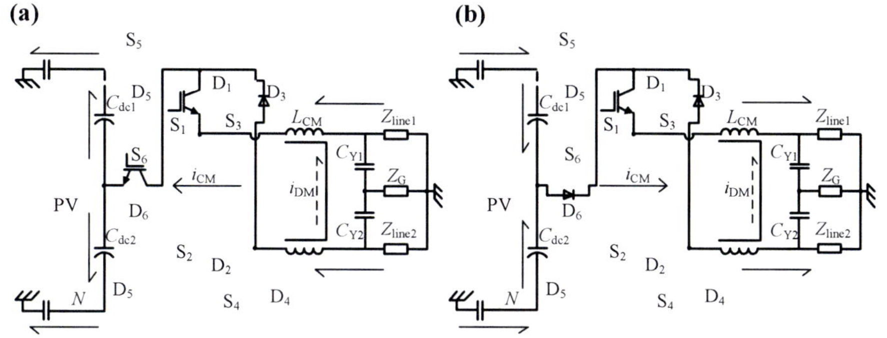

Fig. 3.76 Equivalent circuits when clamping circuit conducting in oH5, **a** potential rises, **b** potential decreases

Table 3.18 DMV and CMV levels of different operating modes in oH5

Modes	u_{1N}	u_{2N}	u_{DM}	u_{CM}
(1)	U_{pv}	0	U_{pv}	$U_{pv}/2$
(2)	$U_{pv}/2$	$U_{pv}/2$	0	$U_{pv}/2$ (uncertain during dead time)
(3)	0	U_{pv}	$-U_{pv}$	$U_{pv}/2$
(4)	$U_{pv}/2$	$U_{pv}/2$	0	$U_{pv}/2$ (uncertain during dead time)

time between S_5 and S_6, CMV can be clamped only when the potential decreases. This is common in the active clamping CMV TLIs.

Based on these modes, the potential of points 1 and 2 against N as well as u_{DM} and u_{CM} is shown in Table 3.18. From this table, it is clear that DMV maintains unipolar SPWM performance. CMV of oH5 is constant except for it during dead time. CMV performance is improved.

In conclusion, one power switch is introduced at DC side in oH5. The PV side and grid side are disconnected during freewheeling stages in oH5. LC will decrease due to the disconnected CMV loop. And CMV is clamped at constant during the whole period except for dead time. The current flows through three power switches in power transmission stages and through two power switches in freewheeling stages, which is the same as H5.

2. Simulation results

A simulation model is built by Matlab/Simulink to verify the performance analysis aforementioned, and the key parameters are shown in Table 3.4.

The simulation waveforms of u_g, grid-in current reference i_{ref}, and grid-in current i_g are shown in Fig. 3.77. The simulation waveforms of DMV u_{DM} in time domain and frequency domain are shown in Fig. 3.78a, b, respectively. u_{DM} is a three-level voltage and contains little harmonics. The simulation waveforms of CMV u_{CM} in time domain and frequency domain are shown in Fig. 3.79a, b, respectively. The amplitude of high-frequency pulses is higher than $U_{pv}/2$, while the amplitude of fundamental

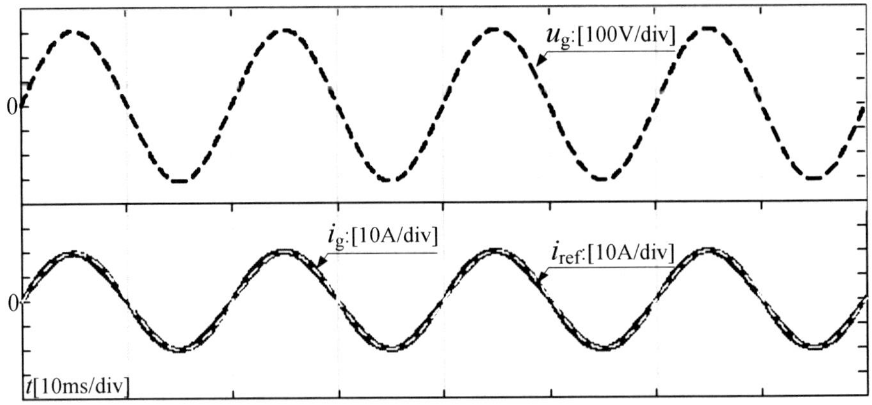

Fig. 3.77 Simulation waveforms of u_g, i_{ref}, and i_g of oH5

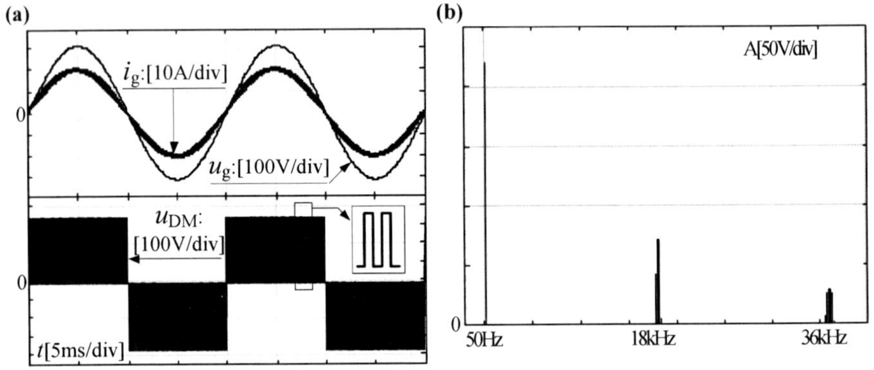

Fig. 3.78 DMV waveforms of oH5, **a** time domain, **b** frequency domain

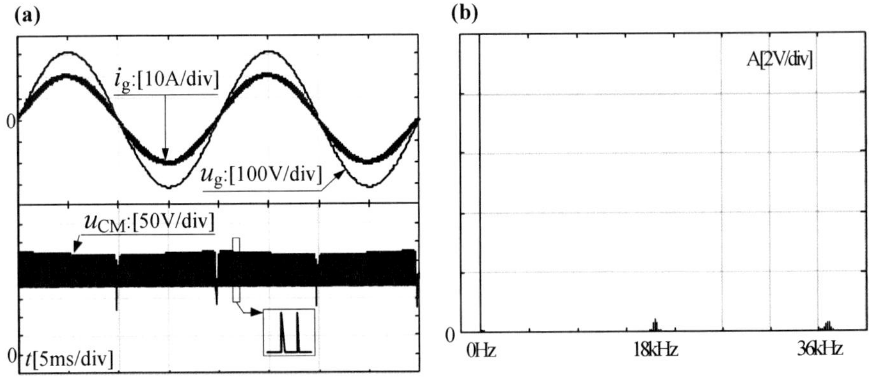

Fig. 3.79 CMV waveforms of oH5, **a** time domain, **b** frequency domain

frequency pulses is lower than $U_{pv}/2$. The former phenomenon is because CMV is uncertain during dead time between power switches, and the latter phenomenon is due to variable CMV during the time when current crossing zero, as shown in Fig. 3.76.

The simulation waveforms of LC i_{CM} in time domain and frequency domain are shown in Fig. 3.80a, b, respectively. From Fig. 3.80, the LC is about 4 mA at switching frequency. LC is reduced by a wide margin because high-frequency components in CMV are suppressed.

3. Experimental results

In order to verify the theoretical analysis and simulation waveforms, a prototype of oH5 is built, and the key parameters are shown in Table 3.19.

The grid voltage u_g, grid current i_g, and inverter output voltage u_{DM} (it is also the differential-mode voltage) from top to bottom are shown in Fig. 3.81a. In this

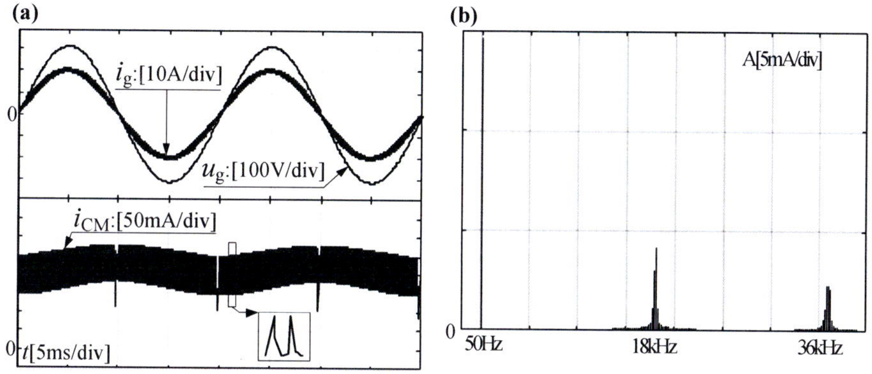

Fig. 3.80 LC waveforms of oH5, **a** time domain, **b** frequency domain

Table 3.19 Key parameters of oH5 prototype

Parameters	Value	Parameters	Value
Input voltage U_{pv}/V	340–700	Filter inductors L_1, L_2/mH	4
Grid voltage u_g/V	240	Filter capacitor C/μF	6.6
Grid frequency/Hz	50	Common-mode inductor L_{CM}	2 × 2 W-43615-TC
Rated power P/W	1000	Wire/mm	Wire: 2 mm
Switching frequency f_s/kHz	20	Turns	Turns: 10 + 10
DC side capacitors C_{dc1}, C_{dc2}/μF	470 μF/400 V	Common-mode capacitor C_{Y1}, C_{Y2}/nF	2.2
MOSFET S_1–S_6	IXFN36N100	PV parasitic capacitor C_{pv1}, C_{pv2}/μF	0.1

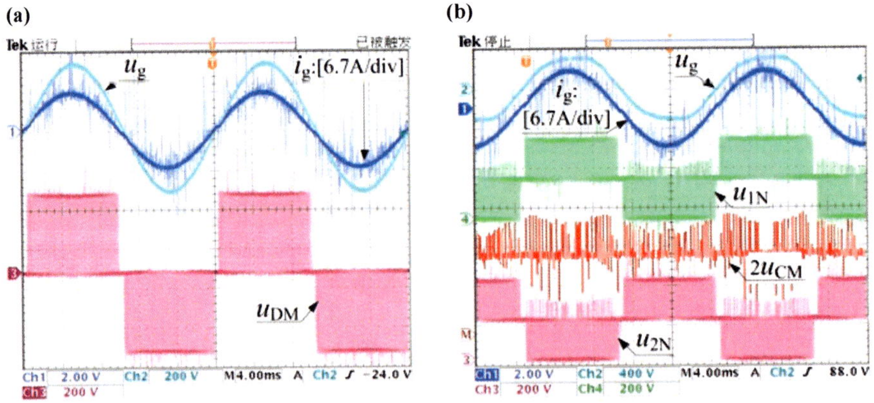

Fig. 3.81 DMV and CMV waveforms of oH5, **a** DMV, **b** CMV

figure, the output voltage u_{DM} has three levels as U_{pv}, 0, and $-U_{pv}$, so a good differential-mode characteristic has been achieved like the unipolar SPWM full-bridge grid-connected inverter with galvanic isolation. The grid voltage u_g, grid current i_g, voltage u_{1N} of midpoint 1 to terminal N, common-mode voltage $u_{CM} = u_{1N} + u_{2N}$, and voltage u_{2N} of middle point 2 to terminal N are shown in Fig. 3.81b from top to down, respectively.

Both of the detailed common-mode voltage waveforms at the peak and the value of grid current are shown in Fig. 3.82a, b, respectively. It can be seen that the narrow pulse width in the u_{1N} turning to zero ring of Fig. 3.82a corresponds to the dead time between the switches S_5 and S_6. In this span, the freewheeling path potential rises because the clamp switch S_5 is still off. When S_6 is on, the freewheeling path potential will be quickly clamped to the midpoint potential of the capacitor divider. The experimental results are in agreement with the theoretical analysis well.

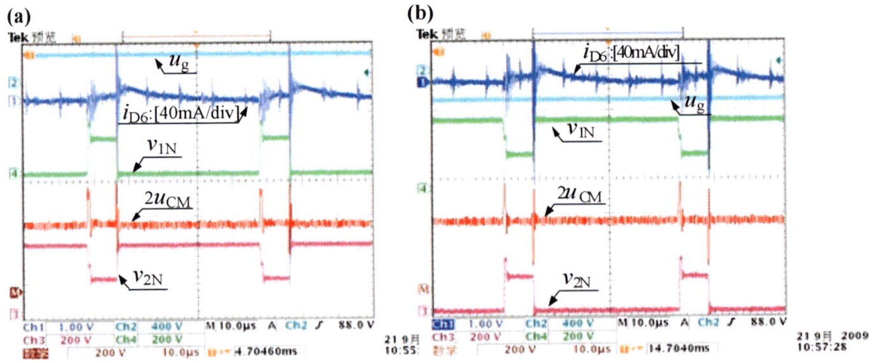

Fig. 3.82 Detailed CMV waveforms of oH5, **a** at the peak of grid current, **b** at the vale of grid current

3.3.3 Diode Clamping H5 Inverter

1. Operating principle

Another active power switch is needed in oH5 inverter. And dead time should be considered between the switch in positive DC bus and clamping switch. On the contrary, diode clamping H5 (H5-D) was proposed in [22], the main circuit of which is shown in Fig. 3.83a. A diode is needed instead of a power switch compared to oH5. This diode connects to the DC side neural point to clamp the potential naturally without dead time. H5-D is still DC side decoupling TLI. The modulation strategy is shown in Fig. 3.83b. S_1–S_4 work in fundamental or high frequency depending on the grid voltage. S_5 works in high frequency. The output voltage is a three-level voltage and H5-D maintains high efficiency as H5.

There are four operating modes based on the directions of grid voltage and grid-in current when H5-D works in unity power factor, as shown in Fig. 3.84. Figure 3.84a, b belongs to the positive grid cycle, while Fig. 3.84c, d belongs to the negative grid cycle.

Mode 1: Refer to Fig. 3.84a; in this stage, the energy transfers from the PV side to the grid through S_1, S_4, S_5, and the filter.

Mode 2: Refer to Fig. 3.84b, S_2–S_5 are all closed while S_1 is conducting from Fig. 3.83b. In this stage, the freewheeling current flows through S_4, D_2, and the filter. The freewheeling circuit is disconnected with PV side completely and the LC loop is cut off compared to conventional full-bridge TLI.

Mode 3: Refer to Fig. 3.84c; in this stage, the energy transfers from the PV side to the grid through S_2, S_3, S_5, and the filter.

Mode 4: Refer to Fig. 3.84d; only S_3 is conducting from Fig. 3.83b. In this stage, the freewheeling current flows through S_2, D_4, and the filter. The freewheeling circuit is disconnected with PV side and the LC loop is also cut off.

The working principles of clamping circuit are shown in Fig. 3.85. In freewheeling stages, LC will flow into DC side through D_c if its potential rises. Thus, CMV is clamped at half of DC input voltage. It is pointed out that CMV cannot be clamped

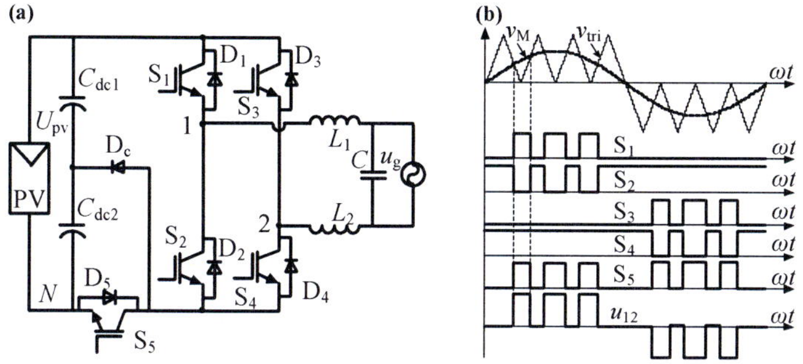

Fig. 3.83 Diode clamping H5 inverter, **a** main circuit, **b** modulation strategy

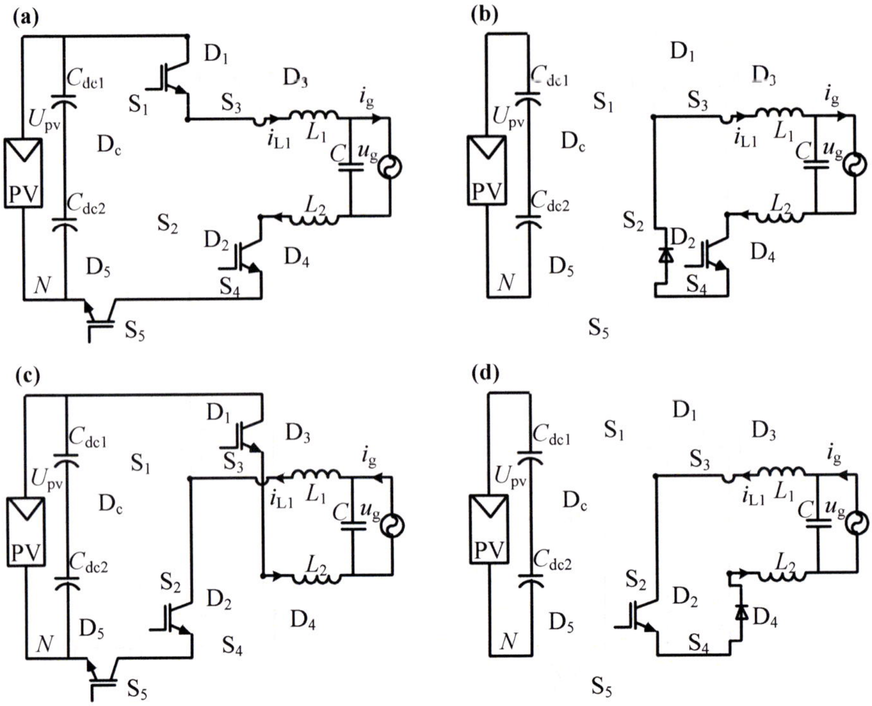

Fig. 3.84 Equivalent circuits of H5-D, **a** mode 1, **b** mode 2, **c** mode 3, **d** mode 4

Fig. 3.85 Equivalent circuit when clamping circuit conducting in H5-D

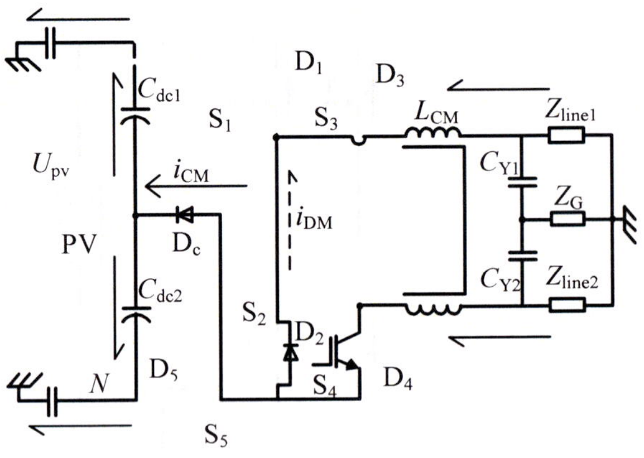

when freewheeling circuit potential decreases. Thus, H5-D is the unidirectional clamping TLI.

Based on these modes, the potential of points 1 and 2 against N as well as u_{DM} and u_{CM} is shown in Table 3.20. From this table, it is clear that DC side and AC side are disconnected in freewheeling stages. DMV maintains unipolar SPWM performance.

Table 3.20 DMV and CMV levels of different operating modes in H5-D

Modes	u_{1N}	u_{2N}	u_{DM}	u_{CM}
(1)	U_{pv}	0	U_{pv}	$U_{pv}/2$
(2)	$U_{pv}/2$	$U_{pv}/2$	0	$\leq U_{pv}/2$
(3)	0	U_{pv}	$-U_{pv}$	$U_{pv}/2$
(4)	$U_{pv}/2$	$U_{pv}/2$	0	$\leq U_{pv}/2$

The uncertain state of CMV is suppressed partly. CMV of H5-D is smaller than $U_{pv}/2$.

In conclusion, one diode is introduced at DC side in H5-D. The PV side and grid side are disconnected during freewheeling stages in oH5. CMV is clamped by the diode. LC will decrease a lot. The current flows through three power switches in power transmission stages and through two power switches in freewheeling stages, which is the same as H5. But, it can only clamp the CMV at $U_{pv}/2$ when the potential of freewheeling circuit increases.

2. Simulations results

A simulation model is built by Matlab/Simulink to verify the performance analysis aforementioned, and the key parameters of H5-D are shown in Table 3.4. The simulation waveforms of u_g, grid-in current reference i_{ref}, and grid-in current i_g are shown in Fig. 3.86.

The simulation waveforms of DMV u_{DM} in time domain and frequency domain are shown in Fig. 3.87a, b, respectively. u_{DM} is a three-level voltage and the RMS of it is about 75 V at switching frequency.

The simulation waveforms of CMV u_{CM} in time domain and frequency domain are shown in Fig. 3.88a, b, respectively. u_{CM} only contains the components whose amplitude is smaller than $U_{pv}/2$, as discussed in Fig. 3.85.

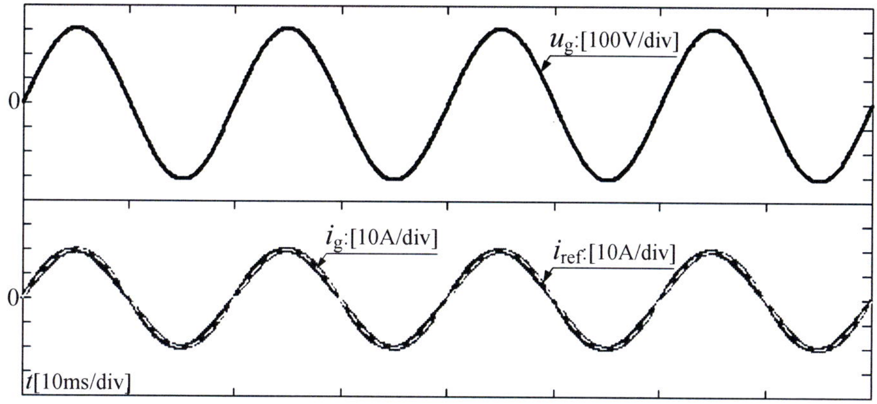

Fig. 3.86 Simulation waveforms of u_g, i_{ref}, and i_g of H5-D

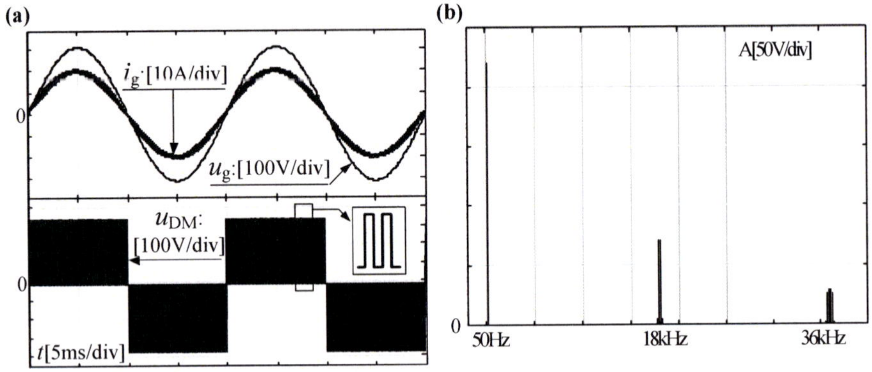

Fig. 3.87 DMV waveforms of H5-D, **a** time domain, **b** frequency domain

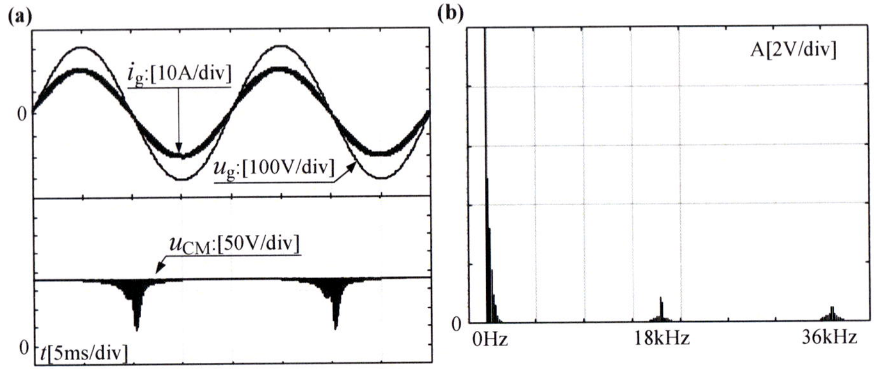

Fig. 3.88 CMV waveforms of H5-D, **a** time domain, **b** frequency domain

The simulation waveforms of LC i_{CM} in time domain and frequency domain are shown in Fig. 3.89a, b, respectively. From Fig. 3.89, the LC is about 8 mA at switching frequency. It is greatly reduced compared to H5 because components whose amplitude higher than $U_{pv}/2$ are removed.

3.3.4 Full-Bridge DC Bypassed Inverter

1. Operating principle

Full-bridge DC bypassed inverter (FB-DCBP) was proposed in [23], the main circuit of which is shown in Fig. 3.90a. It is the diode clamping TLI and can also be called a H6-D. The H6-D is a DC side decoupling TLI and has been commercially used in Inecon Sun TL series (2.5/3.3/6 kW) by Ingeteam Inc. Its European efficiency is

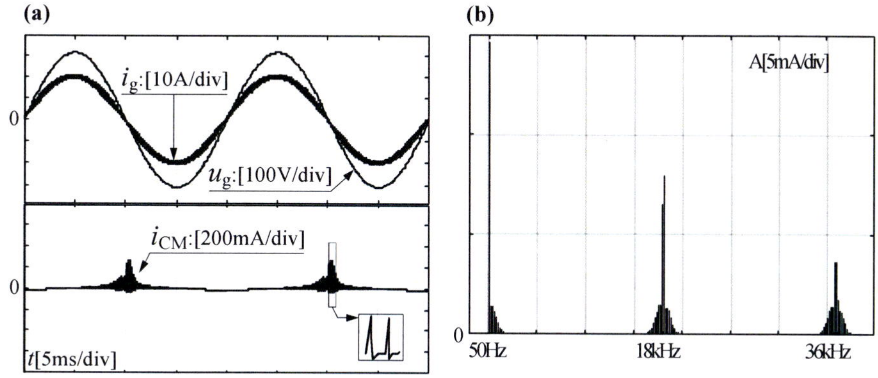

Fig. 3.89 LC waveforms of H5-D, **a** time domain, **b** frequency domain

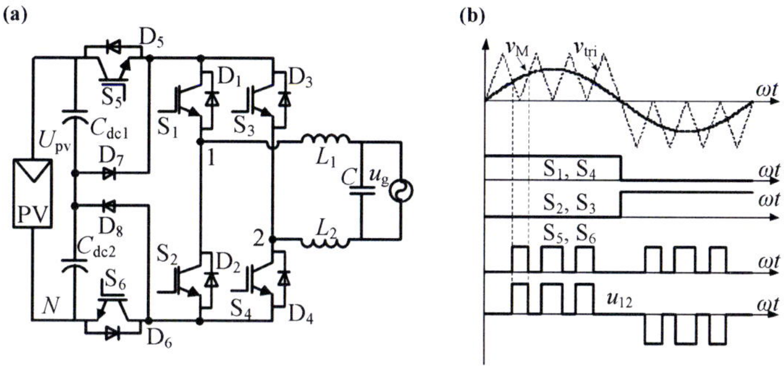

Fig. 3.90 Full-bridge DC bypassed inverter, **a** main circuit, **b** modulation strategy

96.5% as Photon International reported in August, 2007. The modulation strategy is shown in Fig. 3.90b. S_5–S_6 work in high frequency and S_1–S_4 work in fundamental frequency. The output voltage is a three-level voltage.

There are four operating modes based on the directions of grid voltage and grid-in current when H6-D works in unity power factor, as shown in Fig. 3.91. Figure 3.91a, b belongs to the positive grid cycle, while Fig. 3.91c, d belongs to the negative grid cycle.

Mode 1: Refer to Fig. 3.91a; in this stage, the energy transfers from the PV side to the grid through S_1, S_4, S_5, S_6, and the filter.

Mode 2: Refer to Fig. 3.91b; S_1 and S_4 are conducting from Fig. 3.90b. When S_5 and S_6 are turned off, the current splits into two paths: S_1 and the freewheeling diode D_3, and S_4 and the freewheeling diode D_2. Freewheeling circuit is disconnected with PV side and the LC loop is broken.

Mode 3: Refer to Fig. 3.91c; in this stage, the energy transfers from the PV side to the grid through S_2, S_3, S_5, S_6, and the filter.

Mode 4: Refer to Fig. 3.91d; S_2 and S_3 are conducting from Fig. 3.90b. When S_5 and S_6 are turned off, the current splits into two paths: S_3 and the freewheeling diode

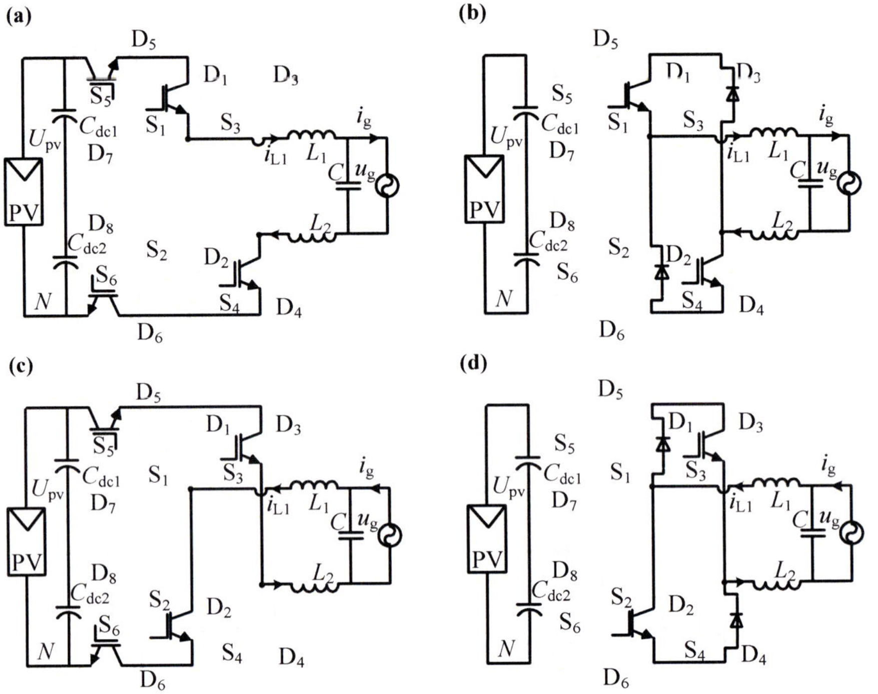

Fig. 3.91 Equivalent circuits of operating modes of H6-D, **a** mode 1, **b** mode 2, **c** mode 3, **d** mode 4

D_1, and S_2 and the freewheeling diode D_4. Freewheeling circuit is disconnected with PV side and the same effect is gained just like Mode 2.

The working principles of clamping circuit during the positive grid cycle are shown in Fig. 3.92. The clamping mode when freewheeling circuit potential rises is shown

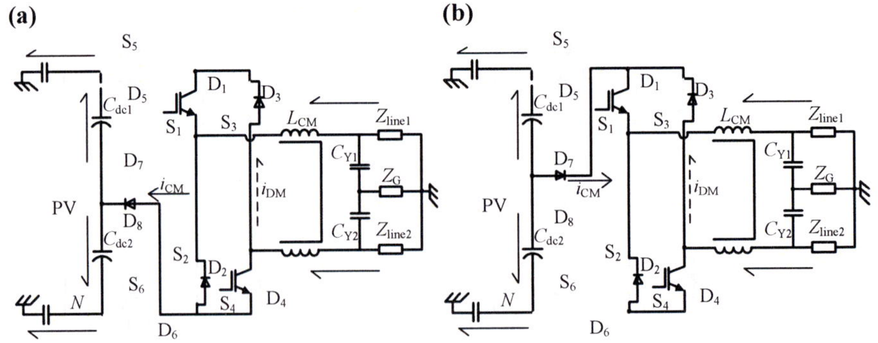

Fig. 3.92 Equivalent circuits when clamping circuit conducting in H6-D, **a** potential rises, **b** potential decreases

Table 3.21 DMV and CMV levels of different operating modes in H6-D

Modes	u_{1N}	u_{2N}	u_{DM}	u_{CM}
(1)	U_{pv}	0	U_{pv}	$U_{pv}/2$
(2)	$U_{pv}/2$	$U_{pv}/2$	0	$U_{pv}/2$
(3)	0	U_{pv}	$-U_{pv}$	$U_{pv}/2$
(4)	$U_{pv}/2$	$U_{pv}/2$	0	$U_{pv}/2$

in Fig. 3.92a. The clamping mode when freewheeling circuit potential decreases is shown in Fig. 3.92b. In freewheeling stages, LC will flow into DC side through D_8 if its potential rises and it will flow from DC side through D_7 if its potential decreases. Thus, CMV is clamped at half of DC input voltage.

Based on these modes, the potential of points 1 and 2 against N as well as u_{DM} and u_{CM} is shown in Table 3.21. From this table, it is clear that DC side and AC side are disconnected in freewheeling stages. DMV maintains unipolar SPWM performance. CMV of H6-D is constant. Its common-mode performance is well.

In conclusion, two power switches S_5 and S_6 are introduced at positive and negative DC buses in H6-D, respectively. The PV and grid are disconnected during freewheeling stages in H6-D. LC is greatly decreased because CMV is clamped at $U_{pv}/2$ by D_7 and D_8. The current flows through four power switches during energy transmission stage, which has negative influence on efficiency.

2. Simulation results

A simulation model is built by Matlab/Simulink to verify the performance analysis aforementioned, and the key parameters of H6-D are shown in Table 3.4.

The simulation waveforms of u_g, grid-in current reference i_{ref}, and grid-in current i_g are shown in Fig. 3.93.

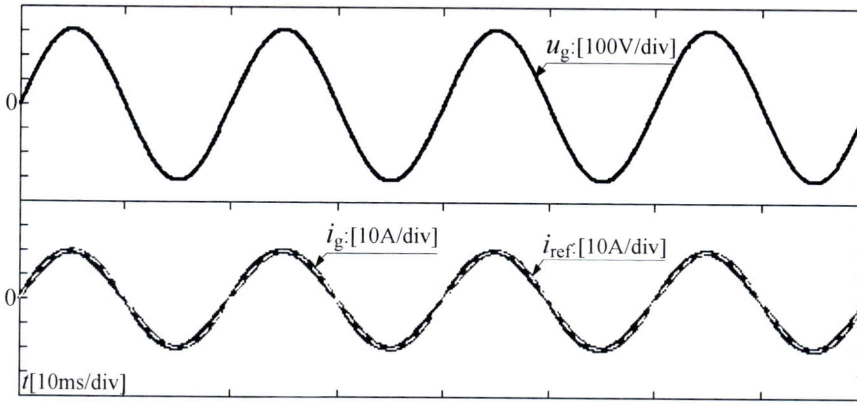

Fig. 3.93 Simulation waveforms of u_g, i_{ref}, and i_g of H6-D

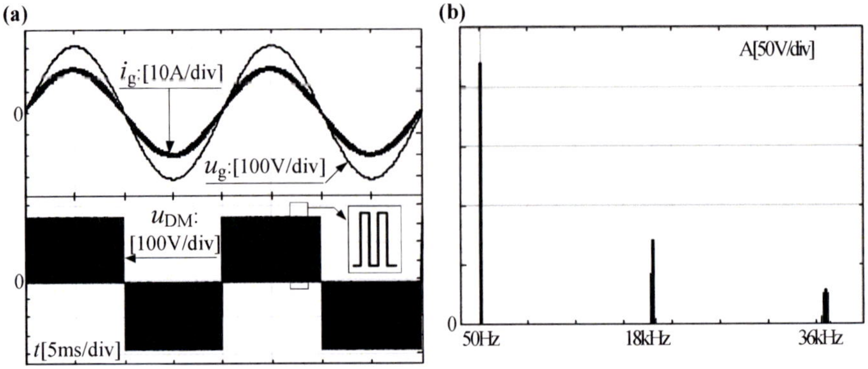

Fig. 3.94 DMV waveforms of H6-D, **a** time domain, **b** frequency domain

The simulation waveforms of DMV u_{DM} in time domain and frequency domain are shown in Fig. 3.94a, b, respectively. u_{DM} is a three-level voltage and the RMS of it is about 75 V at switching frequency.

The simulation waveforms of CMV u_{CM} in time domain and frequency domain are shown in Fig. 3.95a, b, respectively. No high-frequency components are contained in u_{CM} like the discussion in Fig. 3.92.

The simulation waveforms of LC i_{CM} in time domain and frequency domain are shown in Fig. 3.96a, b, respectively. From Fig. 3.96, the LC is about 0.6 mA at switching frequency. It is largely reduced compared to H6-V.

3. Experimental results

In order to verify the theoretical analysis and simulation waveforms, a prototype of H6-D is built, and the key parameters are shown in Table 3.22.

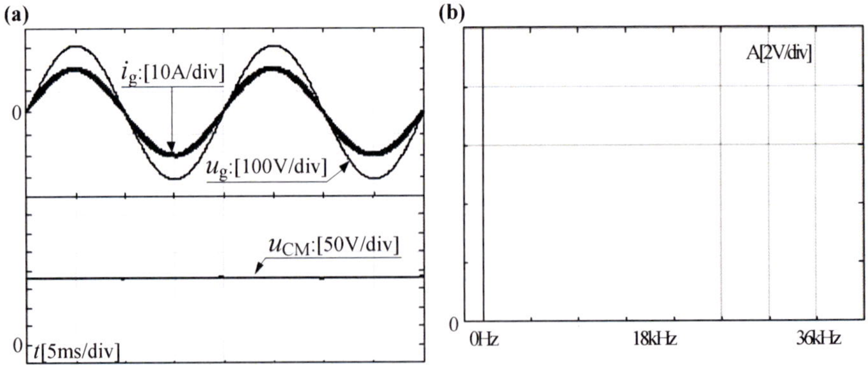

Fig. 3.95 CMV waveforms of H6-D, **a** time domain, **b** frequency domain

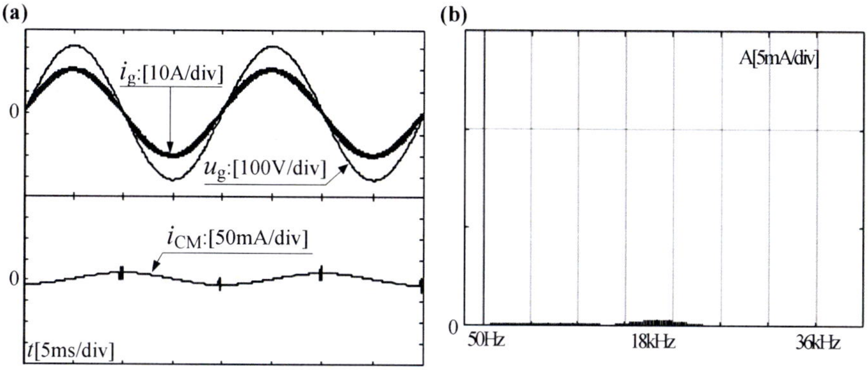

Fig. 3.96 LC waveforms of H6-D, **a** time domain, **b** frequency domain

Table 3.22 Key parameters of H6-D prototype

Parameters	Value	Parameters	Value
Input voltage U_{pv}/V	340–700	Filter inductors L_1, L_2/mH	4
Grid voltage u_g/V	240	Filter capacitor C/μF	6.6
Grid frequency/Hz	50	Common-mode inductor L_{CM}	2 × 2 W-43615-TC
Rated power P/W	1000	Wire/mm	Wire: 2 mm
Switching frequency f_s/kHz	20	Turns	Turns: 10 + 10
DC side capacitors C_{dc1}, C_{dc2}/μF	470 μF/400 V	Common-mode capacitor C_{Y1}, C_{Y2}/nF	2.2
MOSFET S_1–S_6	IXFN36N100	PV parasitic capacitor C_{pv1}, C_{pv2}/μF	0.1

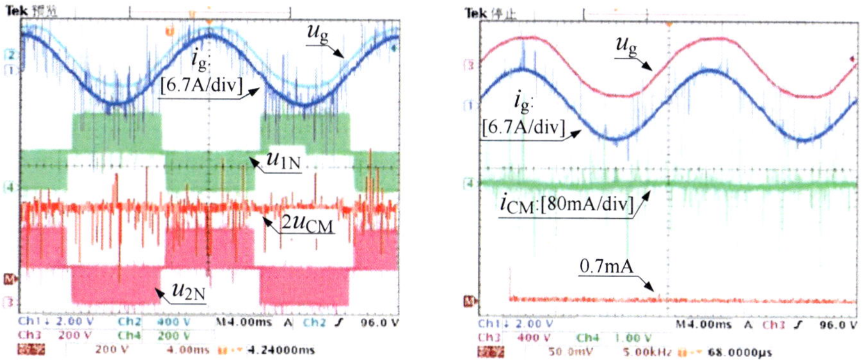

Fig. 3.97 DMV and CMV waveforms of H6-D in grid scale, **a** CMV, **b** LC

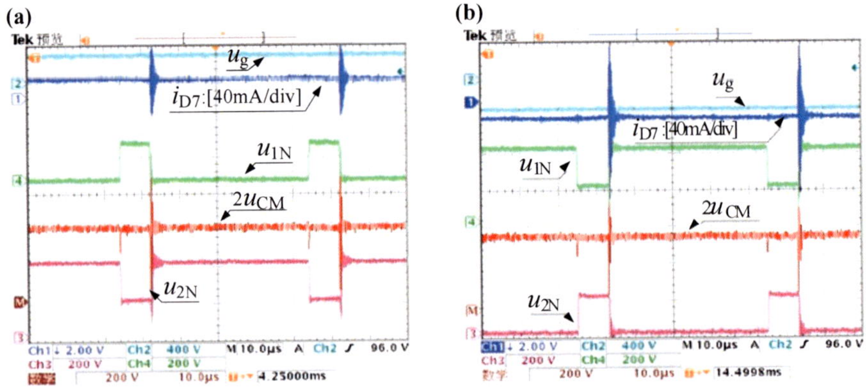

Fig. 3.98 Detailed CMV waveforms of H6-D, **a** at the peak of grid current, **b** at the vale of grid current

The experimental waveforms of u_g, i_g, u_{CM}, and i_{CM} are shown in Fig. 3.97a, b. The grid voltage u_g, grid current i_g, voltage u_{1N} of midpoint 1 to terminal N, common-mode voltage $u_{CM} = u_{1N} + u_{2N}$, and voltage u_{2N} of middle point 2 to terminal N are shown in Fig. 3.98a, b, from top to down, respectively. Both of the detailed common-mode voltage waveforms are also shown in Fig. 3.98a, b, respectively. It can be seen that u_{CM} is constant and LC is about 0.7 mA. The experimental results are in agreement with the theoretical analysis as well.

3.3.5 Active Clamping H6 Inverter I

1. Operating principle

The active clamping H6 inverter I (AH6-I) was proposed in [24], whose main circuit is shown in Fig. 3.99a. AH6-I is the clamping TLI based on TLI proposed in [10].

A new power switch S_7 is added to connect the DC bus and the neutral point of DC input based on TLI proposed in [10]. The AH6-I is an AC side decoupling TLI. The modulation strategy is shown in Fig. 3.99b. S_7 works in high frequency. S_1–S_6 work in fundamental or high frequency depending on the grid voltage. The output voltage is a three-level voltage.

There are six operating modes based on the directions of grid voltage and grid-in current when AH6-I works in unity power factor, as shown in Fig. 3.100. Figure 3.100a–c belongs to the positive grid cycle, while Fig. 3.100d–f belongs to the negative grid cycle.

Mode 1: Refer to Fig. 3.100a; in this stage, the energy transfers from the PV side to the grid through S_2, S_3, and the filter.

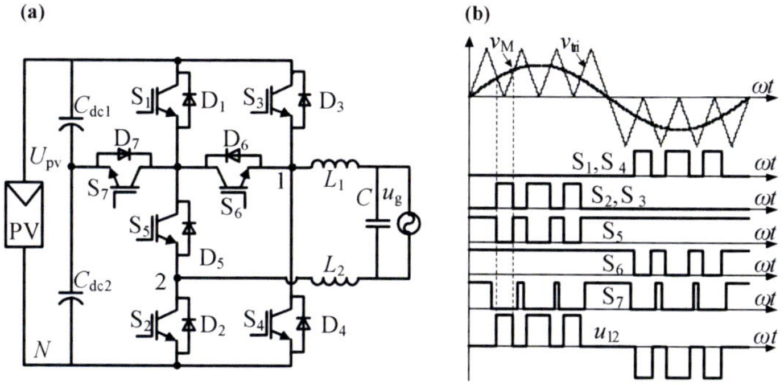

Fig. 3.99 Active clamping H6 inverter I, **a** main circuit, **b** modulation strategy

Mode 2: Refer to Fig. 3.100b; S_5–S_6 are conducting, while S_7 is still closed. In this stage, the freewheeling current flows through S_6, D_5, and the filter. In this mode, the potential of freewheeling circuit is not clamped by power switches or diodes, and narrow pulses may appear in CMV.

Mode 3: Refer to Fig. 3.100c; S_5–S_7 are conducting from Fig. 3.99b. In this stage, the freewheeling current flows through S_6, D_5, and the filter. The freewheeling circuit is disconnected with PV side and the LC loop is open compared to conventional full-bridge TLI.

Mode 4: Refer to Fig. 3.100d; in this stage, the energy transfers from the PV side to the grid through S_1, S_4, S_5, and the filter.

Mode 5: Refer to Fig. 3.100e; S_5 is conducting while S_7 is still closed. In this stage, the freewheeling current flows through S_5, D_6, and the filter. In this mode, the potential of freewheeling circuit is not clamped by power switches or diodes, and narrow pulses may appear in CMV.

Mode 6: Refer to Fig. 3.100f; S_5–S_7 are conducting from Fig. 3.99b. In this stage, the freewheeling current flows through S_5, D_6, and the filter. The freewheeling circuit is disconnected with PV side and the LC loop is cut off.

The working principles of clamping circuit are shown in Fig. 3.101. The clamping mode when freewheeling circuit potential rises is shown in Fig. 3.101a. The clamping mode when freewheeling circuit potential decreases is shown in Fig. 3.101b. In freewheeling stages, LC will flow into DC side through S_7 if its potential rises and it will flow from DC side through D_7 if its potential decreases. Thus, CMV is clamped at half of DC input voltage. It is pointed out that during dead time between S_7 and other switches, CMV can be clamped only when the potential decreases.

Based on these modes, the potential of points 1 and 2 against N as well as u_{DM} and u_{CM} is shown in Table 3.23. From this table, it is clear that DMV maintains unipolar SPWM performance. CMV of AH6-I is constant except for it during dead time.

In conclusion, two power switches are introduced at inverter leg in AH6-I. A new power switch S_7 is connected to the neutral point of DC input. Dead time is also needed. The PV side and grid side are disconnected during freewheeling stages. LC will decrease as a result of almost constant CMV. The current flows through two

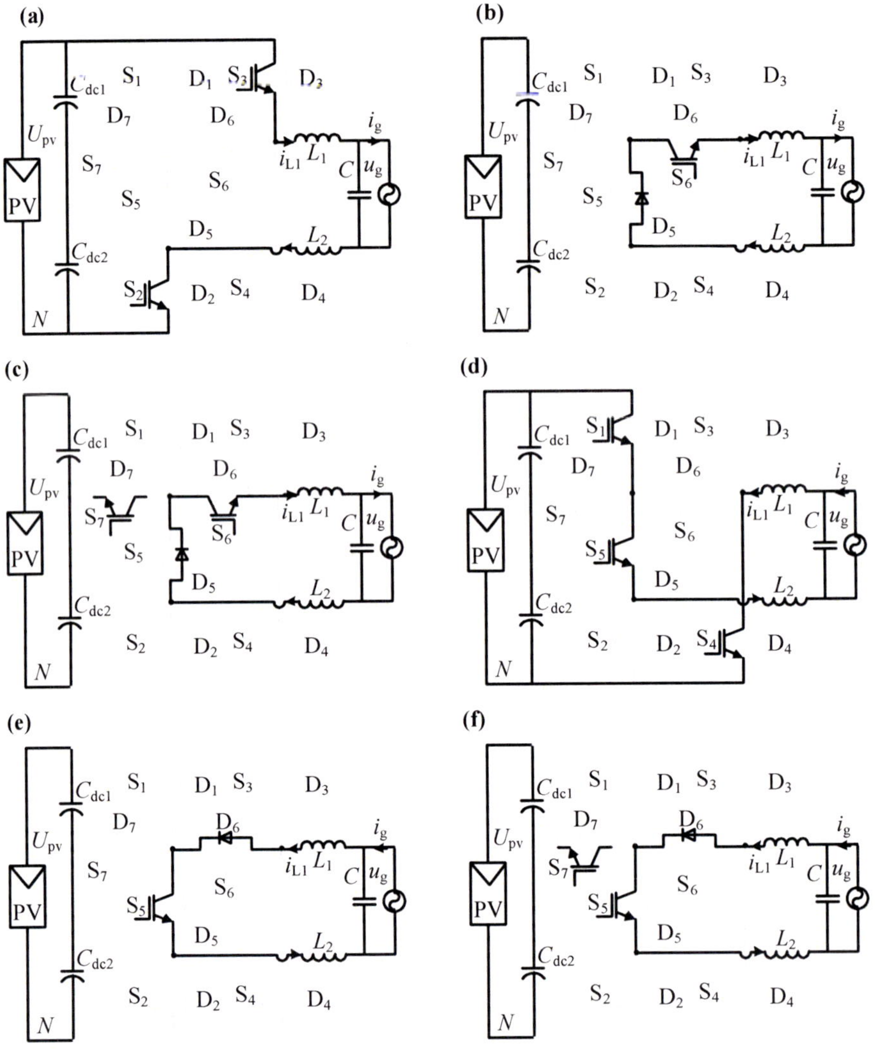

Fig. 3.100 Equivalent circuits of operating modes of AH6-I, **a** mode 1, **b** mode 2, **c** mode 3, **d** mode 4, **e** mode 5, **f** mode 6

power switches in power transmission stages in positive grid cycle. It flows through three power switches in power transmission stages in negative grid cycle. The power loss may be saved compared to H6-D.

2. Simulation results

A simulation model is built by Matlab/Simulink to verify the performance analysis aforementioned, and the key parameters of AH6-I are shown in Table 3.4.

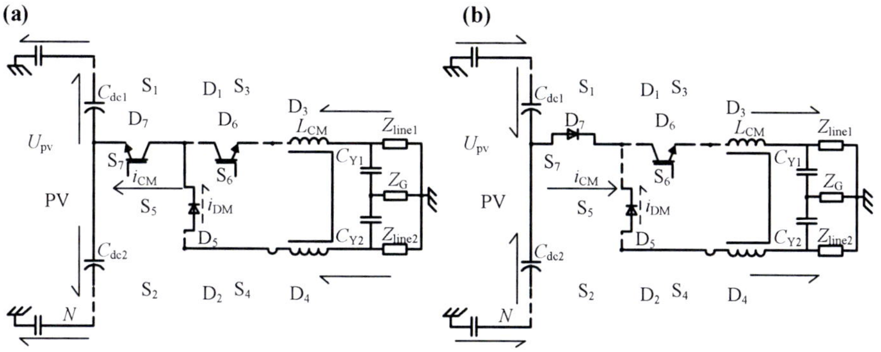

Fig. 3.101 Equivalent circuits when clamping circuit conducting in AH6-I, **a** potential rises, **b** potential decreases

Table 3.23 DMV and CMV levels of different operating modes in AH6-I

Modes	u_{1N}	u_{2N}	u_{DM}	u_{CM}
(1), (3)	U_{pv}	0	U_{pv}	$U_{pv}/2$
(2)	$U_{pv}/2$	$U_{pv}/2$	0	$U_{pv}/2$ (uncertain during dead time)
(4), (6)	0	U_{pv}	$-U_{pv}$	$U_{pv}/2$
(5)	$U_{pv}/2$	$U_{pv}/2$	0	$U_{pv}/2$ (uncertain during dead time)

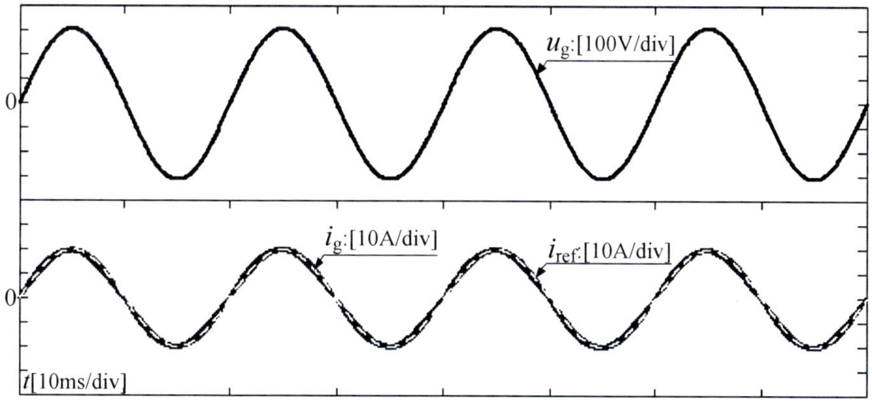

Fig. 3.102 Simulation waveforms of u_g, i_{ref}, and i_g of AH6-I

The simulation waveforms of u_g, grid-in current reference i_{ref}, and grid-in current i_g are shown in Fig. 3.102.

The simulation waveforms of DMV u_{DM} in time domain and frequency domain are shown in Fig. 3.103a, b, respectively. u_{DM} is a three-level voltage and contains little harmonics.

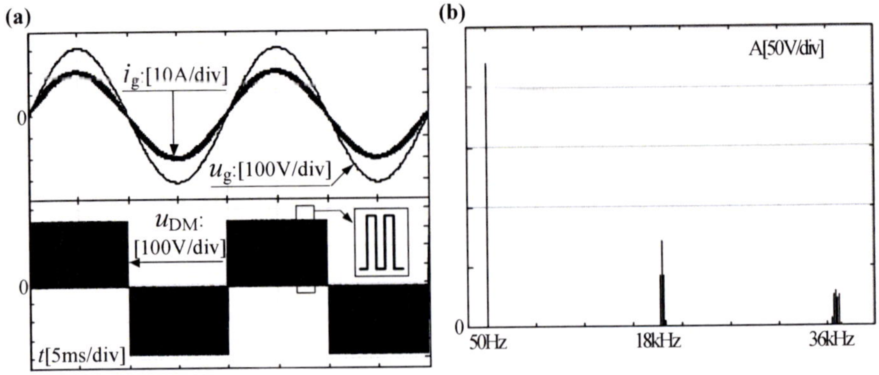

Fig. 3.103 DMV waveforms of AH6-I, **a** time domain, **b** frequency domain

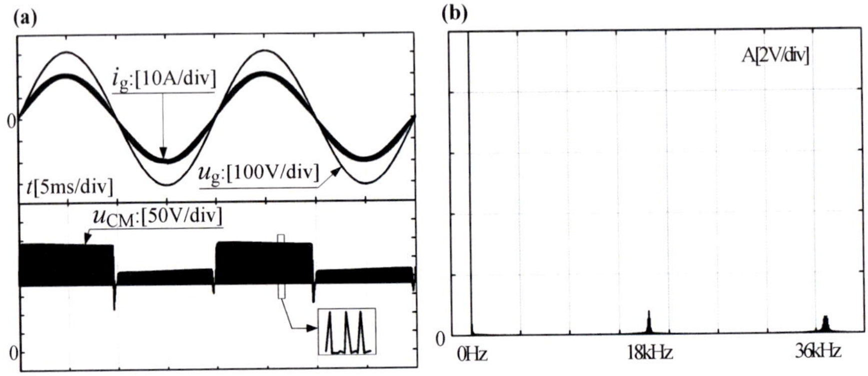

Fig. 3.104 CMV waveforms of AH6-I, **a** time domain, **b** frequency domain

The simulation waveforms of CMV u_{CM} in time domain and frequency domain are shown in Fig. 3.104a, b, respectively. It contains the components whose amplitude is higher than $U_{pv}/2$. This is because CMV is uncertain during dead time between power switches. While the amplitude of fundamental frequency pulses lower than $U_{pv}/2$ also exists, this phenomenon is due to variable CMV when current crossing zero, as discussed in Fig. 3.101.

The simulation waveforms of LC i_{CM} in time domain and frequency domain are shown in Fig. 3.105a, b, respectively. From Fig. 3.105, the LC is about 7 mA at switching frequency. LC is greatly reduced compared to the TLI in [10].

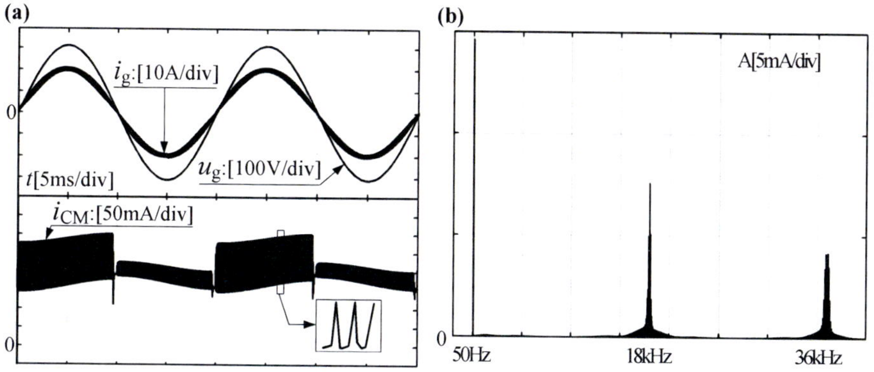

Fig. 3.105 LC waveforms of AH6-I, **a** time domain, **b** frequency domain

3.3.6 *Active Clamping H6 Inverter II*

1. Operating principle

The active clamping H6 inverter II (AH6-II) was proposed in [25], the main circuit of which is shown in Fig. 3.106a. The main circuit of AH6-II is similar to H6-II. Two power switches S_7 and S_8 are added as the clamping circuit. The AH6-II is also AC side decoupling TLI. The modulation strategy is shown in Fig. 3.106b. S_2, S_5, S_7, and S_8 work in fundamental frequency, while others work in fundamental or high frequency depending on the grid voltage. The output voltage is a three-level voltage.

There are four operating modes based on the directions of grid voltage and grid-in current when AH6-II works in unity power factor, as shown in Fig. 3.107. Figure 3.107a, b belongs to the positive grid cycle, while Fig. 3.19c, d belongs to the negative grid cycle.

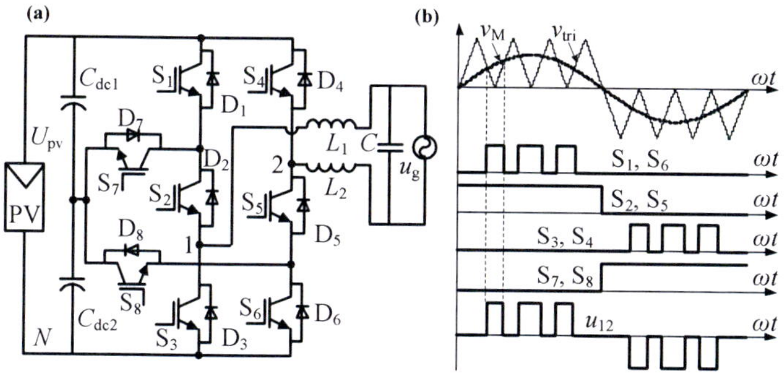

Fig. 3.106 Active clamping H6 inverter II, **a** main circuit, **b** modulation strategy

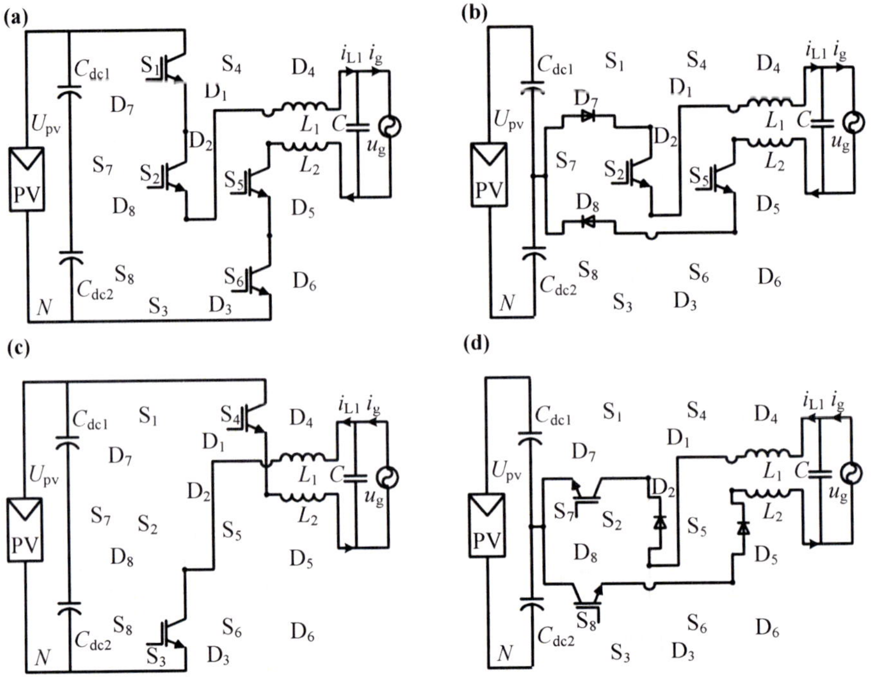

Fig. 3.107 Equivalent circuits of operating modes of AH6-II, **a** mode 1, **b** mode 2, **c** mode 3, **d** mode 4

Mode 1: Refer to Fig. 3.107a; in this stage, the energy transfers from the PV side to the grid through S_1, S_2, S_5, S_6, and the filter. The operating principle is the same as in conventional full-bridge TLI using unipolar SPWM modulation.

Mode 2: Refer to Fig. 3.107b; S_2 and S_5 are all conducting from Fig. 3.106b. In this stage, the freewheeling current flows through S_2, S_5, D_7, D_8, and the filter. The freewheeling circuit is disconnected with PV side and the LC loop is broken.

Mode 3: Refer to Fig. 3.107c; in this stage, the energy transfers from the PV side to the grid through S_3, S_4, and the filter.

Mode 4: Refer to Fig. 3.107d; only S_3 is conducting from Fig. 3.106b. In this stage, the freewheeling current flows through S_7, S_8, D_2, D_5, and the filter. The freewheeling circuit is disconnected with PV side and the LC loop is also cut off.

The working principles of clamping circuit are shown in Fig. 3.108. Clamping mode when freewheeling circuit potential rises is shown in Fig. 3.108a. Clamping mode when freewheeling circuit potential decreases is shown in Fig. 3.108b. In freewheeling stages, LC will flow into DC side through D_8 if its potential rises and it will flow from DC side through D_7 if its potential decreases. Thus, CMV is clamped at half of DC input voltage. Besides, the working principles of clamping circuits are different from clamping type TLIs discussed above according to Fig. 3.108. The clamping circuit in TLIs above only used to clamp the freewheeling potential. However, not only common mode current but also differential mode current flows

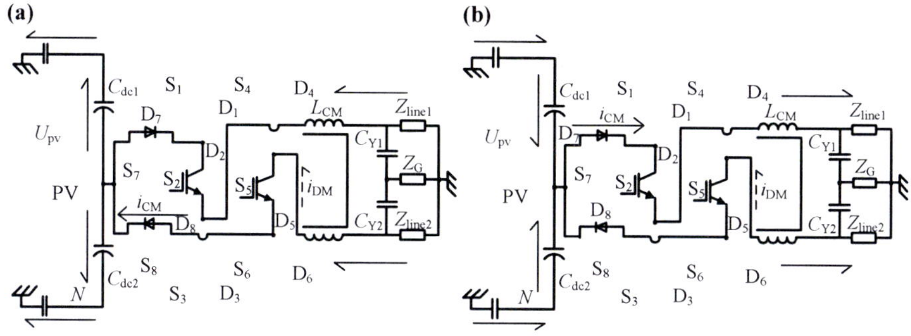

Fig. 3.108 Equivalent circuits when clamping circuit conducting in AH6-II, **a** potential rises, **b** potential decreases

Table 3.24 DMV and CMV levels of different operating modes in AH6-II

Modes	u_{1N}	u_{2N}	u_{DM}	u_{CM}
(1)	U_{pv}	0	U_{pv}	$U_{pv}/2$
(2)	$U_{pv}/2$	$U_{pv}/2$	0	$U_{pv}/2$
(3)	0	U_{pv}	$-U_{pv}$	$U_{pv}/2$
(4)	$U_{pv}/2$	$U_{pv}/2$	0	$U_{pv}/2$

through clamping circuit in AH6-II. Thus, clamping circuit is part of differential mode loop in AH6-II, and the current flows through four power switching components.

Based on these modes, the potential of points 1 and 2 against N as well as u_{DM} and u_{CM} is shown in Table 3.24. From this table, it is clear that DMV maintains unipolar performance. CMV of AH6-II is constant and has the perfect CMV performance.

In conclusion, two power switches S_7, S_8 are introduced in AH6-II. The PV and grid are disconnected during freewheeling stages in AH6-II. LC will decrease due to the constant CMV. The current flows through four power switches in energy transmission stage during positive grid cycle and two in negative cycle. It flows through two power switches and two diodes in freewheeling stages during whole grid period. The power loss may be higher compared to H6-D.

2. Simulation results

A simulation model is built by Matlab/Simulink to verify the performance analysis aforementioned, and the key parameters of AH6-II are shown in Table 3.4.

The simulation waveforms of u_g, grid-in current reference i_{ref}, and grid-in current i_g are shown in Fig. 3.109.

The simulation waveforms of DMV u_{DM} in time domain and frequency domain are shown in Fig. 3.110a, b, respectively. u_{DM} is a three-level voltage and contains little harmonics.

The simulation waveforms of CMV u_{CM} in time domain and frequency domain are shown in Fig. 3.111a, b, respectively. u_{CM} contains no high-frequency components and keeps constant.

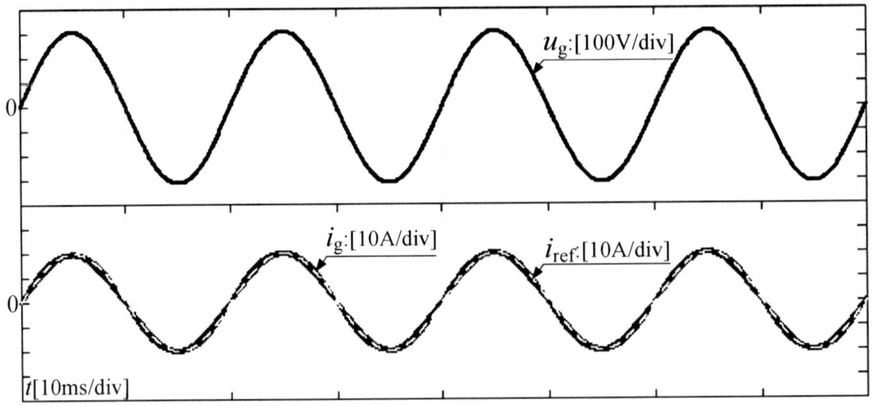

Fig. 3.109 Simulation waveforms of u_g, i_{ref}, and i_g of AH6-II

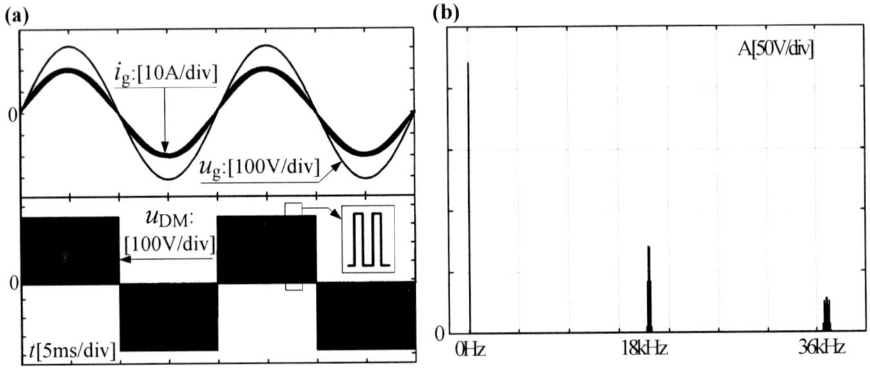

Fig. 3.110 DMV waveforms of AH6-II, **a** time domain, **b** frequency domain

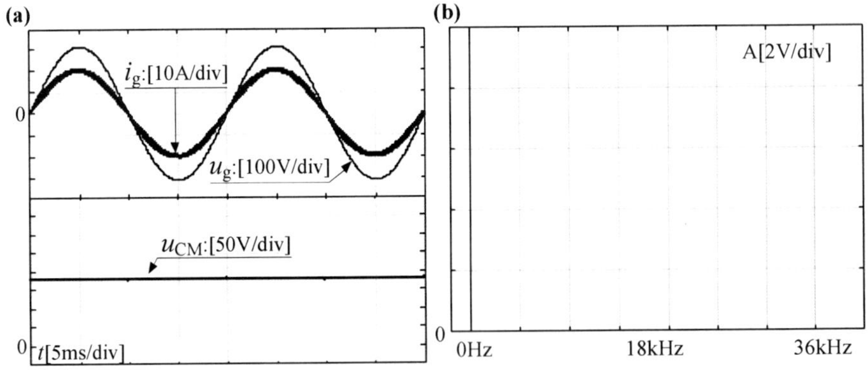

Fig. 3.111 CMV waveforms of AH6-II, **a** time domain, **b** frequency domain

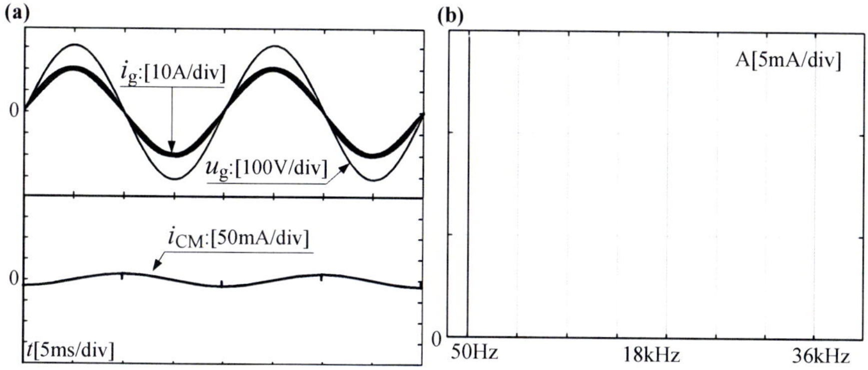

Fig. 3.112 LC waveforms of AH6-II, **a** time domain, **b** frequency domain

The simulation waveforms of LC i_{CM} in time domain and frequency domain are shown in Fig. 3.112a, b, respectively. From Fig. 3.112, the LC is about 0.3 mA at switching frequency. It meets the standard and greatly reduced compared to H6-II.

Although CMV is constant in the DC side clamping TLIs and two active clamping H6 TLIs, current flows through three or four power switches in energy transmission stages. Power loss is increased obviously. On the contrary, current flows through few power switches in AC side clamping TLIs based on HERIC. The high efficiency characteristic is maintained. Thus, researchers have proposed many AC side clamping TLIs [26–30].

3.3.7 Full-Bridge Zero-Voltage Rectifier Inverter

1. Operating principle

Full-bridge zero-voltage rectifier (FB-ZVR) [26] was proposed by T. Kerekes and R. Teodorescu from Aalborg University in Denmark. Its main circuit is shown in Fig. 3.113a. Power switch S_5/D_5 and a rectifier D_{rect} are added to AC side in FB-ZVR. Besides, switch S_5 is connected to the neutral point of DC side through D_6. Thus, the clamping structure is composed of those components. The modulation strategy is shown in Fig. 3.113b. S_1–S_4 work in fundamental or high frequency depending on the grid voltage. S_5 works in high frequency. Dead time is needed between S_5 and S_1–S_4 to avoid short circuit. Thus, the output voltage is the quasi-unipolar voltage.

There are six operating modes according to the directions of grid voltage and grid-in current when FB-ZVR works in unity power factor, as shown in Fig. 3.114. Figure 3.114a–c belongs to the positive grid cycle, while Fig. 3.114d–f belongs to the negative grid cycle.

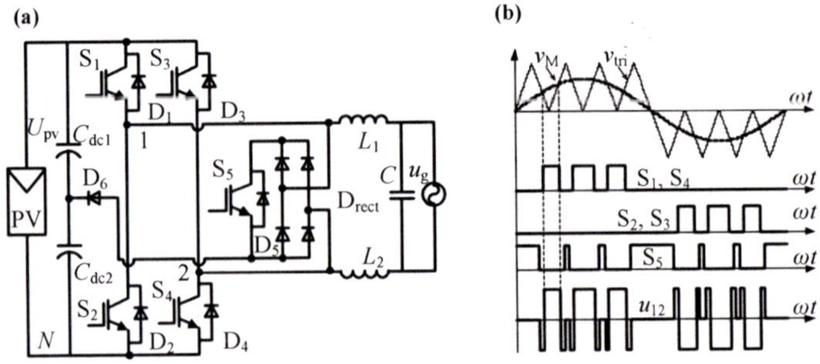

Fig. 3.113 Full-bridge zero voltage rectifier, **a** main circuit, **b** modulation strategy

Mode 1: Refer to Fig. 3.114a; in this stage, the energy transfers from the PV side to the grid through S_1, S_4, and the filter. The operating principle is the same as in conventional full-bridge TLI using unipolar SPWM modulation.

Mode 2: Refer to Fig. 3.114b; in this stage, the mode gets into the dead time. Power switch S_1 and S_4 are closed, but switch S_5 is not triggered. The current flows through D_2, D_3, and the filter to DC side. Thus, the narrow reverse pulse will appear in the output voltage.

Mode 3: Refer to Fig. 3.114c; S_1–S_4 are all closed while S_5 is conducting from Fig. 3.113b. In this stage, the freewheeling current flows through S_5, D_{rect}, and the filter. The freewheeling circuit is disconnected with PV side completely and the LC loop is broken. The potential of freewheeling circuit is clamped at constant because it is connected to the neutral point of DC side by D_6. Besides, short circuit will not happen because current can only flow through diode D_6.

Mode 4: Refer to Fig. 3.114d; in this stage, the energy transfers from the PV side to the grid through S_2, S_3, and the filter. The operating principle is the same as in conventional full-bridge TLI using unipolar SPWM modulation.

Mode 5: Refer to Fig. 3.114e; in this stage, the mode gets into the dead time. Power S_1 and S_4 are closed, and switch S_5 is not triggered. The current flows through D_1, D_4, and the filter to DC side. Thus, narrow reverse pulse will appear in the output voltage.

Mode 6: Refer to Fig. 3.114f, S_1–S_4 are all closed while S_5 is conducting from Fig. 3.113b. In this stage, the freewheeling current flows through S_5, D_{rect}, and the filter. The freewheeling circuit is disconnected with PV side. The potential of freewheeling circuit is clamped at constant by D_6.

The working principles of clamping circuit are shown in Fig. 3.115. In freewheeling stages, LC will flow into DC side through S_5, D_{rect}, and D_6 if its potential rises. Thus, CMV is clamped at half of DC input voltage. It is pointed out that CMV is uncertain and cannot be clamped when freewheeling circuit potential decreases. Thus, FB-ZVR is the unidirectional clamping TLI.

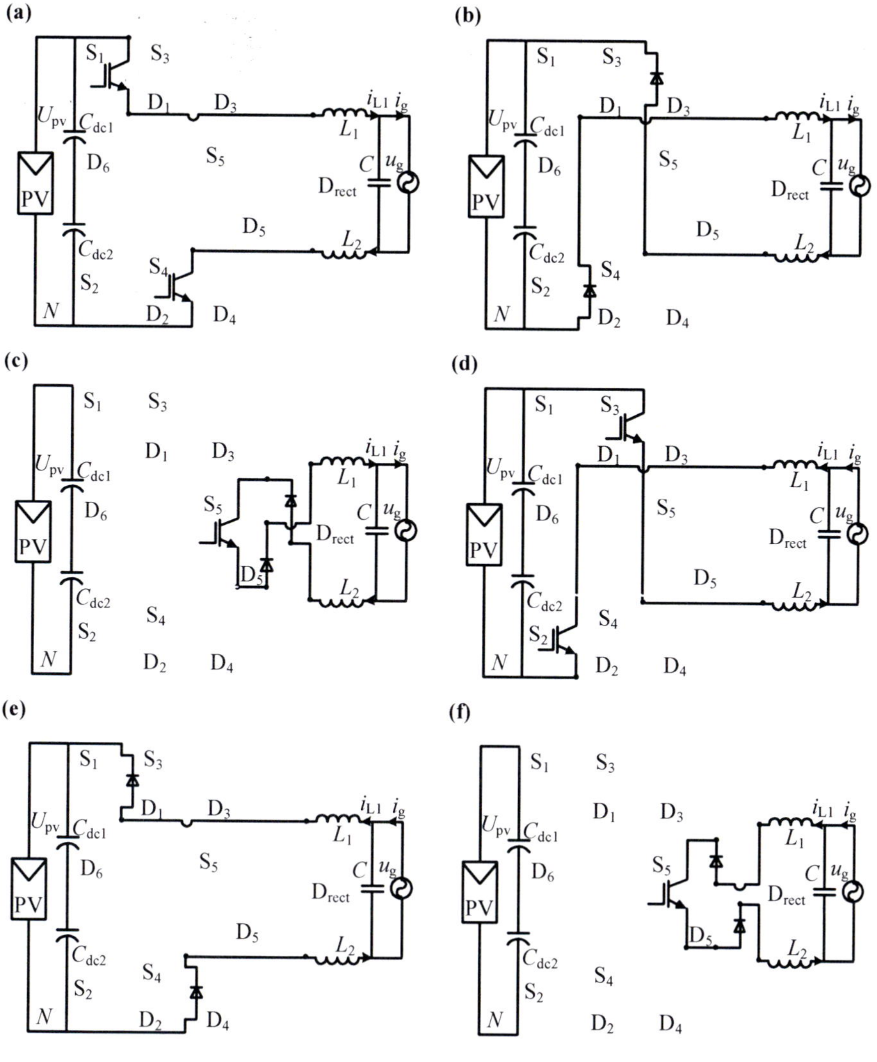

Fig. 3.114 Equivalent circuits of operating modes of FB-ZVR, **a** mode 1, **b** mode 2, **c** mode 3, **d** mode 4, **e** mode 5, **f** mode 6

Based on these modes, the potential of points 1 and 2 against N as well as u_{DM} and u_{CM} is shown in Table 3.25. From this table, it is clear that DMV is the quasi-unipolar SPWM voltage. CMV of FB-ZVR is constant except for the freewheeling circuit potential decreases.

In conclusion, many components are introduced in FB-ZVR. The PV and grid are disconnected during freewheeling stages in this TLI. CMV is clamped by D_6 and LC will be reduced. The current flows through two power switches in power

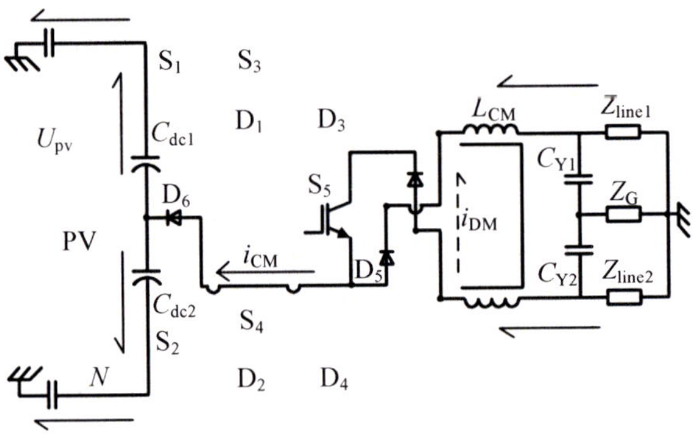

Fig. 3.115 Equivalent circuit when clamping circuit conducting in FB-ZVR

Table 3.25 DMV and CMV of different operating modes in FB-ZVR

Modes	u_{1N}	u_{2N}	u_{DM}	u_{CM}
(1), (5)	U_{pv}	0	U_{pv}	$U_{pv}/2$
(3)	$U_{pv}/2$	$U_{pv}/2$	0	$\le U_{pv}/2$
(2), (4)	0	U_{pv}	$-U_{pv}$	$U_{pv}/2$
(6)	$U_{pv}/2$	$U_{pv}/2$	0	$\le U_{pv}/2$

transmission stages and one power switch with two diodes in freewheeling stages. But, it can only clamp the CMV at $U_{pv}/2$ when the freewheeling circuit rises.

2. Simulation results

A simulation model is built by Matlab/Simulink to verify the performance analysis aforementioned, and the key parameters of FB-ZVR are shown in Table 3.4.

The simulation waveforms of u_g, grid-in current reference i_{ref}, and grid-in current i_g are shown in Fig. 3.116.

The simulation waveforms of DMV u_{DM} in time domain and frequency domain are shown in Fig. 3.117a, b, respectively. From this figure, it is clear that the reverse pulse exists in the output voltage. u_{DM} is the quasi-unipolar three-level voltage and contains little harmonics. The RMS of u_{DM} is about 75 V at switching frequency.

The simulation waveforms of CMV u_{CM} in time domain and frequency domain are shown in Fig. 3.118a, b, respectively. u_{CM} only contains the components whose amplitude is smaller than $U_{pv}/2$, as discussed in Fig. 3.115.

The simulation waveforms of LC i_{CM} in time domain and frequency domain are shown in Fig. 3.119a, b, respectively. From Fig. 3.119, the LC is about 10 mA at switching frequency. It is greatly reduced compared to HERIC because components whose amplitude higher than $U_{pv}/2$ are removed.

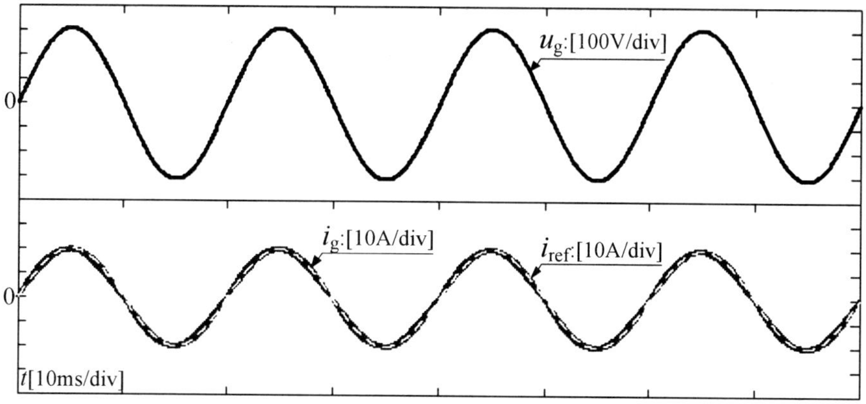

Fig. 3.116 Simulation waveforms of u_g, i_{ref}, and i_g of FB-ZVR

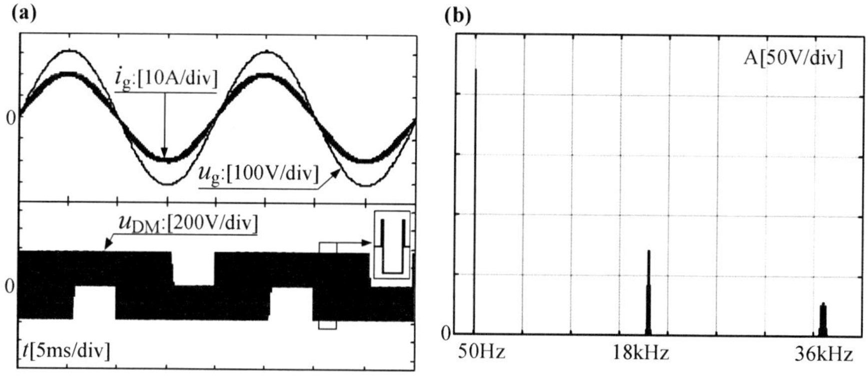

Fig. 3.117 DMV waveforms of FB-ZVR, **a** time domain, **b** frequency domain

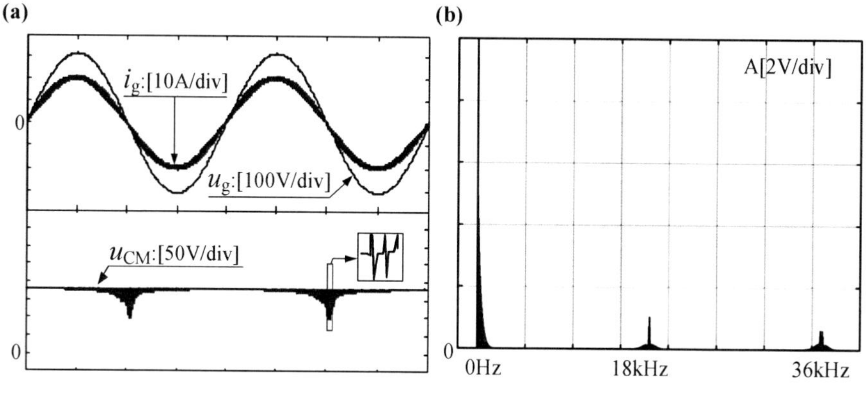

Fig. 3.118 CMV waveforms of FB-ZVR, **a** time domain, **b** frequency domain

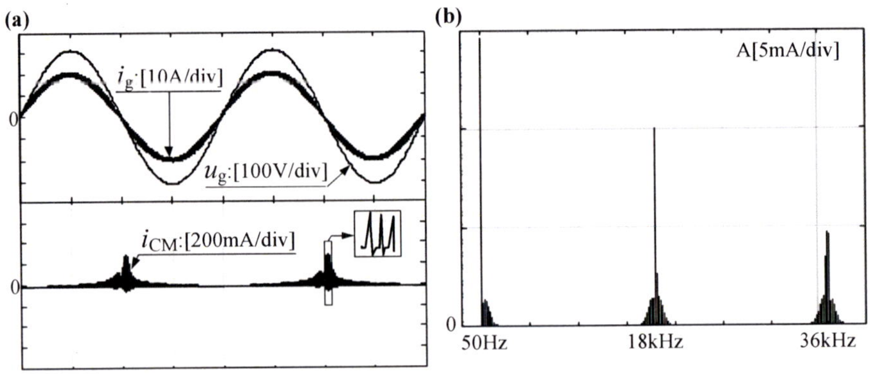

Fig. 3.119 LC waveforms of FB-ZVR, **a** time domain, **b** frequency domain

3.3.8 Diode Clamping Full-Bridge Zero-Voltage Rectifier Inverter

1. Operating principle

The diode clamping full-bridge zero-voltage rectifier (FB-ZVR-D) inverter was proposed in [27], whose main circuit is shown in Fig. 3.120a. Another diode is added in FB-ZVR-D compared to FB-ZVR. FB-ZVR-D has bidirectional clamping ability. The modulation strategy is shown in Fig. 3.120b. S_5 works in high frequency. S_1–S_4 work in fundamental or high frequency depending on the grid voltage. Dead time is also needed between S_5 and S_1–S_4 to avoid short circuit. Thus, the output voltage is the quasi-unipolar voltage.

The modulation strategy and differential mode performance are similar to FB-ZVR. In unity power factor, there are also six operating modes which can refer to Fig. 3.114. However, the common-mode performance is improved. The CMV is clamped bidirectional.

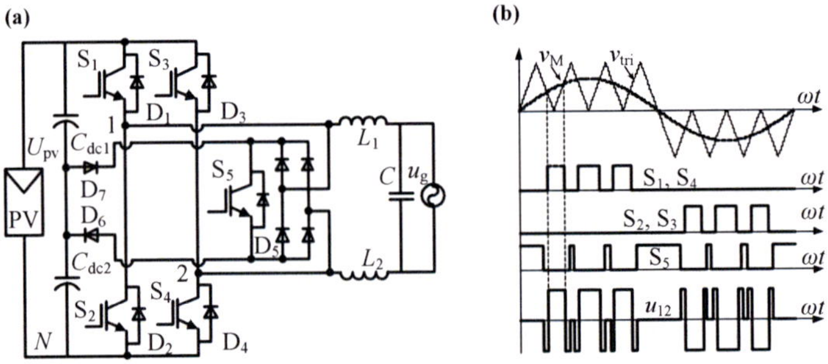

Fig. 3.120 Diode clamping full-bridge zero-voltage rectifier, **a** main circuit, **b** modulation strategy

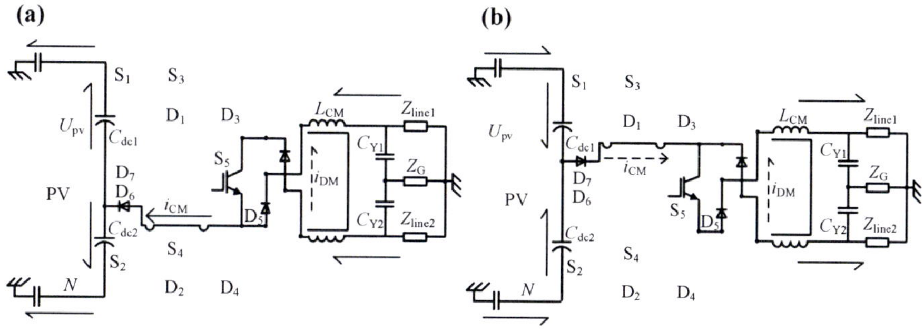

Fig. 3.121 Equivalent circuits when clamping circuit conducting in FB-ZVR-D, **a** potential rises, **b** potential decreases

The working principles of clamping circuit are shown in Fig. 3.121. The clamping mode when freewheeling circuit potential rises is shown in Fig. 3.121a. The clamping mode when freewheeling circuit potential decreases is shown in Fig. 3.121b. In freewheeling stages, LC will flow into DC side through D_6 if its potential rises and it will flow from DC side through D_7 if its potential decreases. Thus, CMV is clamped at half of DC input voltage. The clamping ability is dependent on the properties of the diode and CMV can still be clamped at constant during dead time.

Based on these modes, the potential of points 1 and 2 against N as well as u_{DM} and u_{CM} is shown in Table 3.26. From this table, it is clear that the DMV maintains quasi-unipolar SPWM performance. The CMV of FB-ZVR-D is constant. The common-mode performance is optimal.

In conclusion, freewheeling and clamping structures are introduced in FB-ZVR-D. The PV side and grid side are disconnected during freewheeling stages. CMV is clamped at constant in all modes. The current flows through two power switches in power transmission stages and one power switch and two diodes in freewheeling stages, like FB-ZVR. But, the CMV is clamped at $U_{pv}/2$ whenever the potential freewheeling circuit rises or decreases.

2. Simulation results

A simulation model is built by Matlab/Simulink to verify the performance analysis aforementioned, and the key parameters of FB-ZVR-D are shown in Table 3.4.

Table 3.26 DMV and CMV levels of different operating modes in FB-ZVR-D

Modes	u_{1N}	u_{2N}	u_{DM}	u_{CM}
(1), (5)	U_{pv}	0	U_{pv}	$U_{pv}/2$
(3)	$U_{pv}/2$	$U_{pv}/2$	0	$U_{pv}/2$
(2), (4)	0	U_{pv}	$-U_{pv}$	$U_{pv}/2$
(6)	$U_{pv}/2$	$U_{pv}/2$	0	$U_{pv}/2$

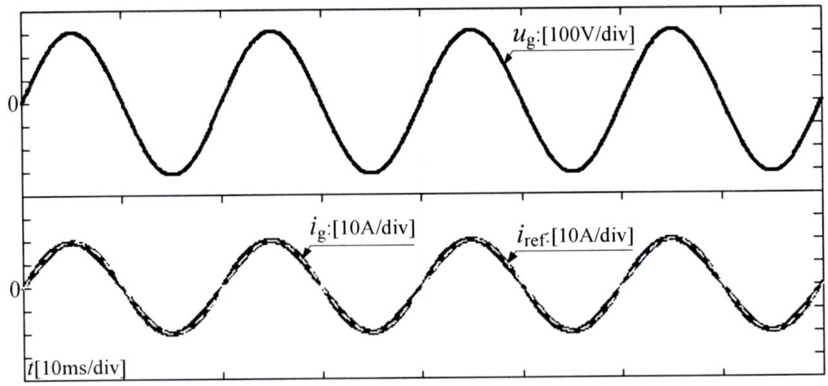

Fig. 3.122 Simulation waveforms of u_g, i_{ref}, and i_g of FB-ZVR-D

The simulation waveforms of u_g, grid-in current reference i_{ref}, and grid-in current i_g are shown in Fig. 3.122.

The simulation waveforms of DMV u_{DM} in time domain and frequency domain are shown in Fig. 3.123a, b, respectively. From this figure, it is clear that reverse pulse exists in the output voltage. u_{DM} is a quasi-unipolar three-level voltage and contains little harmonics. The RMS of u_{DM} is about 75 V at switching frequency.

The simulation waveforms of CMV u_{CM} in time domain and frequency domain are shown in Fig. 3.124a, b, respectively. u_{CM} is constant and only contains fundamental component.

The simulation waveforms of LC i_{CM} in time domain and frequency domain are shown in Fig. 3.125a, b, respectively. From Fig. 3.125, the LC is decreased to 0.3 mA compared to 10 mA in FB-ZVR at switching frequency.

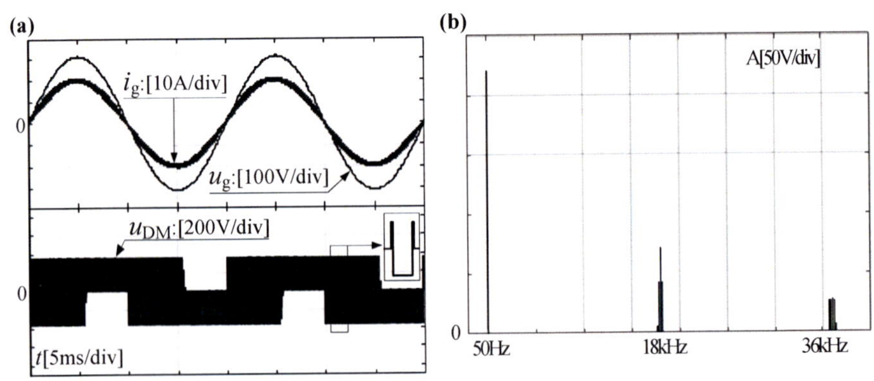

Fig. 3.123 DMV waveforms of FB-ZVR-D, **a** time domain, **b** frequency domain

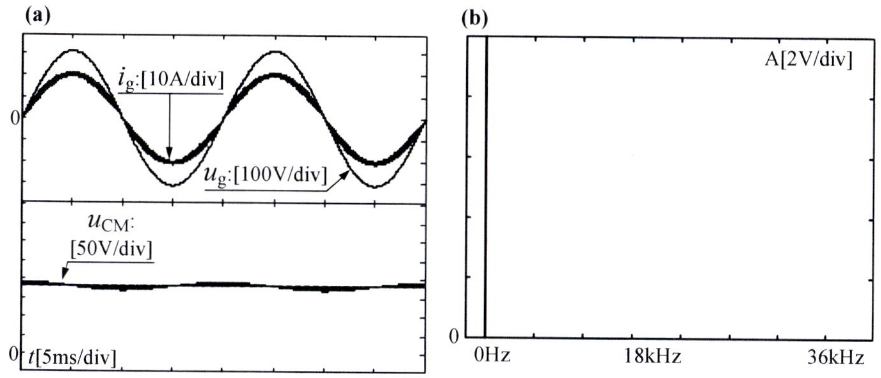

Fig. 3.124 CMV waveforms of FB-ZVR-D, **a** time domain, **b** frequency domain

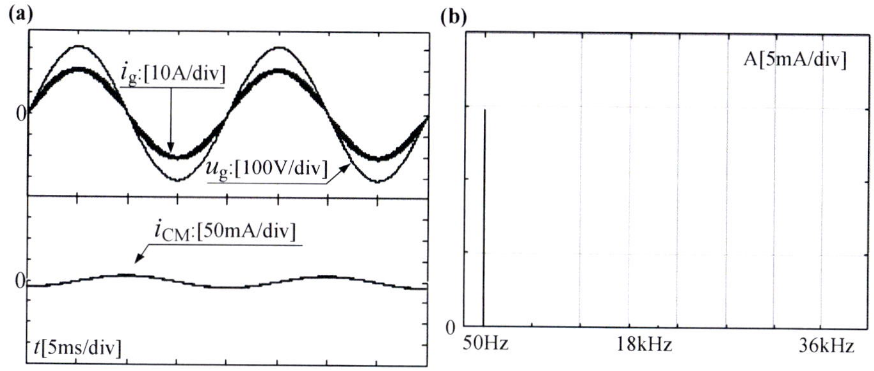

Fig. 3.125 LC waveforms of FB-ZVR-D, **a** time domain, **b** frequency domain

3.3.9 HERIC with Tri-direction Clamping Cell Inverter

1. Operating principle

HERIC with tri-direction clamping cell (HERIC-TDCC) was proposed in [28], whose main circuit is shown in Fig. 3.126a. A new power switch S_7 is added to connect the neutral point of DC input and output compared to HERIC. HERIC-TDCC is also an AC side decoupling TLI. The modulation strategy is shown in Fig. 3.126b. S_7 works in high frequency. S_1–S_6 work in fundamental or high frequency depending on the grid voltage. Dead time is needed between S_7 and S_1–S_4 to avoid short circuit. The output voltage is still three-level voltage.

The modulation strategy and differential mode performance are similar to HERIC. In unity power factor, there are also four operating modes which can refer to Fig. 3.9. However, the common-mode performance is improved. The CMV is clamped.

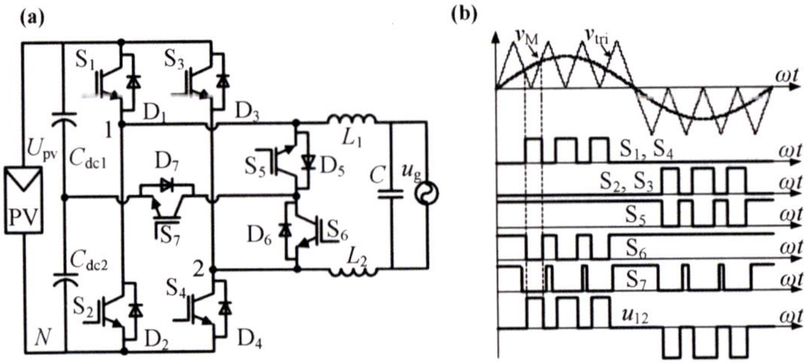

Fig. 3.126 HERIC with tri-direction clamping cell inverter, **a** main circuit, **b** modulation strategy

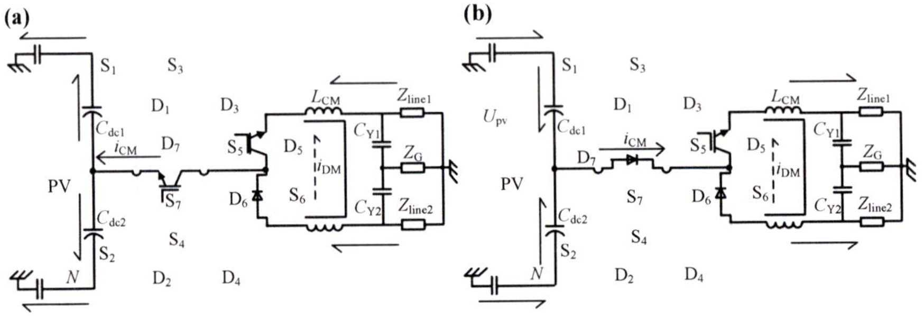

Fig. 3.127 Equivalent circuits when clamping circuit conducting in HERIC-TDCC, **a** potential rises, **b** potential decreases

The working principles of the clamping circuit are shown in Fig. 3.127. The clamping mode when freewheeling circuit potential rises is shown in Fig. 3.127a. The clamping mode when freewheeling circuit potential decreases is shown in Fig. 3.127b. In freewheeling stages, LC will flow into DC side through S_7 if its potential rises. LC will flow from DC side through D_7 if its potential decreases. CMV is clamped naturally. Furthermore, CMV cannot be clamped when freewheeling circuit potential rises during dead time.

Based on these modes, the potential of points 1 and 2 against N as well as u_{DM} and u_{CM} is shown in Table 3.27. From this table, it is clear that DMV maintains unipolar

Table 3.27 DMV and CMV of different operating modes in HERIC-TDCC

Modes	u_{1N}	u_{2N}	u_{DM}	u_{CM}
(1)	U_{pv}	0	U_{pv}	$U_{pv}/2$
(2)	$U_{pv}/2$	$U_{pv}/2$	0	$U_{pv}/2$ (uncertain during dead time)
(3)	0	U_{pv}	$-U_{pv}$	$U_{pv}/2$
(4)	$U_{pv}/2$	$U_{pv}/2$	0	$U_{pv}/2$ (uncertain during dead time)

SPWM performance. CMV of HERIC-TDCC is constant except for it during dead time.

In conclusion, freewheeling and clamping structures are introduced at AC side in HERIC-TDCC. The PV side and grid side are disconnected during freewheeling stages. The output voltage is also unipolar three-level voltage. CMV is clamped at constant in all modes except for dead time. The current flows through two power switches in power transmission or freewheeling stages, like HERIC. Power loss is decreased compared to FB-ZVR or FB-ZVR-D.

2. Simulation results

A simulation model is built by Matlab/Simulink to verify the performance analysis aforementioned, and the key parameters of HERIC-TDCC are shown in Table 3.4.

The simulation waveforms of u_g, grid-in current reference i_{ref}, and grid-in current i_g are shown in Fig. 3.128.

The simulation waveforms of DMV u_{DM} in time domain and frequency domain are shown in Fig. 3.129a, b, respectively. u_{DM} is a three-level voltage and contains little harmonics. The RMS of u_{DM} is about 75 V at switching frequency.

The simulation waveforms of CMV u_{CM} in time domain and frequency domain are shown in Fig. 3.130a, b, respectively. It contains the components that its amplitude higher than $U_{pv}/2$; this is because CMV is uncertain during the dead time between power switches. While the amplitude of fundamental frequency pulses is lower than $U_{pv}/2$ also exists, this phenomenon is due to variable CMV when current crossing zero, as discussed in Fig. 3.127.

The simulation waveforms of LC i_{CM} in time domain and frequency domain are shown in Fig. 3.131a, b, respectively. From Fig. 3.131, the LC is about 4 mA at switching frequency. LC is greatly reduced compared to HERIC.

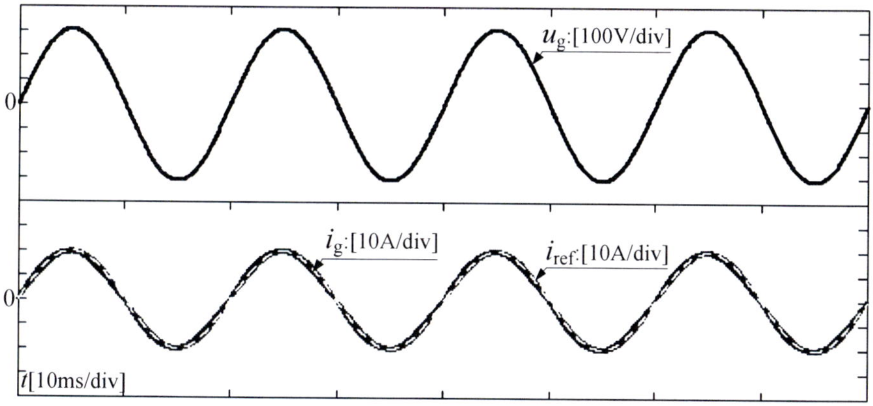

Fig. 3.128 Simulation waveforms of u_g, i_{ref}, and i_g of HERIC-TDCC

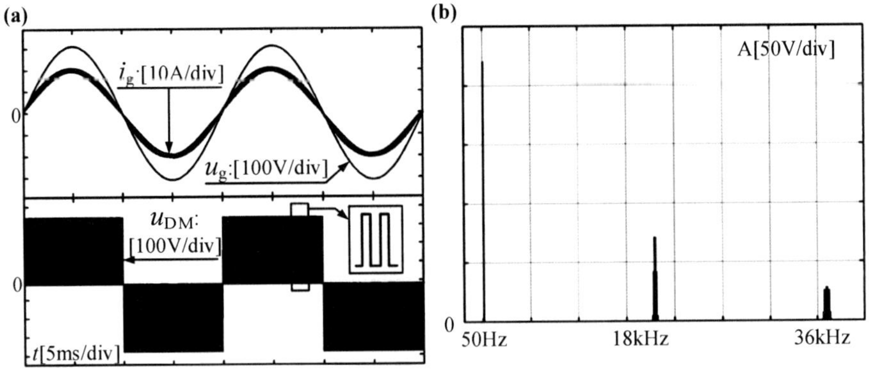

Fig. 3.129 DMV waveforms of HERIC-TDCC, **a** time domain, **b** frequency domain

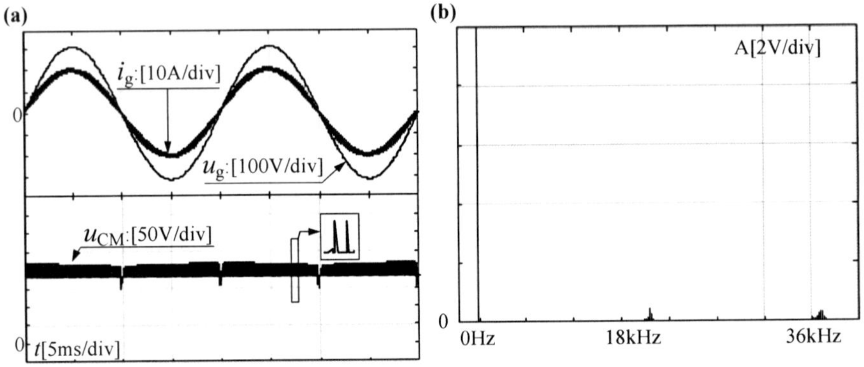

Fig. 3.130 CMV waveforms of HERIC-TDCC, **a** time domain, **b** frequency domain

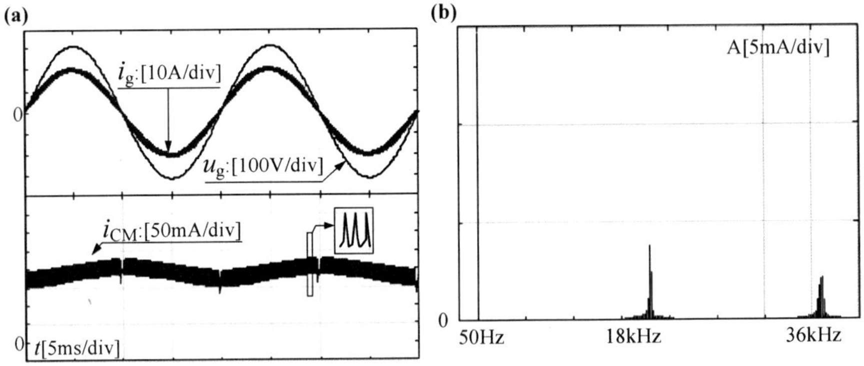

Fig. 3.131 LC waveforms of HERIC-TDCC, **a** time domain, **b** frequency domain

3.3.10 Optimized HERIC Inverter

1. Operating principle

The optimized HERIC transformerless inverter (oHERIC) was proposed in [29], whose main circuit is shown in Fig. 3.132a. Two new power switches S_7 and S_8 are added to connect the neutral point of DC input and output compared to HERIC. The oHERIC is AC side decoupling TLI. The modulation strategy is shown in Fig. 3.132b. S_7 and S_8 work in high frequency. S_1–S_6 work in fundamental or high frequency depending on the grid voltage. Dead time is needed between S_7, S_8 and S_1–S_4 to avoid short circuit. The output voltage is still three-level voltage.

The modulation strategy and differential mode performance are similar to HERIC. In unity power factor, there are also four operating modes which can refer to Fig. 3.9. However, the common-mode performance improved. The CMV is clamped. Dead time is also needed at the switching scale.

The working principles of the clamping circuit are shown in Fig. 3.133. The clamping mode when freewheeling circuit potential rises is shown in Fig. 3.133a. The clamping mode when freewheeling circuit potential decreases is shown in Fig. 3.133b. In freewheeling stages, LC will flow into DC side through S_7 and D_8 if

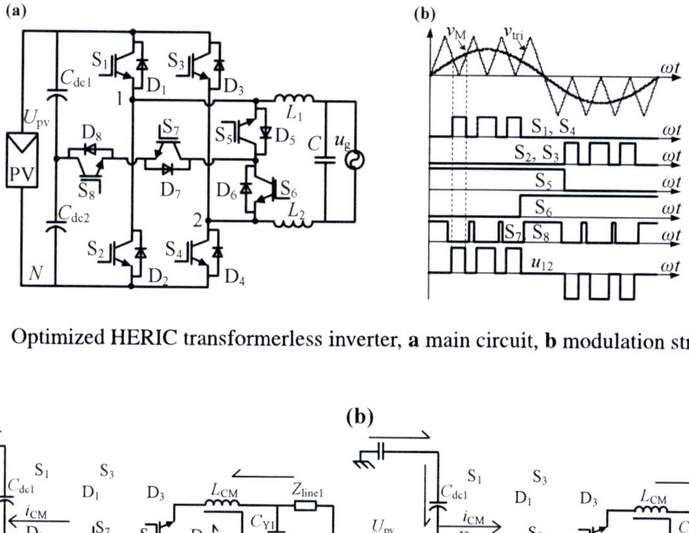

Fig. 3.132 Optimized HERIC transformerless inverter, a main circuit, b modulation strategy

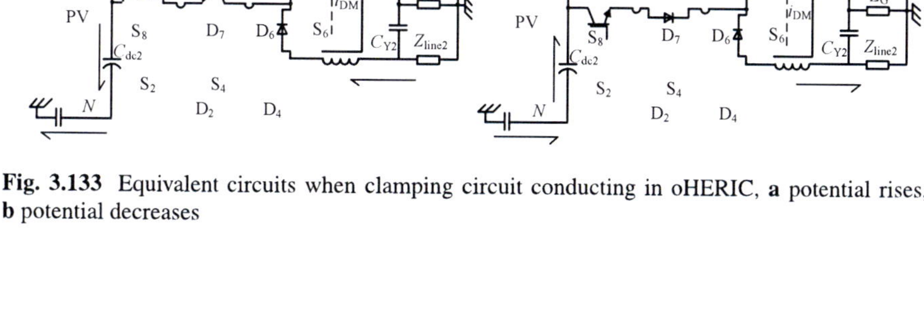

Fig. 3.133 Equivalent circuits when clamping circuit conducting in oHERIC, a potential rises, b potential decreases

Table 3.28 DMV and CMV levels of different operating modes in oHERIC

Modes	u_{1N}	u_{2N}	u_{DM}	u_{CM}
(1)	U_{pv}	0	U_{pv}	$U_{pv}/2$
(2)	$U_{pv}/2$	$U_{pv}/2$	0	$U_{pv}/2$ (uncertain during dead time)
(3)	0	U_{pv}	$-U_{pv}$	$U_{pv}/2$
(4)	$U_{pv}/2$	$U_{pv}/2$	0	$U_{pv}/2$ (uncertain during dead time)

potential rises. LC will flow from DC side through S_8 and D_7 if its potential decreases. Furthermore, CMV cannot be clamped during dead time.

Based on these modes, the potential of points 1 and 2 against N as well as u_{DM} and u_{CM} is shown in Table 3.28. From this table, it is clear that DMV is a three-level voltage. CMV of oHERIC is constant except for it during dead time.

In conclusion, freewheeling and clamping structures are introduced at the AC side in oHERIC. The PV and grid are disconnected during the freewheeling stages. The output voltage is also unipolar three-level voltage. CMV is clamped at constant in all modes except for short dead time. It is more flexible using two clamping power switches. The current flows through two power switches in power transmission or freewheeling stages, like HERIC.

2. Experimental results

The working principles of oHERIC are the same as HERIC-TDCC. Only experimental results are given for avoiding duplication. Simulation waveforms can refer to Figs. 3.128, 3.129, 3.130, and 3.131. To verify the theoretical analysis, a prototype of oHERIC is built, and the key parameters are shown in Table 3.29.

The prototype is shown in Fig. 3.134a. The experimental waveforms of u_g, i_g, and DMV u_{DM} are shown in Fig. 3.134b. From Fig. 3.134b, u_{DM} is a unipolar three-level voltage.

The experimental waveforms of u_g, i_g, u_{1N}, u_{2N}, u_{CM} in the grid frequency scale are shown in Fig. 3.135a. The experimental waveforms of u_{CM} in positive and negative grid cycles at switching frequency scales are shown in Fig. 3.135b and c, respectively. In Fig. 3.135a, CMV is oscillating in large amplitude. This is because clamping switches S_7 and S_8 are not triggered during current crossing zero range to commuting reliable. The CMV oscillating is owing to the dead time in Fig. 3.135b, c. In other time intervals, CMV is constant as the theoretical analysis above.

3.3.11 Full-Bridge with Constant Common-Mode Voltage Inverter

Although CMV can be clamped in FB-ZVR and its improved structure, power loss in freewheeling is improved because conducting components are increasing from two to three components. In HERIC-TDCC and oHERIC, the cost is also increased for using

Table 3.29 Key parameters of oHERIC prototype

Parameters	Value	Parameters	Value
Input voltage U_{pv}/V	400	Diodes D_5, D_6	IDD12SG60C
Grid voltage u_g/V, Hz	220, 50	Filter inductors L_1, L_2/mH	0.5
Rated power P/W	1000	Filter capacitor C/μF	2.2
Switching frequency f_s/kHz	20	Common-mode inductor L_{CM}	2 × B82726-S6223-N40
DC side capacitors C_{dc1}, C_{dc2}/μF	470μF/350 V	Inductor L_{CM} value/mH	L: 2 × 1.6
IGBT S_5, S_6	IRG4BC30S-S	Turns	Turns: 11 + 11
MOSFET S_1–S_4	FZ06BIA070FS-P894E-14	Common-mode capacitor C_{Y1}, C_{Y2}/nF	22
MOSFET S_7, S_8	FQB5N50C	PV parasitic capacitor C_{pv1}, C_{pv2}/μF	0.1

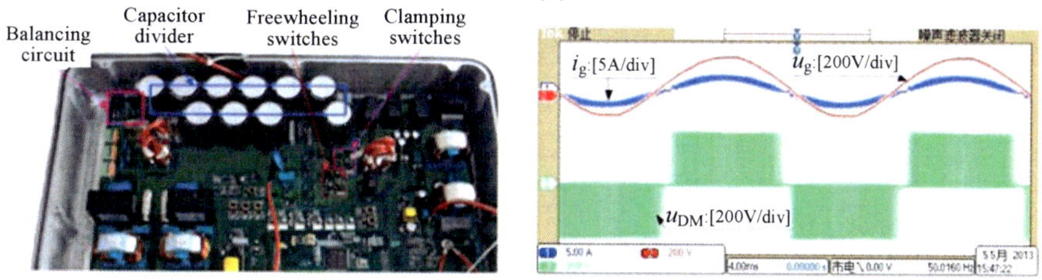

Fig. 3.134 oHERIC prototype and DMV waveforms, **a** oHERIC prototype, **b** DMV waveforms of oHERIC

more active power switches. Besides, CMV cannot be clamped during dead time. Thus, researchers proposed a TLI based on HERIC while CMV is clamped only using diodes, which is called full-bridge inverter topology with constant common-mode voltage (FB-CCV) [30].

1. Operating principle

The FB-CCV was proposed in [30], whose main circuit is shown in Fig. 3.136a. Two diodes D_7 and D_8 are added based on Fig. 3.71c. D_7 and D_8 connect the neutral point of DC side and S_5 and S_6, respectively. The modulation strategy is shown in Fig. 3.136b. S_5–S_6 work in high frequency. The output voltage is a three-level

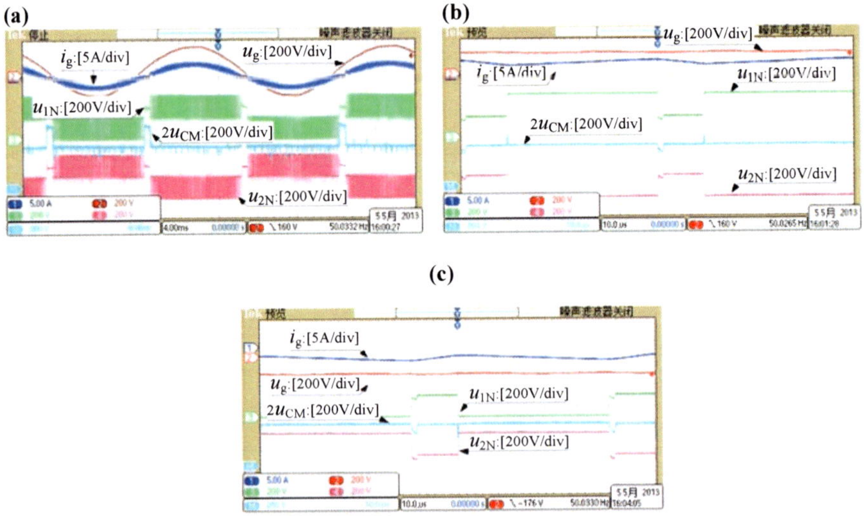

Fig. 3.135 CMV waveforms of oHERIC, **a** grid frequency scale, **b** positive (switching frequency scale), **c** negative (switching frequency scale)

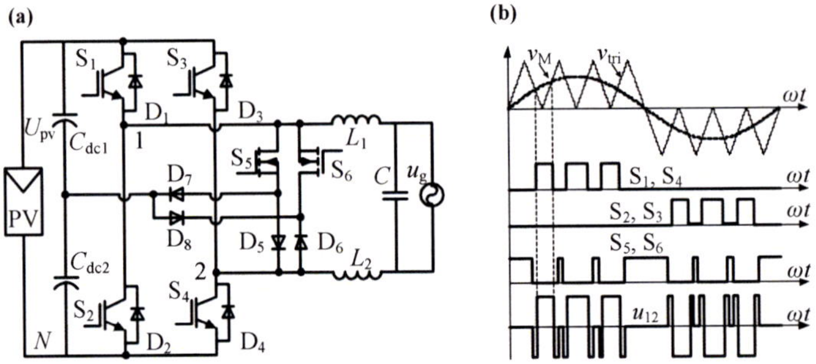

Fig. 3.136 Full-bridge with constant common-mode voltage inverter, **a** main circuit, **b** modulation strategy

unipolar voltage. But actually, dead time is needed between S_5, S_6, and S_1–S_4. Thus, the output voltage u_{12} is quasi-unipolar three-level voltage, as shown in Fig. 3.136b.

There are six operating modes based on the directions of grid voltage and grid-in current when FB-CCV works in unity power factor, as shown in Fig. 3.137. Figure 3.137a–c belongs to the positive grid cycle, while Fig. 3.137d–f belongs to the negative grid cycle.

Mode 1: Refer to Fig. 3.137a; in this stage, the energy transfers from the PV side to the grid through S_1, S_4, and the filter. The operating principle is the same as HERIC.

Mode 2: Refer to Fig. 3.137b; in this stage, the mode gets into the dead time. Power S_1 and S_4 are closed, and switches S_5 and S_6 are not triggered. The current

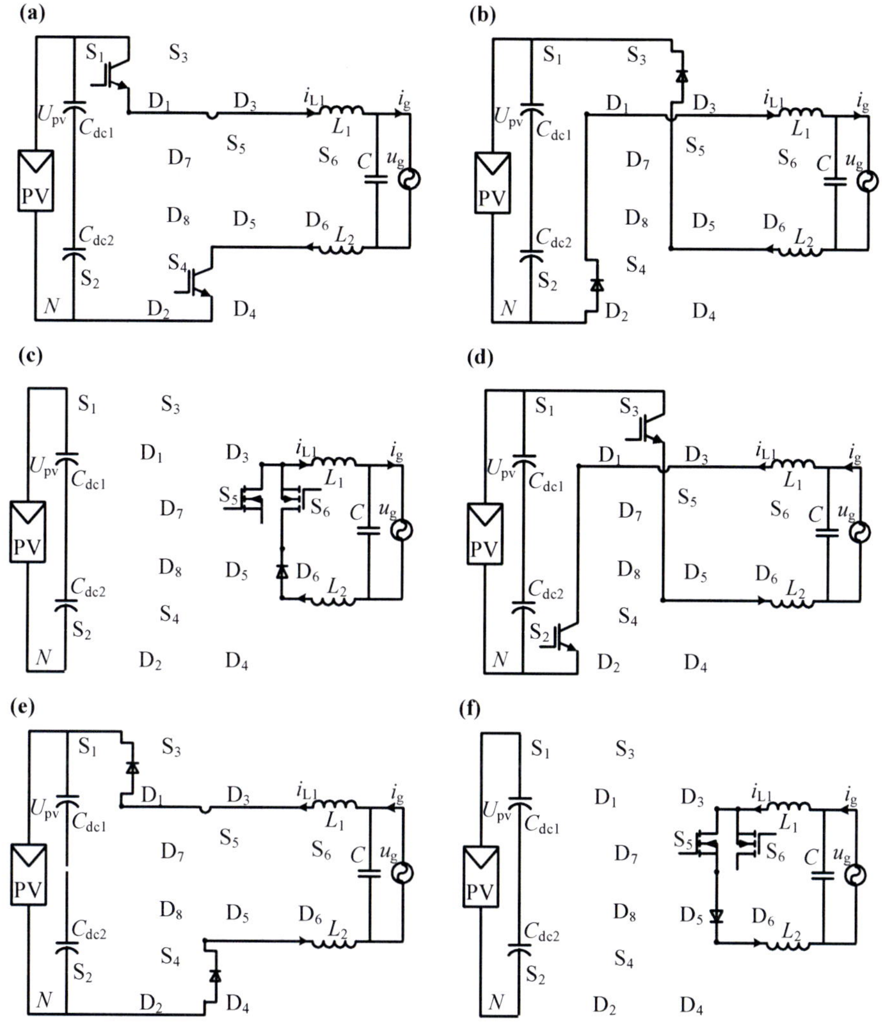

Fig. 3.137 Equivalent circuits of operating modes of FB-CCV, **a** mode 1, **b** mode 2, **c** mode 3, **d** mode 4, **e** mode 5, **f** mode 6

flows through D_2, D_3, and the filter to DC side. Thus, the narrow reverse pulse will appear in the output voltage.

Mode 3: Refer to Fig. 3.137c; S_1–S_4 are all closed while S_5 and S_6 are conducting from Fig. 3.136b. In this stage, the freewheeling current flows through S_6, D_6, and the filter. The freewheeling circuit is disconnected with PV side completely and the LC loop is cut off.

Mode 4: Refer to Fig. 3.137d; in this stage, the energy transfers from the PV side to the grid through S_2, S_3, and the filter. The operating principle is the same as in conventional full-bridge TLI.

Mode 5: Refer to Fig. 3.137e; in this stage, the mode gets into the dead time stage. Power S_1 and S_4 are closed, and switches S_5 and S_6 are not triggered. The current flows through D_1, D_4, and the filter to DC side. Thus, narrow reverse pulse will appear in the output voltage.

Mode 6: Refer to Fig. 3.137f; S_1–S_4 are all closed while S_5 and S_6 are conducting from Fig. 3.136b. In this stage, the freewheeling current flows through S_5, D_5, and the filter. The freewheeling circuit is disconnected with PV side and the LC loop is broken compared to conventional full-bridge TLI. The potential of freewheeling circuit is clamped at constant by diodes naturally.

The working principles of clamping circuit are shown in Fig. 3.138. The clamping mode when freewheeling circuit potential rises is shown in Fig. 3.138a. The clamping mode when freewheeling circuit potential decreases is shown in Fig. 3.138b. In freewheeling stages, LC will flow into DC side through D_7 if its potential rises and it will flow from DC side through D_8 if its potential decreases. Thus, CMV is clamped at half of DC input voltage. The clamping ability is dependent on the properties of the diode.

Based on these modes, the potential of points 1 and 2 against N as well as u_{DM} and u_{CM} is shown in Table 3.30. From this table, it is clear that DC side and AC side are disconnected in both Mode 3 and Mode 6. DMV is a quasi-unipolar voltage. CMV of FB-CCV is constant. It has perfect CMV performance.

In conclusion, freewheeling and clamping structures are introduced at AC side in FB-CCV. The PV side and grid side are disconnected during freewheeling stages. CMV is clamped naturally at constant in all modes, avoiding uncertain state during

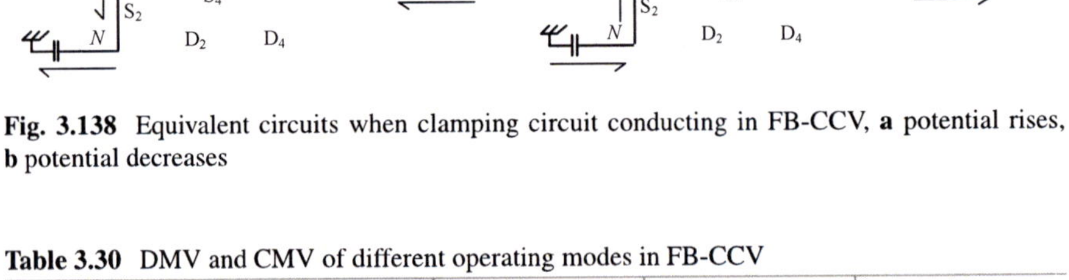

Fig. 3.138 Equivalent circuits when clamping circuit conducting in FB-CCV, **a** potential rises, **b** potential decreases

Table 3.30 DMV and CMV of different operating modes in FB-CCV

Modes	u_{1N}	u_{2N}	u_{DM}	u_{CM}
(1), (5)	U_{pv}	0	U_{pv}	$U_{pv}/2$
(3)	$U_{pv}/2$	$U_{pv}/2$	0	$U_{pv}/2$
(2), (4)	0	U_{pv}	$-U_{pv}$	$U_{pv}/2$
(6)	$U_{pv}/2$	$U_{pv}/2$	0	$U_{pv}/2$

dead time in active clamping TLIs. The current flows through two power switches in power transmission stages and one power switch with two diodes in freewheeling stages. The CMV is clamped at $U_{pv}/2$ whenever the potential freewheeling circuit rises or decreases.

2. Simulation results

A simulation model is built by Matlab/Simulink to verify the performance analysis aforementioned, and the key parameters of FB-CCV are shown in Table 3.4.

The simulation waveforms of u_g, grid-in current reference i_{ref}, and grid-in current i_g are shown in Fig. 3.139.

The simulation waveforms of DMV u_{DM} in time domain and frequency domain are shown in Fig. 3.140a, b, respectively. From these figures, it is clear that reverse pulse exists in the output voltage. u_{DM} is the quasi-unipolar three-level voltage and the switching frequency components are the same with unipolar three-level voltage. Besides, dead time is not needed and eliminating CMV oscillation in active clamping TLIs.

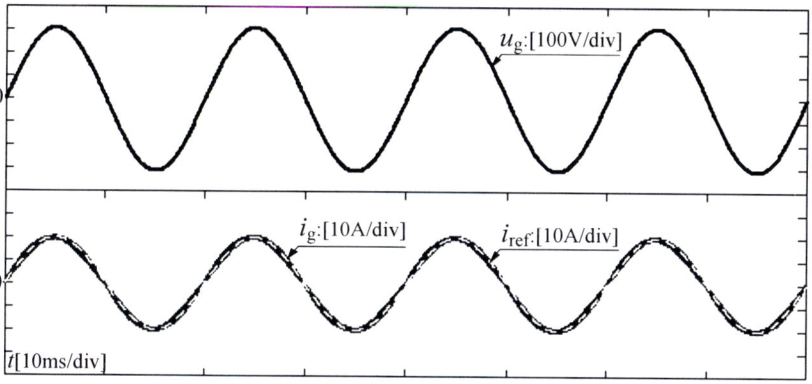

Fig. 3.139 Simulation waveforms of u_g, i_{ref}, and i_g of FB-CCV

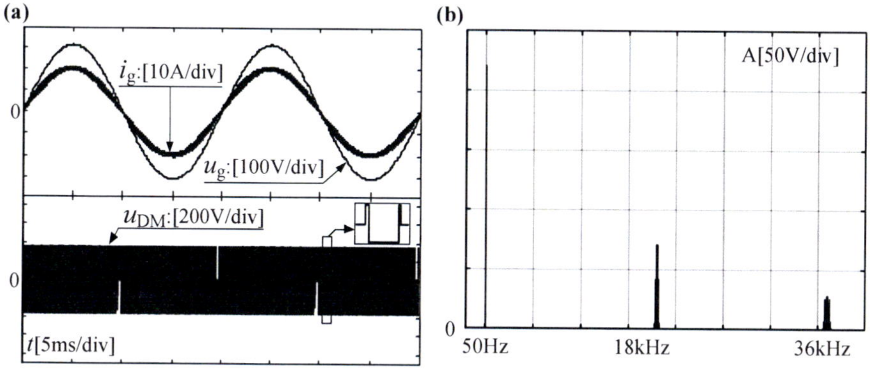

Fig. 3.140 DMV waveforms of FB-CCV, **a** time domain, **b** frequency domain

The simulation waveforms of CMV u_{CM} in time domain and frequency domain are shown in Fig. 3.141a, b, respectively. u_{CM} is constant and only contains fundamental component.

The simulation waveforms of LC i_{CM} in time domain and frequency domain are shown in Fig. 3.142a, b, respectively. From Fig. 3.142, it was observed that the LC decreased to 0.2 mA at switching frequency. It decreased greatly compared to HERIC.

3. Experimental results

In order to verify the theoretical analysis and simulation waveforms, a prototype of FB-CCV is built, and the key parameters are shown in Table 3.31.

The experimental waveforms of u_g, i_g, and u_{CM} in grid frequency and switching frequency scale are shown in Fig. 3.143a, b, respectively. From Fig. 3.143, DMV is quasi-unipolar three-level voltage due to dead time between S_5, S_6 and S_1–S_4. The grid voltage u_g, grid current i_g, voltage u_{1N} of midpoint 1 to terminal N, common-mode voltage $u_{CM} = u_{1N} + u_{2N}$, and voltage u_{2N} of middle point 2 to terminal N in grid

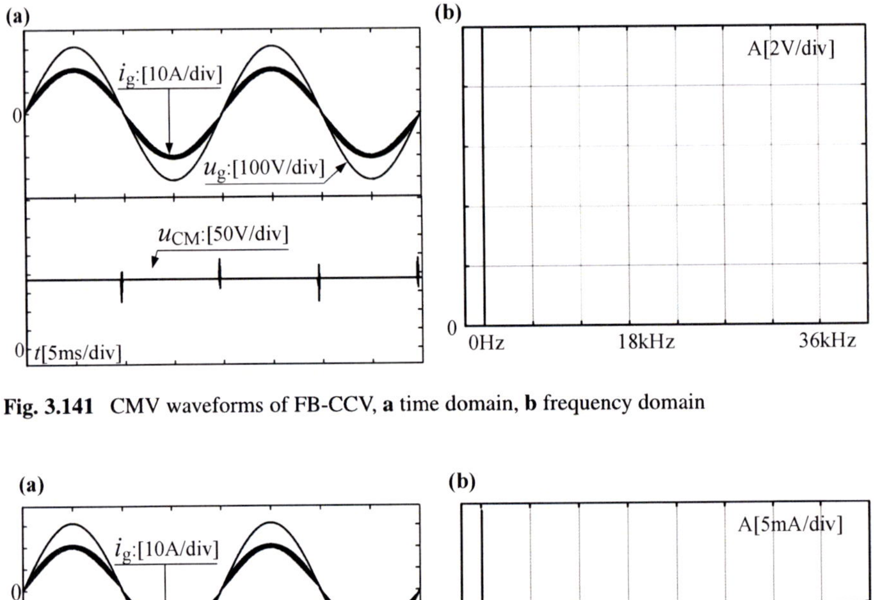

Fig. 3.141 CMV waveforms of FB-CCV, **a** time domain, **b** frequency domain

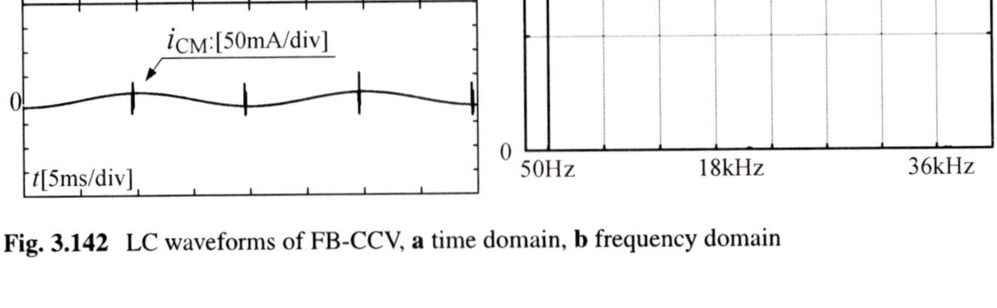

Fig. 3.142 LC waveforms of FB-CCV, **a** time domain, **b** frequency domain

Table 3.31 Key parameters of FB-CCV prototype

Parameters	Value	Parameters	Value
Input voltage U_{pv}/V	400	Freewheeling diodes	APT60DQ60B
Grid voltage u_g/V	220	Clamping diodes	IDP08E65D1
Grid frequency/Hz	50	Filter inductors L_1, L_2/mH	0.5
Rated power P/W	5000	Filter capacitor C/μF	2
Switching frequency f_s/kHz	20	Common-mode inductor L_{CM}	2 × B82726-S6223-N40
DC side capacitors C_{dc1}, C_{dc2}/μF	470 μF/350 V	Inductor value/mH	L: 2 × 1.6 mH
IGBT	IKW75N60T	Turns	Turns: 11 + 11
MOSFET	IPW65R019C7	Common-mode capacitor C_{Y1}, C_{Y2}/nF	22
CoolMOSFET	SPW47N60C3	PV parasitic capacitor C_{pv1}, C_{pv2}/μF	0.1

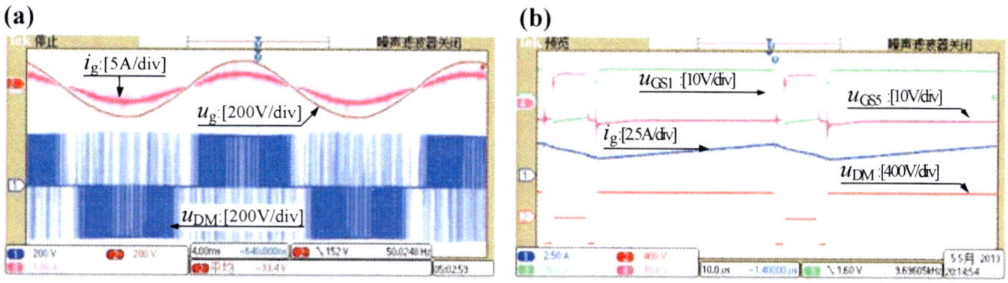

Fig. 3.143 Experimental waveforms of DMV in FB-CCV, **a** grid frequency scale, **b** switching frequency scale

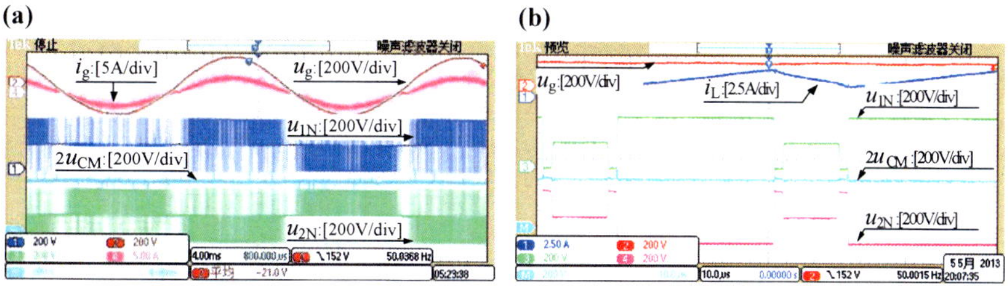

Fig. 3.144 Experimental waveforms of CMV in FB-CCV, **a** grid frequency scale, **b** switching frequency scale

frequency and switching frequency scale are shown in Fig. 3.144a, b, respectively. From Fig. 3.144, CMV is constant and in accordance to theoretical analysis.

3.4 Summary

Various full-bridge TLIs are analyzed in this chapter. It can be classified as decoupling and clamping categories. Besides, it can also be classified as AC side or DC side decoupling type TLIs. Clamping conception is designed to improve the CMV performance of full-bridge TLIs. Unidirectional and bidirectional clamping TLIs are two types based on clamping conception. Moreover, it can also be classified as active clamping or passive clamping TLIs based on the components used in the clamping circuit. Table 3.32 is the comparison of DMV and CMV performance of TLIs analyzed above and can also be used as the guidelines when choosing the proper TLIs.

Table 3.32 DMV and CMV performance of full-bridge TLIs

		Topologies	DMV	CMV
Decoupling	AC side	HERIC [1]	Unipolar	Uncertain when freewheeling
		H6-I [5]	Unipolar	Uncertain when freewheeling
		H6-II [7]	Unipolar	Uncertain when freewheeling
		H6-III [8]	Unipolar	Uncertain when freewheeling
		HERTLI [17]	Unipolar	Uncertain when freewheeling
	DC side	H5 [4]	Unipolar	Uncertain when freewheeling
		H6-IV [12]	Unipolar	Uncertain when freewheeling
		H6-V [13]	Unipolar	Uncertain when freewheeling
Clamping	Unidirectional	H5-D [22]	Unipolar	$\leq U_{pv}/2$
		FB-ZVR [26]	Quasi-unipolar	$\leq U_{pv}/2$
	Bidirectional	oH5 [21]	Unipolar	Uncertain during dead time
		H6-D [23]	Unipolar	Constant
		AH6-I [24]	Unipolar	Uncertain during dead time
		AH6-II [25]	Unipolar	Constant
		FB-ZVR-D [27]	Quasi-unipolar	Constant
		HERIC-TDCC [28]	Unipolar	Uncertain during dead time
		oHERIC [29]	Unipolar	Uncertain during dead time
		FB-CCV [30]	Quasi-unipolar	Constant

References

1. Schmidt H, Siedle C, Ketterer J (2003) Wechselrichter zum Unwandeln einer elektrischen Gleichspannung in einen Wechselstrom oder eine Wechselspannung, EP 2 086 102 A2
2. VDE 0126 1 1, 2006 (2008) Automatic disconnection device between a generator and the public low voltage grid. DIN VDE Normo
3. Technical requirements for connecting photovoltaic power station to power system (2012) Standardization Administration of the P.R.C, GB/T 19964-2012
4. Victor M, Greizer F, Bremicker S et al (2008) Method of converting a direct current voltage from a source of direct current voltage, more specifically from a photovoltaic source of direct current voltage, into a alternating current voltage. Am US 7411802:B2
5. Yu WS, Lai JS, Qian H et al (2011) High-efficiency MOSFET inverter with H6-type configuration for photovoltaic nonisolated AC-module applications. IEEE Trans Power Electron 26(4):1253–1260
6. Ji BJ, Wang JH, Zhao JF et al (2013) High efficiency single phase transformerless PV H6 inverter with hybrid modulation method. IEEE Trans Industr Electron 60(5):2104–2115
7. Islam M, Mekhilef S (2015) H6-type transformerless single-phase inverter for grid-tied photovoltaic system. IET Power Electron 8(4):636–644
8. Cui WF, Luo HZ, Gu YJ et al (2015) Hybrid-bridge transformerless photovoltaic grid-connected inverter. IET Power Electron 8(3):439–446
9. Cui WF, Yang B, Zhao Yi et al, A novel single-phase transformerless grid-connected inverter In: 2011 Annual conference of the IEEE industrial electronics society, pp 1126–1130
10. San GC, Qi HH, Wu JJ et al, A new three-level six-switch topology for transformerless photovoltaic systems, In: 2012 IEEE international power electronics and motion control conference, pp 163–166
11. Li WH, Gu YJ, Luo HZ et al (2015) Topology review and derivation methodology of single-phase transformerless photovoltaic inverters for leakage current suppression. IEEE Trans Industr Electron 62(7):4537–4551
12. Yang B, Li WH, Gu YJ et al (2012) Improved transformerless inverter with common-mode leakage current elimination for a photovoltaic grid-connected power system. IEEE Trans Power Electron 27(2):752–762
13. Zhang L, Sun K, Xing Y et al (2014) H6 transformerless full-bridge PV grid-tied inverters. IEEE Trans Power Electron 29(3):1229–1238
14. Chen BF, Sun PW, Liu C et al, High efficiency transformerless photovoltaic inverter with wide-range power factor capability In: 2012 IEEE applied power electronics conference and exposition, Orlando, pp 575–582
15. Chen BF, Gu B, Lai J, et al. Current distortion correction in dual buck photovoltaic inverter with a novel PWM modulation and control method. In: 2013 IEEE applied power electronics conference and exposition, Long Beach, pp 727–731
16. Garabandic D (2002) Method and apparatus for reducing switching losses in a switching circuit, American, US 6 847 196
17. Gu B, Dominic J, Lai J et al (2013) High reliability and efficiency single-phase transformerless inverter for grid-connected photovoltaic systems. IEEE Trans Power Electron 28(5):2235–2245
18. Chen BF, Gu B, Zhang L et al (2015) A high-efficiency MOSFET transformerless inverter for nonisolated microinverter applications. IEEE Trans Power Electron 30(7):3610–3622
19. Araujo SV, Zacharias P, Mallwitz R (2010) Highly efficient single phase transformer-less inverters for grid-connected photovoltaic systems. IEEE Trans Power Electron 57(9):3118–3128
20. Zhou LW, Gao F, Xu TY (2017) Implementation of active NPC circuits in transformer-less single-phase inverter with low leakage current. IEEE Trans Ind Appl 53(6):5658–5667
21. Xiao HF, Xie SJ, Yang C et al (2011) An optimized transformerless photovoltaic grid-connected inverter. IEEE Trans Industr Electron 58(5):1887–1895
22. Li H, Zeng Y, Zhang B et al (2019) An improved h5 topology with low common-mode current for transformerless pv grid connected inverter. IEEE Trans Power Electron 34(2):1254–1265

23. Gonzalez R, Lopez J, Sanchis P et al (2007) Transformerless inverter for single- phase photovoltaic systems. IEEE Trans Power Electron 22(2):693–697
24. Ahmad Z, Singh SN (2017) An improved single phase transformerless inverter topology for grid connected PV system with reduce leakage current and reactive power capability. Sol Energy 157:133–146
25. Zhang L, Sun K, Feng LL et al (2013) A family of neutral point clamped full-bridge topologies for transformerless photovoltaic grid-tied inverters. IEEE Trans Power Electron 28(2):730–739
26. Kerekes T, Teodorescu R, Rodriguez P et al (2011) A new high- efficiency single-phase transformerless PV inverter topology. IEEE Trans Industr Electron 58(1):184–191
27. Freddy T, Rahim NA, Hew WP et al (2014) Comparison and analysis of single-phase transformerless grid-connected PV inverters. IEEE Trans Power Electron 29(10):5358–5369
28. Mei Y, Hu S, Lin L et al (2016) Highly efficient and reliable inverter concept-based transformerless photovoltaic inverters with tri-direction clamping cell for leakage current elimination. IET Power Electron 9(8):1675–1683
29. Xiao HF, Liu XP, Lan K (2014) Optimised full-bridge transformerless photovoltaic grid-connected inverter with low conduction loss and low leakage current. IET Power Electron 7(4):1008–1015
30. Xiao HF, Lan K, Zhang L (2014) A quasi-unipolar SPWM full-bridge transformerless PV grid-connected inverter with constant common-mode voltage. IEEE Trans Power Electron 29(6):3122–3132

Chapter 4
Half-Bridge Transformerless PV Grid-Connected Inverters

Check for updates

Abstract *Rule* 1 for eliminating leakage current of full-bridge TLIs has been discussed in Chap. 3. In this chapter, *Rule* 2 to eliminate leakage current of half-bridge TLIs will be discussed in detail. Because the neutral line of the grid is directly connected to the midpoint of DC side, the inductor of output AC filter is asymmetrically placed. Therefore, the total CMV keeps constant.

Keywords Transformerless PV grid-connected inverter · Half-bridge · Three-level · Loss distribution

4.1 Conventional Half-Bridge TLI

A conventional single-phase two-level half-bridge inverter circuit is shown in Fig. 4.1a, U_{pv} is the output voltage of PV arrays, C_{dc1} and C_{dc2} are the DC voltage dividing capacitors, S_1 and S_2 represent power switches, and u_g is the grid voltage. Different from the full-bridge inverter circuit, the filter inductor *L* is only placed in live wire of the grid, so it is an asymmetric structure. *L* and *C* represent inductor and capacitor of the AC filter, respectively. Modulation strategy is shown in Fig. 4.1b, where u_{12} is the output voltage.

When half-bridge inverter works with unity power factor, four operating modes are shown in Fig. 4.2 according to the polarities of the grid voltage and grid-in current.

According to the definitions of DMV u_{DM} and CMV u_{CM} (referring to (2.1), (2.2)), and u_{CM_tot} (referring to (2.14)) in Chap. 2, $u_{CM\text{-}DM}$ and u_{CM_tot} are relisted in (4.1) and (4.2) for convenience of reading.

$$u_{CM-DM} = -\frac{u_{DM}}{2} \tag{4.1}$$

$$u_{CM\_tot} = u_{CM} + u_{CM-DM} = u_{CM} - \frac{u_{DM}}{2} = \frac{u_{PN}}{2} \tag{4.2}$$

H. Xiao and X. Wang, *Transformerless Photovoltaic Grid-Connected Inverters*, CPSS Power Electronics Series, https://doi.org/10.1007/978-981-15-8525-8_4

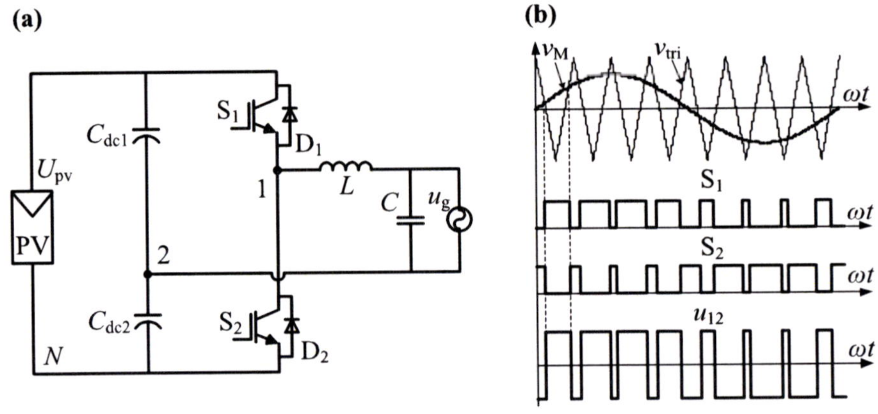

Fig. 4.1 Conventional two-level half-bridge TLI, **a** Main circuit, **b** Modulation strategy

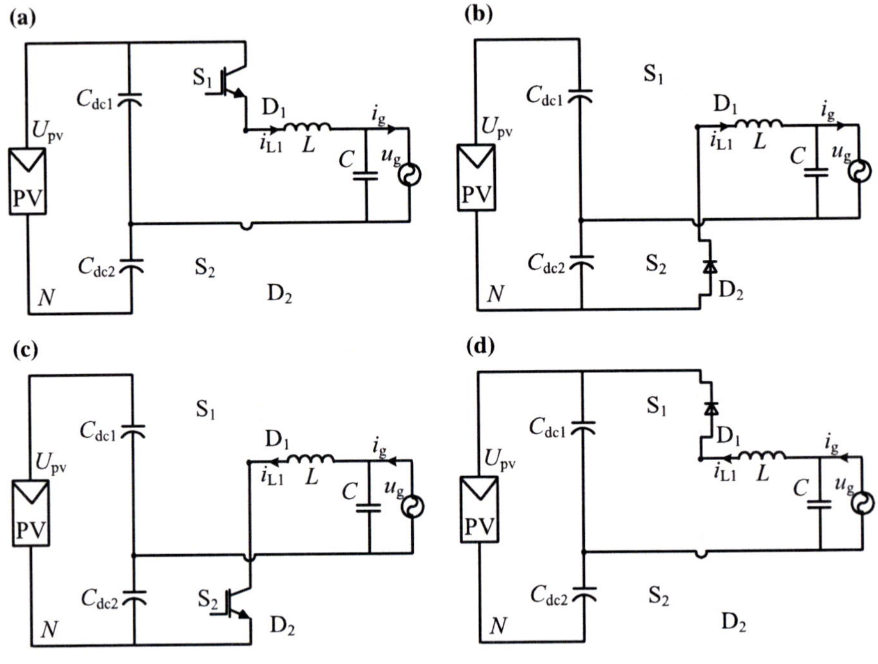

Fig. 4.2 Equivalent circuits of operating modes of bipolar SPWM modulation, **a** Mode 1, **b** Mode 2, **c** Mode 3, **d** Mode 4

Based on the operating modes, the potentials of points 1 and 2 against N, as well as DMV and CMV are shown in Table 4.1. It can be seen that DMV is a two-level voltage, and CMV is a constant value. As discussed in Chap. 3, the constant CMV characteristic is perfect for TLI applications, but two-level DMV characteristic

Table 4.1 DMV and CMV levels at different working stages with bipolar SPWM

Modes	u_{1N}	u_{2N}	u_{DM}	u_{CM}	$u_{CM\text{-}DM}$	u_{CM_tot}
(1)	U_{pv}	$U_{pv}/2$	$U_{pv}/2$	$3U_{pv}/4$	$-U_{pv}/4$	$U_{pv}/2$
(2)	0	$U_{pv}/2$	$-U_{pv}/2$	$U_{pv}/4$	$U_{pv}/4$	$U_{pv}/2$
(3)	0	$U_{pv}/2$	$-U_{pv}/2$	$U_{pv}/4$	$U_{pv}/4$	$U_{pv}/2$
(4)	U_{pv}	$U_{pv}/2$	$U_{pv}/2$	$3U_{pv}/4$	$-U_{pv}/4$	$U_{pv}/2$

has the disadvantages of abundant output harmonics and low conversion efficiency. Therefore, the combination of three-level DMV and constant CMV is developing direction of half-bridge TLIs.

4.2 I-Type Half-Bridge TLIs

I-type half-bridge TLIs mainly include neutral point clamping inverter (NPC) and its improved topologies. NPC was first proposed by Japanese scholars Akira Nabae, Isao Takahashi, and Hirofumi Akagi in 1981 [1], which was used by Danfoss Solar in TripleLynx series products (three-phase TLIs rated at about 10/12.5/15 kW). According to the report of Photon Magazine in July 2010, the European efficiency is 97%, and the peak efficiency is 98%.

4.2.1 NPC Inverter

1. Operating principle

The main circuit of NPC grid-connected inverter is shown in Fig. 4.3a [1]. Compared with the conventional half-bridge circuit, two active switch combinations S_2/D_2 and S_3/D_3 and two clamping diodes D_5 and D_6 are added. At the same time, the diodes are connected to the midpoint of the capacitor bridge to guarantee that the voltage stress of all power devices is reduced to half of the input voltage. In this circuit, a positive voltage level is obtained when switches S_1 and S_2 are on; a zero voltage is outputted while switches S_2 and S_3 are on; when switches S_3 and S_4 are on, the inverter outputs a negative voltage level.

There are four operating modes according to the polarities of the grid voltage and grid current when the power factor is 1, as shown in Fig. 4.4. Mode 1 and Mode 2 in Fig. 4.4a, b, respectively, work in positive half period of the grid voltage; symmetrically, Mode 3 and Mode 4 in Fig. 4.4c, d work in negative half period of the grid voltage.

Mode 1: Refer to Fig. 4.4a; the energy is delivered from the PV side into the grid through S_1, S_2, and the AC filter at this stage. The switches S_3 and S_4 are off, and

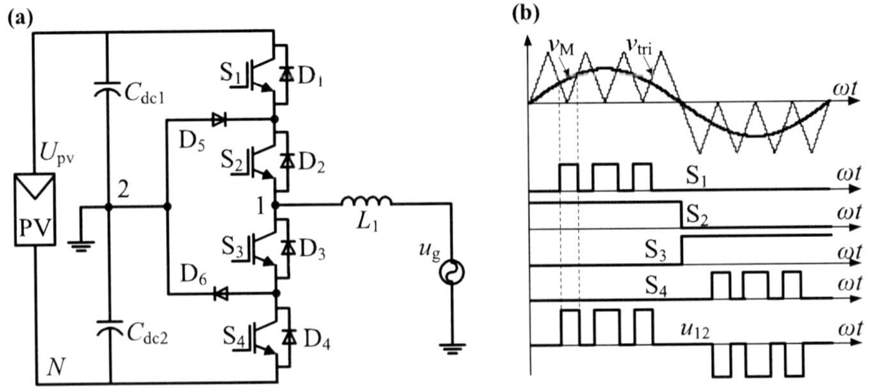

Fig. 4.3 NPC grid-connected inverter, **a** Main circuit, **b** Modulation strategy

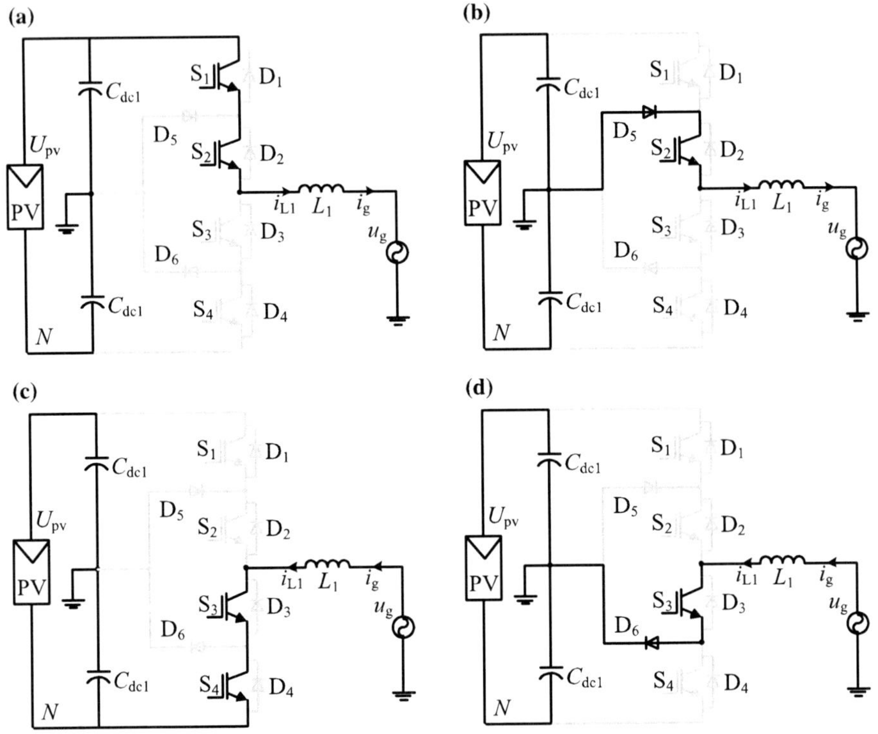

Fig. 4.4 Equivalent circuits of operating modes of NPC, **a** Mode 1, **b** Mode 2, **c** Mode 3, **d** Mode 4

share the input voltage U_{pv} in series, so their turn-off voltage stress is half of the input voltage due to the existence of clamping diode D_6.

Mode 2: Refer to Fig. 4.4b; only S_2 is on and rest of the switches are off according to Fig. 4.3b. At this stage, the freewheeling current flows through D_5, S_2, and the filter, and keeps feeding the grid. It is worth noting that the potential of the freewheeling path is clamped at the ground.

Mode 3: Refer to Fig. 4.4c; the energy is delivered from the PV side into the grid through S_3, S_4, and the AC filter at this stage. The switches S_1 and S_2 are off, and evenly share the input voltage U_{pv} because of clamping diode D_5.

Mode 4: Refer to Fig. 4.4d; only S_3 is on and rest of the switches are off according to Fig. 4.3b. At this stage, the freewheeling current flows through D_6, S_3, and the filter, and keeps feeding the grid. At the same time, the potential of the freewheeling path is clamped at the ground.

According to the aforementioned analysis of operating modes, the potentials of points 1 and 2 against N as well as related to DMV and CMV are shown in Table 4.2. It can be seen that the output voltage u_{DM} is a three-level voltage with unipolar SPWM, and the total CMV u_{CM_tot} remains constant. As a result, NPC inverter can be adopted in distributed TLI-PVPG system with low LC and high efficiency. Besides, the circuit structure of NPC can block the DC component in the grid-in current, and this issue will be further discussed in Chap. 6.

2. Simulation results

In order to verify the aforementioned analysis, a simulation model of NPC grid-connected inverter is established in Matlab/Simulink, and the key parameters are listed in Table 4.3.

Table 4.2 DMV and CMV levels of different operating modes for NPC

Modes	u_{1N}	u_{2N}	u_{DM}	u_{CM}	$u_{CM\text{-}DM}$	u_{CM_tot}
(1)	U_{pv}	$U_{pv}/2$	$U_{pv}/2$	$3U_{pv}/4$	$-U_{pv}/4$	$U_{pv}/2$
(2)	$U_{pv}/2$	$U_{pv}/2$	0	$U_{pv}/2$	0	$U_{pv}/2$
(3)	0	$U_{pv}/2$	$-U_{pv}/2$	$U_{pv}/4$	$U_{pv}/4$	$U_{pv}/2$
(4)	$U_{pv}/2$	$U_{pv}/2$	0	$U_{pv}/2$	0	$U_{pv}/2$

Table 4.3 Key parameters of NPC

Parameters	Value	Parameters	Value
Input voltage U_{pv}/V	720	Switching frequency f_s/kHz	18
Grid voltage u_g/V, frequency/Hz	220/50	Filter inductor L_1/mH	1.72
Rated power P/W	3000	Parasitic capacitance of PV panel C_{pv}/nF	300

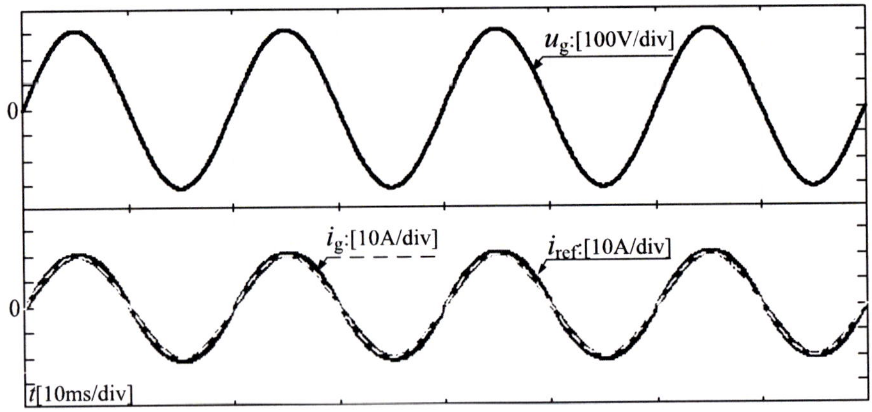

Fig. 4.5 Simulation waveforms of u_g, i_{ref}, and i_g of NPC

When the NPC operates with unity power factor, the steady-state waveforms of the grid voltage u_g, reference signal of grid-in current i_{ref}, and actual grid-in current i_g are shown in Fig. 4.5.

The simulation waveforms of DMV u_{DM} in time domain and frequency domain are shown in Fig. 4.6a, b, respectively. It can be observed that the RMS of DMV is about 75 V at switching frequency, which is equivalent to unipolar SPWM full-bridge (referring to Fig. 3.5b in Chap 3).

The simulation waveforms of total CMV u_{CM_tot} in time domain and frequency domain are shown in Fig. 4.7a, b, respectively. It can be observed that the amplitude of CMV is zero at switching frequency, but the DC component of CMV is significant, and the simulation waveforms are in accordance with theoretical analysis. The simulation waveforms of LC i_{CM} in time domain and frequency domain are shown in Fig. 4.8a, b, respectively. It is outstanding that no LC appears among whole frequency span. It proves that the DC component of CMV does not induce LC.

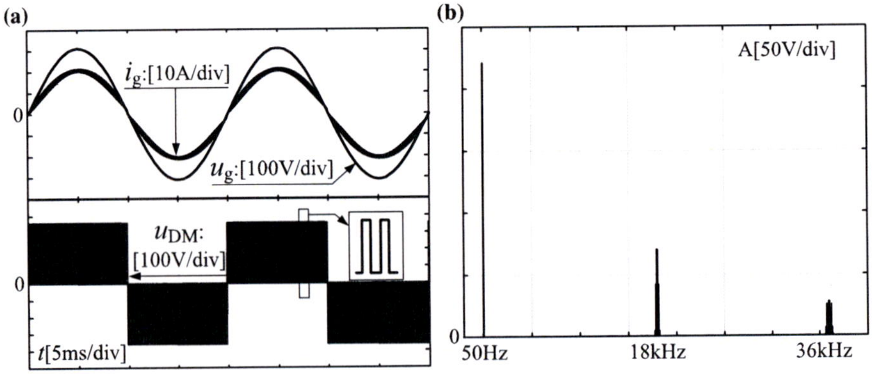

Fig. 4.6 DMV waveforms of NPC, **a** Time domain, **b** Frequency domain

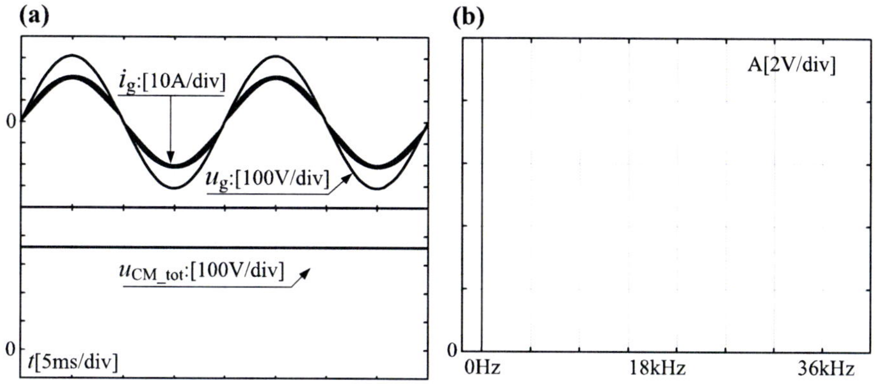

Fig. 4.7 CMV waveforms of NPC, **a** Time domain, **b** Frequency domain

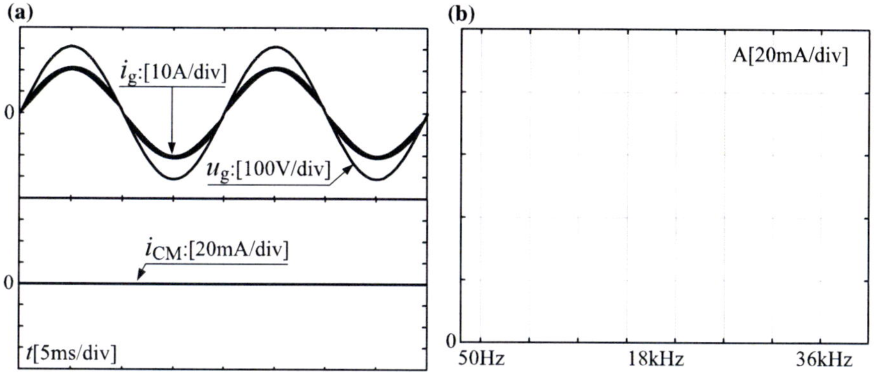

Fig. 4.8 LC waveforms of NPC, **a** Time domain, **b** Frequency domain

3. Experimental results

In order to further verify the effectiveness of operating principle and simulation waveforms, a grid-connected experimental platform with NPC is established in lab, and the key component parameters are shown in Table 4.4.

The experimental waveforms of the grid voltage u_g, grid-in current i_g, bridge voltages u_{1N} and u_{2N} are shown in Fig. 4.9. It can be seen that the middle point voltage of input capacitors u_{2N} has a line-frequency fluctuation because of output power pulse, and it is consistent with the practice.

The waveforms of the LC i_{CM} are shown in Fig. 4.10. It is necessary to point out that the PV output was simulated by using a DC source with solar array simulation in this experiment, and this is the reason that some LC components appear in Fig. 4.10a when the circuit breaker is closed and the NPC inverter has not started yet. After the NPC inverter starts working to deliver the energy into the grid, a new LC component appears in Fig. 4.10b, and its frequency is the switching frequency of the NPC inverter

Table 4.4 Key parameters of NPC prototype

Parameters	Value	Parameters	Value
Input voltage U_{pv}/V	800	Filter inductor L_1/mH	8
Grid voltage u_g/V, frequency/Hz	240, 50	Common-mode inductor L_{CM}	2 × 2 W-43615-TC
Rated power P/W	1000	Wire diameter	2 mm
Switching frequency f_s/kHz	20	Number of turns	10 + 10
Capacitance of DC side C_{dc1}, C_{dc2}/μF	235 μF/400 V	Common-mode capacitor C_{Y1}, C_{Y2}/nF	2.2
Switches S_1–S_2	IXFN36N100	Parasitic capacitance of PV panel C_{pv1}, C_{pv2}/μF	0.1

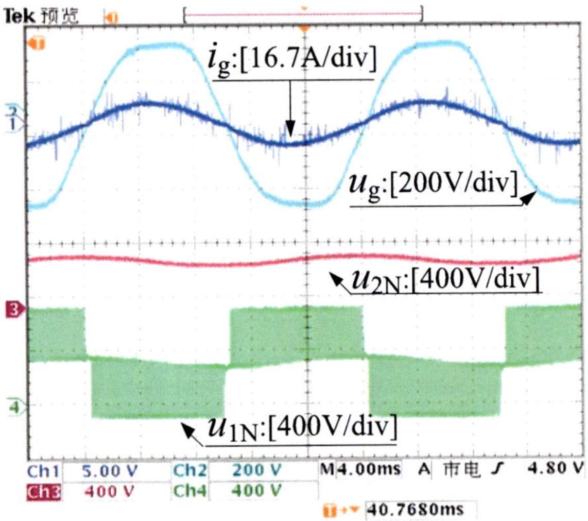

Fig. 4.9 Output voltage waveforms of NPC

(i.e., 20 kHz). By equivalent calculation, the value of LC is about 1 mA, which meets the limitation of DIN VDE 0126-1-1 and NB/T 32004-2013 of 30 mA.

4.2.2 Active NPC Inverter

The aforementioned NPC inverter features high efficiency as well as low LC and DC current in the grid-in current, but the active switches in the bridge arm have uneven loss distribution, and this will jeopardize the reliability of distributed PVPG system. In order to solve this issue, active neutral point clamped three-level inverter (A-NPC) is proposed.

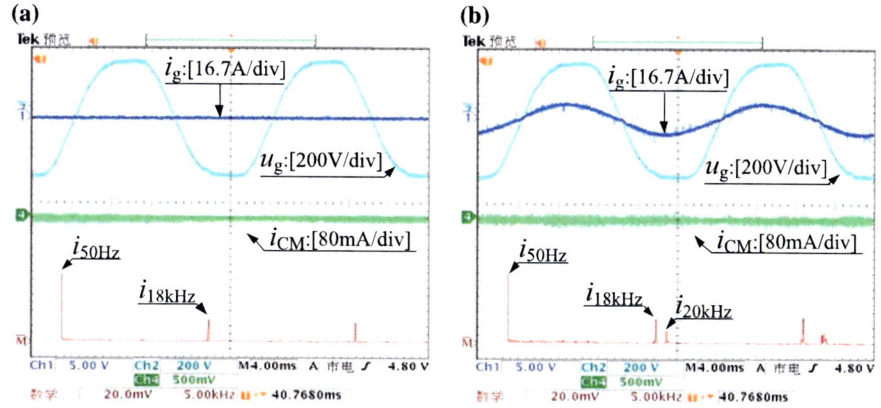

Fig. 4.10 LC waveforms of NPC, **a** NPC inverter not working, **b** NPC inverter working

1. Operating principle

Compared with NPC, the diodes of A-NPC are replaced by active switches S_5/D_5 and S_6/D_6 to realize active clamping, as shown in Fig. 4.11a. The voltage of neutral point is clamped when upper switches S_2 and S_5 are on, as well as the case that lower switches S_3 and S_6 are on. And thus the freewheeling current flows through the upper clamp branch or the lower clamp branch. A variety of modulation strategies can be adopted in A-NPC [3–6], among which two conventional modulation strategies are discussed in this book, as shown in Fig. 4.11b, c.

There are four operating modes according to the polarities of the grid voltage and grid current when A-NPC works in unity power factor case with modulation strategy I, as shown in Fig. 4.12. Mode 1 and Mode 2 in Fig. 4.12a, b, respectively, work in positive half period of the grid voltage; symmetrically, Mode 3 and Mode 4 work in negative half period of the grid voltage.

Mode 1: Refer to Fig. 4.12a; the energy is delivered from the PV side into the grid through S_1, S_2, and the AC filter. The output voltage of the bridge is $U_{pv}/2$ at this stage.

Mode 2: Refer to Fig. 4.12b; S_1, S_6 are on and rest of the switches are off according to Fig. 4.11b. The freewheeling current flows through S_6, D_3, and the filter, and keeps feeding the grid at this stage. It is worth noting that the potential of the freewheeling path is clamped at the ground.

Mode 3: Refer to Fig. 4.12c; the energy is delivered from the PV side into the grid through S_3, S_4, and the AC filter. The output voltage of the bridge is $-U_{pv}/2$ at this stage.

Mode 4: Refer to Fig. 4.12d; S_2, S_5 are on and rest of the switches are off according to Fig. 4.11b. Consequently, the freewheeling current flows through S_5, D_2, and the filter at this stage. It is worth noting that the potential of the freewheeling path is clamped at the ground.

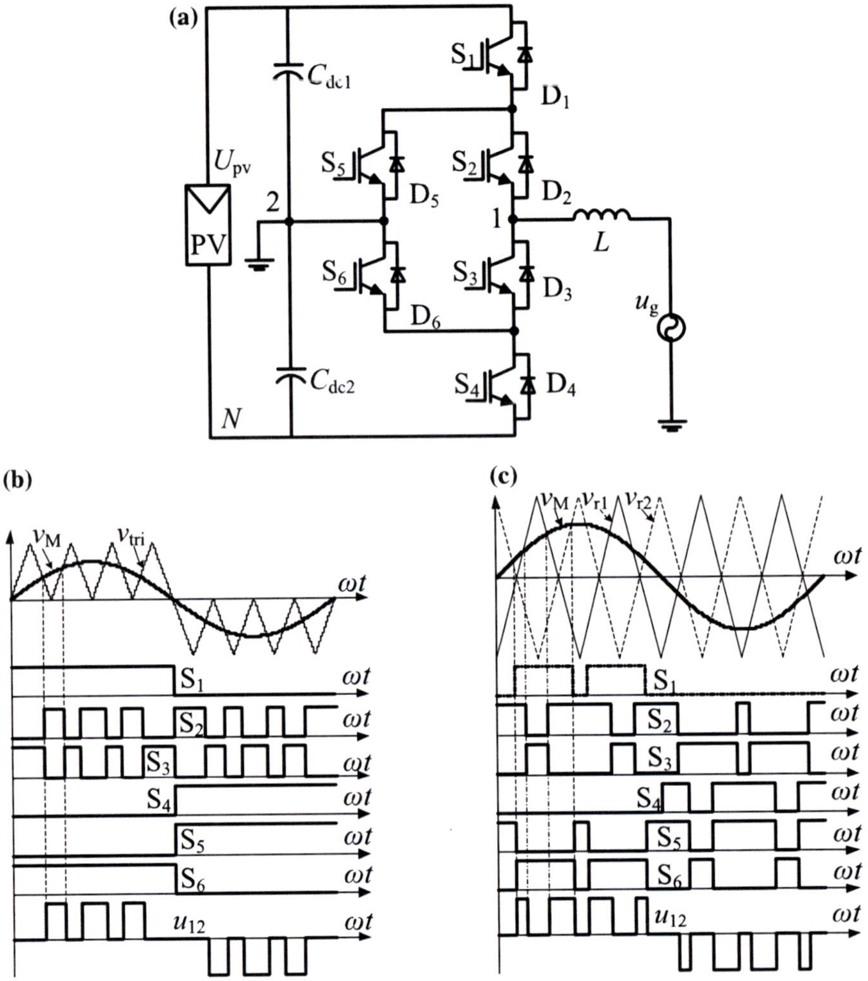

Fig. 4.11 Acitve neutral point clamping inverter, **a** Main circuit, **b** Modulation strategy I, **c** Modulation strategy II

Based on the operating modes, the potentials of points 1 and 2 against N, as well as DMV and CMV are shown in Table 4.5. Different from operating characteristics of NPC, it can be noticed that opposite clamping branch in the freewheeling stage will optimize the distribution of switching loss. Subsequently, the operating modes of A-NPC with modulation strategy II will be discussed.

There are six operating modes according to the polarities of the grid voltage and grid current when A-NPC works in unity power factor case with modulation strategy II, as shown in Fig. 4.13. Mode 1, Mode 2, and Mode 3 in Fig. 4.13a–c, respectively, work in positive half period of the grid voltage; symmetrically, Mode 3, Mode 4,

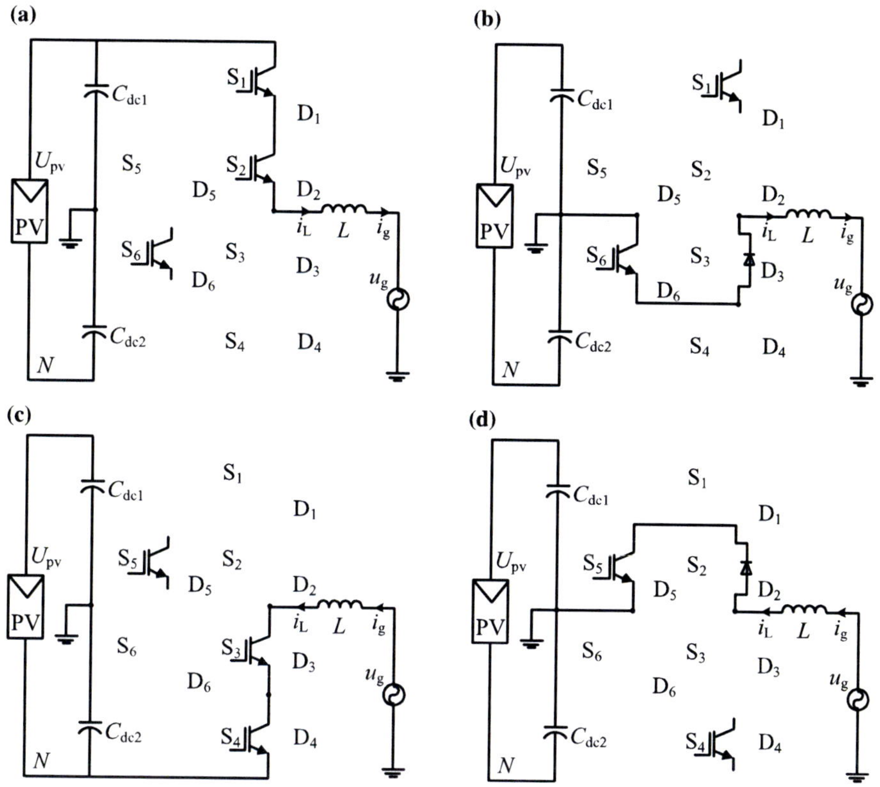

Fig. 4.12 Equivalent circuits of operating modes of A-NPC with modulation strategy I, **a** Mode 1, **b** Mode 2, **c** Mode 3, **d** Mode 4

Table 4.5 DMV and CMV levels of different operating modes for A-NPC with modulation strategy I

Modes	u_{1N}	u_{2N}	u_{DM}	u_{CM}	$u_{CM\text{-}DM}$	u_{CM_tot}
(1)	U_{pv}	$U_{pv}/2$	$U_{pv}/2$	$3U_{pv}/4$	$-U_{pv}/4$	$U_{pv}/2$
(2)	$U_{pv}/2$	$U_{pv}/2$	0	$U_{pv}/2$	0	$U_{pv}/2$
(3)	0	$U_{pv}/2$	$-U_{pv}/2$	$U_{pv}/4$	$U_{pv}/4$	$U_{pv}/2$
(4)	$U_{pv}/2$	$U_{pv}/2$	0	$U_{pv}/2$	0	$U_{pv}/2$

and Mode 5 in Fig. 4.13d–f, respectively, work in negative half period of the grid voltage.

Mode 1: Refer to Fig. 4.13a; S_2, S_5 are on and rest of the switches are off according to Fig. 4.11c. At this stage, the freewheeling current flows through D_5, S_2, and the filter, and keeps feeding the grid. At the same time, the potential of the freewheeling path is clamped at the ground.

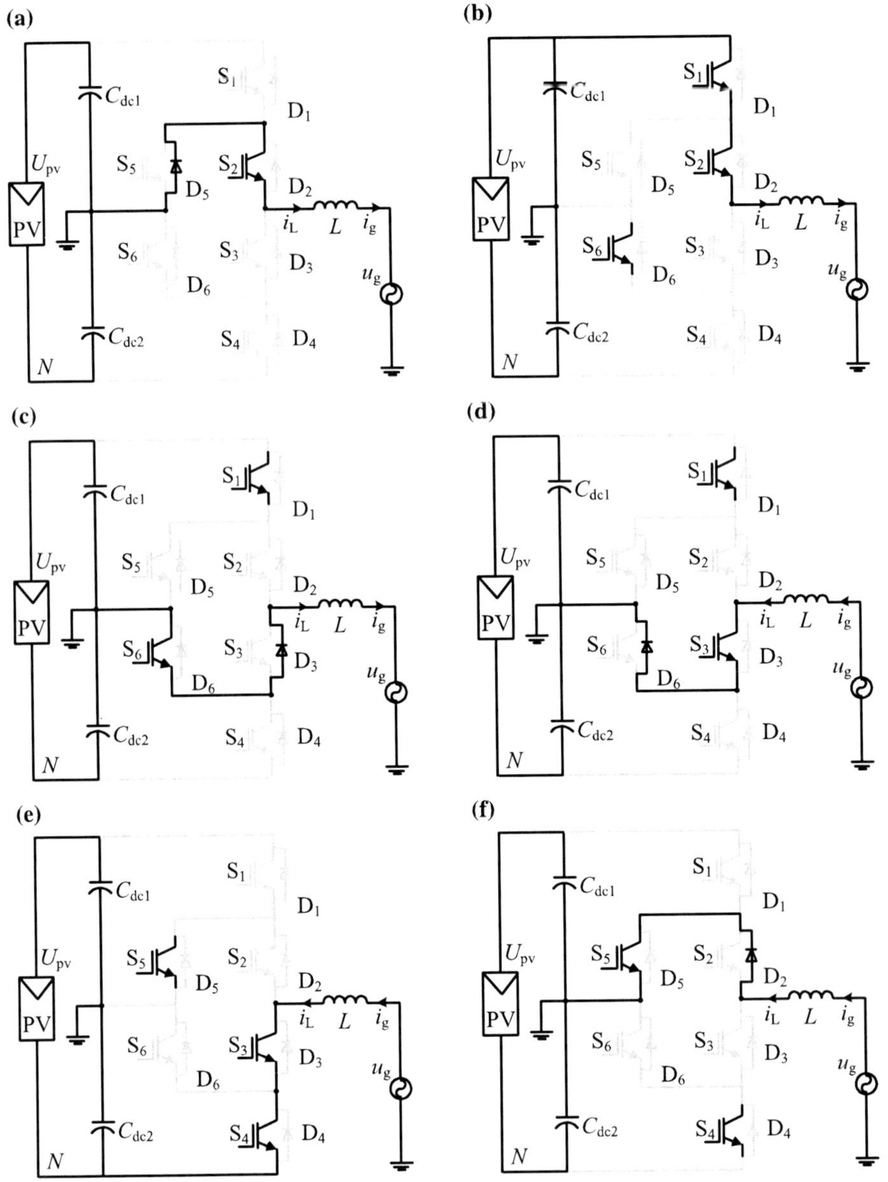

Fig. 4.13 Equivalent circuits of operating modes of A-NPC with modulation strategy II, **a** Mode 1, **b** Mode 2, **c** Mode 3, **d** Mode 4, **e** Mode 5, **f** Mode 6

Mode 2: Refer to Fig. 4.13b; the energy is delivered from the PV side into the grid through S_1, S_2, and the AC filter at this stage. The switches S_3 and S_4 are off according to Fig. 4.11c, and share the input voltage U_{pv} in series, so their turn-off voltage stress is half of the input voltage.

Mode 3: Refer to Fig. 4.13c; S_1, S_3, S_6 are on and rest of the switches are off according to Fig. 4.11c. At this stage, the freewheeling current flows through S_6, D_3, and the filter, and keeps feeding the grid. At the same time, the potential of the freewheeling path is also clamped at the ground.

Mode 4: Refer to Fig. 4.13d; S_3, S_6 are on and rest of the switches are off according to Fig. 4.11c. The freewheeling current flows through D_6, S_3, and the filter at this stage. Meanwhile, the potential of the freewheeling path is clamped at the ground.

Mode 5: Refer to Fig. 4.13e; the energy is delivered from the PV side into the grid through S_4, S_3, and the AC filter at this stage. The switches S_1 and S_2 are off according to Fig. 4.11c, and share the input voltage U_{pv} in series, so their turn-off voltage stress is half of the input voltage.

Mode 6: Refer to Fig. 4.13f; S_2, S_4, S_5 are on and rest of the switches are off according to Fig. 4.11c. The freewheeling current flows through S_5, D_2, and the filter at this stage. Meanwhile, the potential of the freewheeling path is clamped at the ground.

Based on the operating modes, the potentials of points 1 and 2 against N, as well as DMV and CMV are shown in Table 4.6.

Furthermore, the loss distribution of NPC and A-NPC in unity power factor case is shown in Figs. 4.14 and 4.15 [4]. Loss of the outermost switch S_1 and the innermost switch S_2 varies significantly when modulation strategy I is used for A-NPC, as shown in Fig. 4.15a. Consequently, the life and output power of the inverter will be limited by individual switch. Meanwhile, the total loss of NPC is highly in accordance with that of A-NPC with modulation strategy I, because they have a consistent model in the energy-delivering stage and an equivalent model in the freewheeling stage.

The loss distribution of A-NPC with modulation strategy II is shown in Fig. 4.15b. It is worth mentioning that the difference between power loss of the outermost switch S_1 and the innermost switch S_2 is significantly reduced, which will bring advantage for heat dissipation and improvement of inverter life.

Table 4.6 DMV and CMV levels of different operating modes for A-NPC with modulation strategy II

Modes	u_{1N}	u_{2N}	u_{DM}	u_{CM}	$u_{CM\text{-}DM}$	u_{CM_tot}
(1)	$U_{pv}/2$	$U_{pv}/2$	0	$U_{pv}/2$	0	$U_{pv}/2$
(2)	U_{pv}	$U_{pv}/2$	$U_{pv}/2$	$3U_{pv}/4$	$-U_{pv}/4$	$U_{pv}/2$
(3)	$U_{pv}/2$	$U_{pv}/2$	0	$U_{pv}/2$	0	$U_{pv}/2$
(4)	$U_{pv}/2$	$U_{pv}/2$	0	$U_{pv}/2$	0	$U_{pv}/2$
(5)	0	$U_{pv}/2$	$-U_{pv}/2$	$U_{pv}/4$	$U_{pv}/4$	$U_{pv}/2$
(6)	$U_{pv}/2$	$U_{pv}/2$	0	$U_{pv}/2$	0	$U_{pv}/2$

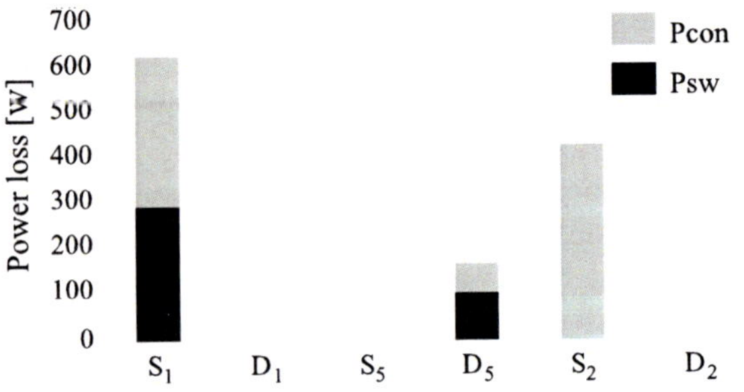

Fig. 4.14 Loss distribution of NPC

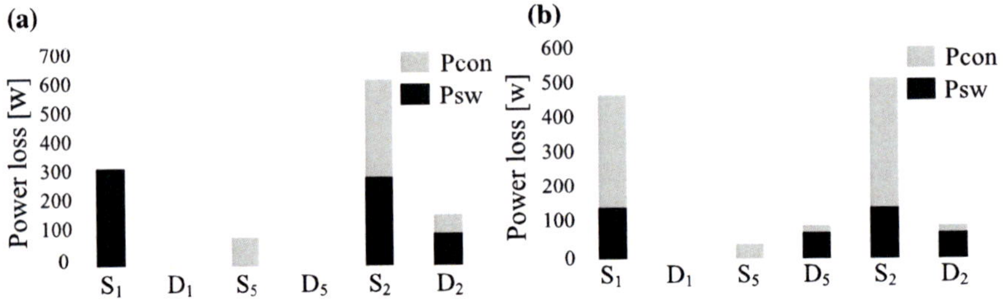

Fig. 4.15 Loss distributions of A-NPC with different modulation strategies, **a** Loss distribution with modulation strategy I, **b** Loss distribution with modulation strategy II

As indicated by aforementioned discussion, the existence of two freewheeling stages in each switching cycle yields more promising loss distribution. Besides, there are two voltage pulses in every switching cycle when modulation strategy II is used, which breeds an equivalent effect of double frequency in A-NPC.

2. Simulation results

In order to verify the aforementioned operating principle and property analysis, a simulation model of A-NPC grid-connected inverter is established in Matlab/Simulink, and the key parameters are listed in Table 4.3.

When modulation strategy I is used in A-NPC, the simulation waveforms of DMV u_{DM} in time domain and frequency domain are shown in Fig. 4.16a, b, respectively. In addition to the case that modulation strategy II is used in A-NPC, the associated simulation waveforms are shown in Fig. 4.17. It can be seen that the amplitude of DMV is zero at switching frequency, but there exists double frequency component. An equivalent effect of double frequency appears due to the adoption of modulation strategy II, which is consistent with the theoretical analysis. Besides, the CMV characteristics of A-NPC are in accordance with that of NPC, as shown in the simulation waveforms (Figs. 4.5, 4.6, 4.7, and 4.8).

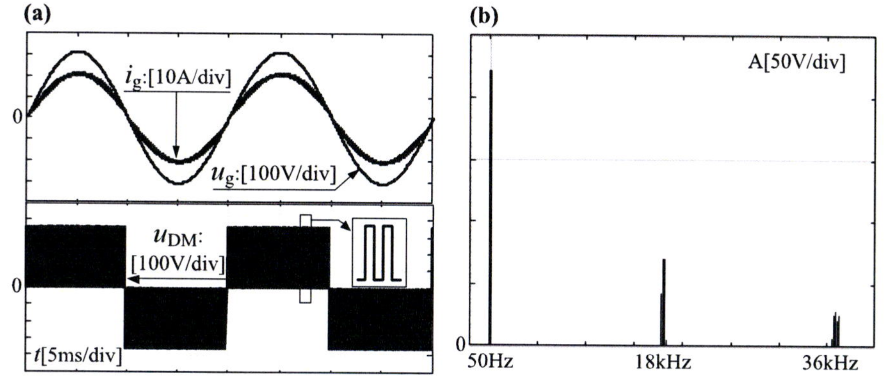

Fig. 4.16 DMV waveforms of A-NPC with modulation strategy I, **a** Time domain, **b** Frequency domain

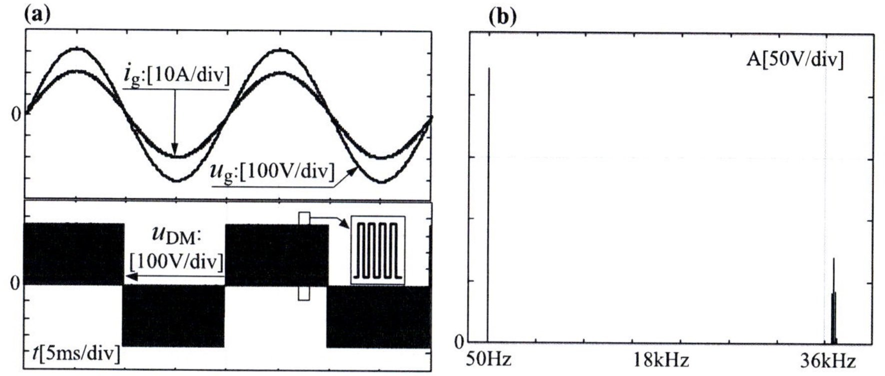

Fig. 4.17 DMV waveforms of A-NPC with modulation strategy II, **a** Time domain, **b** Frequency domain

4.2.3 Double Capacitors Legs NPC Inverter

1. Operating principle

The DC component of the grid-in current can be reduced by the shunt capacitor of DC side in NPC, but the grid-in current does not flow through the capacitor during freewheeling stages, which will introduce DC component to a certain extent. As shown in Fig. 4.18a, double capacitors legs neutral point clamped three-level inverter was proposed to further reduce DC component of grid-in current [7]. Compared with NPC, a shunt capacitor bridge composed of capacitors C_{dc3} and C_{dc4} is added to the circuit while the midpoint is connected to the neutral point of the grid, as main circuit shown in Fig. 4.18a. In accordance with the modulation strategy of NPC, the modulation strategy of NPC-C is shown in Fig. 4.18b.

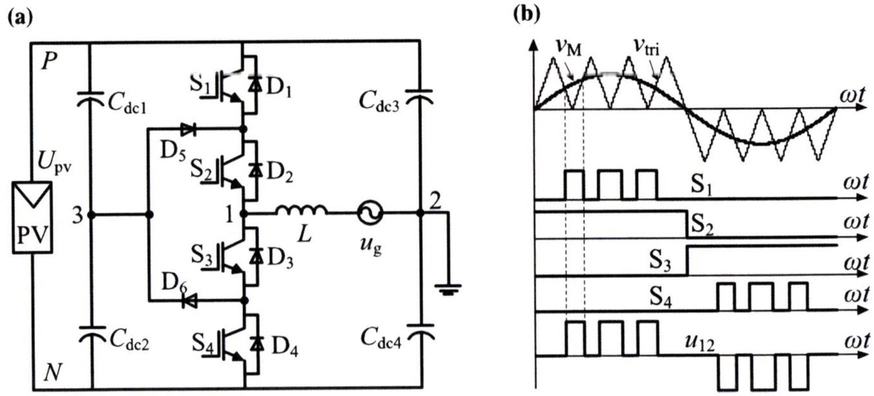

Fig. 4.18 NPC-C grid-connected inverter, **a** Main circuit, **b** Modulation strategy

There are four operating modes according to the polarities of the grid voltage and grid current when NPC-C works in unity power factor case, as shown in Fig. 4.19. Mode 1 and Mode 2 in Fig. 4.19a, b, respectively, work in positive half period of the

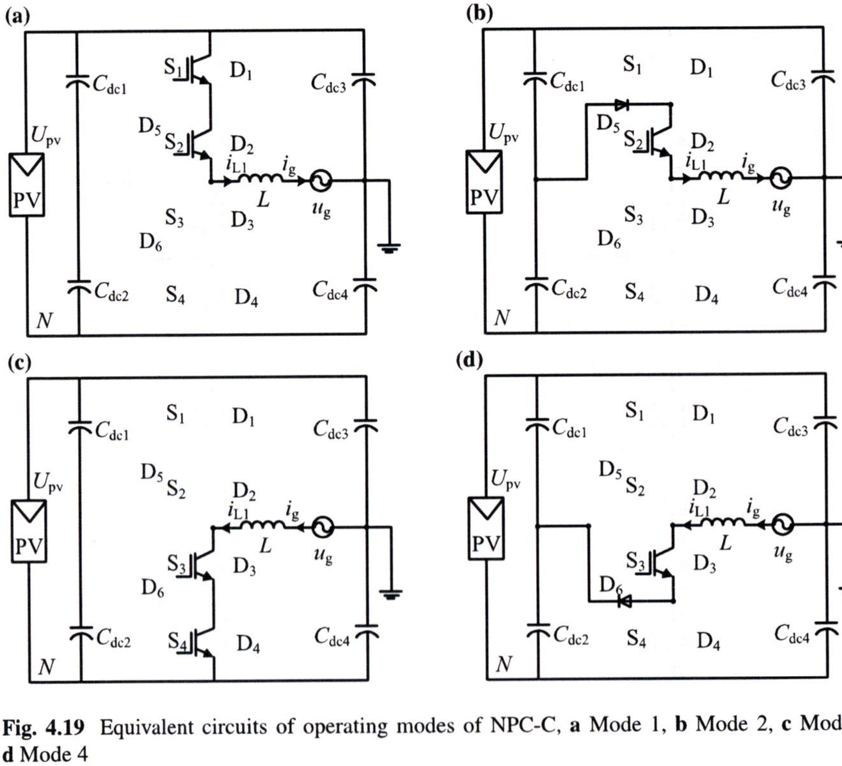

Fig. 4.19 Equivalent circuits of operating modes of NPC-C, **a** Mode 1, **b** Mode 2, **c** Mode 3, **d** Mode 4

grid voltage; symmetrically, Mode 3 and Mode 4 in Fig. 4.19c, d, respectively, work in negative half period of the grid voltage.

Mode 1: Refer to Fig. 4.19a; the energy is delivered from the PV side into the grid through S_1, S_2, and the AC filter at this stage. The switches S_3 and S_4 are off according to Fig. 4.18b, and share the input voltage U_{pv} in series, so their turn-off voltage stress is half of the input voltage.

Mode 2: Refer to Fig. 4.19b; only S_2 is on and rest of the switches are off according to Fig. 4.18b. At this stage, the freewheeling current flows through D_5, S_2, and the filter, and keeps feeding the grid. It is worth noting that the potential of the freewheeling path is clamped at the ground.

Mode 3: Refer to Fig. 4.19c; the energy is delivered from the PV side into the grid through S_4, S_3, and the AC filter at this stage. The switches S_1 and S_2 are off according to Fig. 4.18b, and share the input voltage U_{pv} in series, so their turn-off voltage stress is half of the input voltage.

Mode 4: Refer to Fig. 4.19d; only S_3 is on and rest of the switches are off according to Fig. 4.18b. At this stage, the freewheeling current flows through D_6, S_3, and the filter, and keeps feeding the grid. It is worth noting that the potential of the freewheeling path is clamped at the ground.

As indicated by aforementioned operating principle, the switches S_1 and S_2 are simultaneously conducted at the energy-delivering stage with positive grid voltage, and then the value of u_{12} can be derived as $U_{pv}-u_{Cdc4}$. In addition to the freewheeling stage with positive grid voltage, u_{12} can be expressed as $u_{Cdc2}-u_{Cdc4}$. At the energy-delivering stage and the freewheeling stage with negative grid voltage, the value of u_{12} can be obtained by $u_{Cdc2}-U_{pv}$ and $u_{Cdc2}-u_{Cdc4}$, respectively. If u_{Cdc2} and u_{Cdc4} have an equivalent value, the DMV of NPC-C is shaped as unipolar and three-level waveform, which is basically consistent with NPC. Therefore, the voltage of capacitor C_{dc4} should be controlled as a constant value to effectively balance the input voltage and reduce DC component of the grid-in current.

Based on the operating modes, the potentials of points 1 and 2 against N, as well as DMV and CMV are shown in Table 4.7. It can be noticed that the total CMV u_{CM_tot} of NPC-C is kept in a constant value. The voltage balance of the input-side capacitor, as well as that of the output-side capacitor is crucial in the controlling of NPC-C.

Table 4.7 DMV and CMV levels of different operating modes for NPC-C

Modes	u_{1N}	u_{2N}	u_{DM}	u_{CM}	$u_{CM\text{-}DM}$	u_{CM_tot}
(1)	U_{pv}	$U_{pv}/2$	$U_{pv}/2$	$3U_{pv}/4$	$-U_{pv}/4$	$U_{pv}/2$
(2)	$U_{pv}/2$	$U_{pv}/2$	0	$U_{pv}/2$	0	$U_{pv}/2$
(3)	0	$U_{pv}/2$	$-U_{pv}/2$	$U_{pv}/4$	$U_{pv}/4$	$U_{pv}/2$
(4)	$U_{pv}/2$	$U_{pv}/2$	0	$U_{pv}/2$	0	$U_{pv}/2$

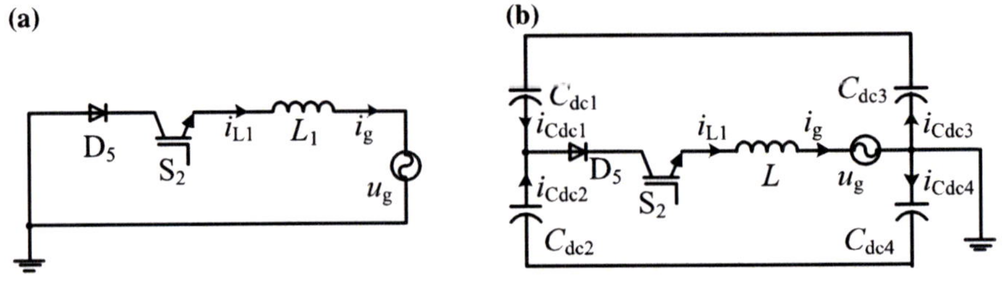

Fig. 4.20 Equivalent freewheeling circuits of NPC and NPC-C, **a** NPC, **b** NPC-C

2. Simulation results

In order to verify the aforementioned operating principle and property analysis, a simulation model of NPC-C inverter is established in Matlab/Simulink, and the key parameters in this circuit are listed in Table 4.3. Subsequently, the CMV and DMV characteristics are consistent with that of NPC, as simulation waveforms are shown in Figs. 4.5, 4.6, 4.7, and 4.8.

In positive half period of the grid voltage, equivalent circuits of the freewheeling stage of NPC and NPC-C are shown in Fig. 4.20. It can be noticed that the grid current flows through D_5, S_2, and the filter in Fig. 4.20a, while the grid current flows through D_5, S_2, the filter, and the capacitors in Fig. 4.20b. As a result, the DC component of grid-in current is significantly reduced with the isolation of capacitors.

When the reference of grid-in current i_{ref} is configured as $20\sin\omega t + 1$, the simulation waveforms of grid-in current of HERIC, NPC, and NPC-C in time domain and frequency domain are shown in Fig. 4.21. Furthermore, the spectrum of grid-in current is listed in Table 4.8, in which 0 Hz represents the DC component and 50 Hz is the fundamental current. In accordance with the theoretical analysis, NPC and NPC-C have better DC-suppression effect than HERIC, while the DC component NPC-C is eliminated the most significantly.

NPCs aforementioned are generally used in occasions with high-input DC voltage. However, there still exists the shoot through risk of bridge, as indicated by analysis of operating modes in NPC, A-NPC, and NPC-C. Therefore, the NPC with split-inductor was proposed to improve operational reliability.

4.2.4 *Split-Inductor NPC Inverter*

1. Operating principle

The main circuit of split-inductor neutral point camped three-level inverter (SI-NPC) is shown in Fig. 4.22a [8], where output point of NPC circuit is separated by two inductors connected to the grid voltage. The switches S_1 and S_2 are on in positive half period of the grid voltage, as shown in Fig. 4.22b. Symmetrically, the switches

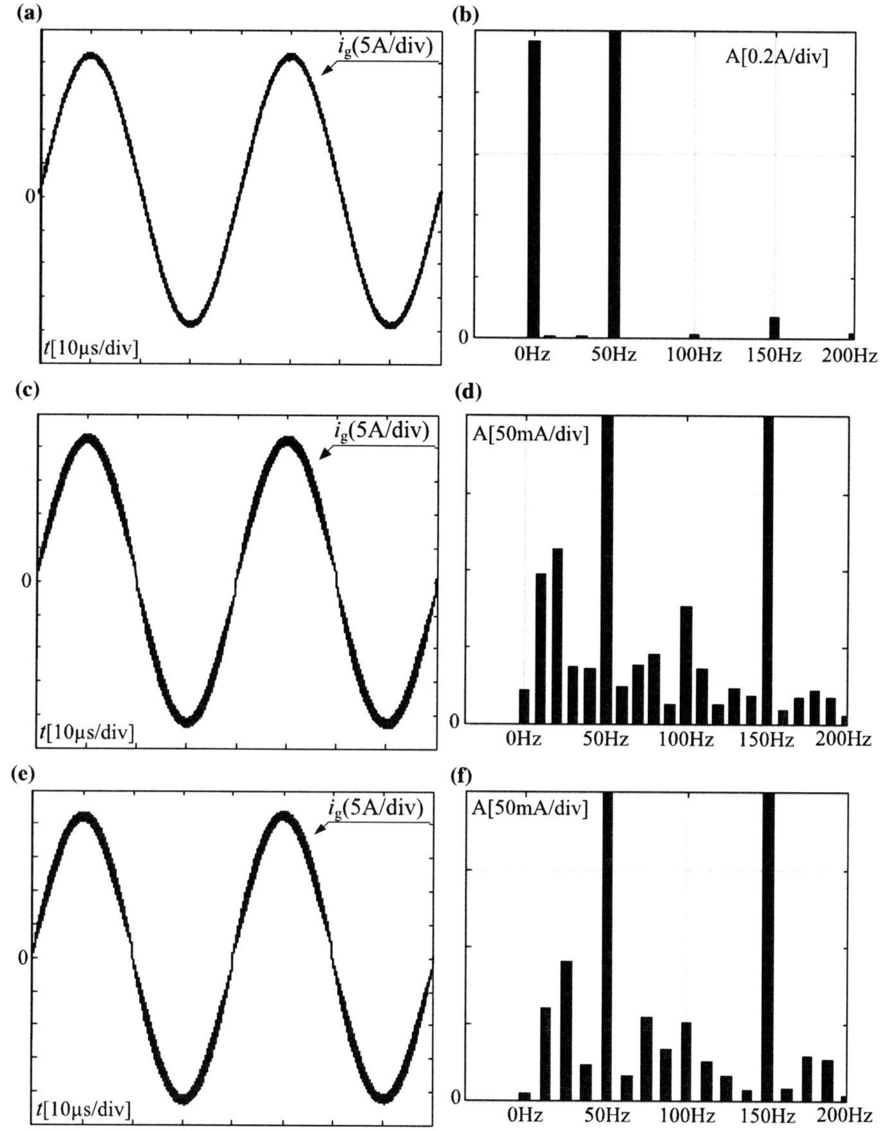

Fig. 4.21 Grid-in current waveforms of different TLIs, **a** HERIC in time domain, **b** HERIC in frequency domain, **c** NPC in time domain, **d** NPC in frequency domain, **e** NPC-C in time domain, **f** NPC-C in frequency domain

Table 4.8 Components of grid-in current at different frequencies with different TLIs

Frequency (Hz)	HERIC (A)	NPC (A)	NPC-C (A)
0	0.976	0.023	0.006
50	20.07	21.33	21.34
100	0.02	0.08	0.01
150	0.08	0.23	0.23
200	0.01	0.01	0.01

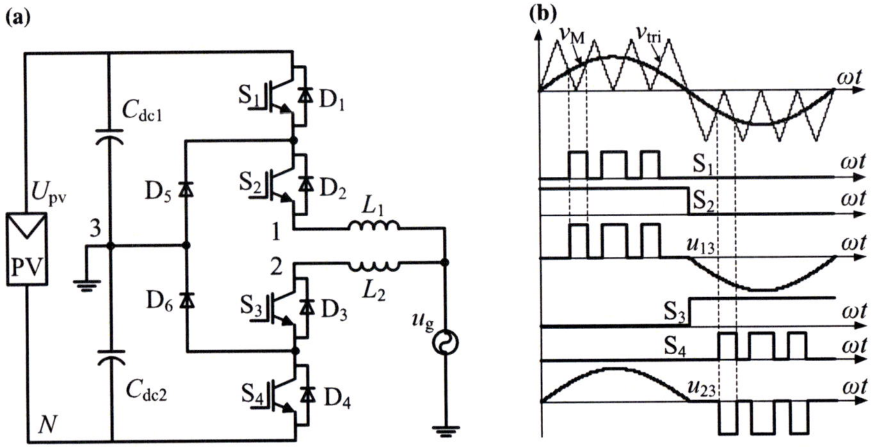

Fig. 4.22 SI-NPC grid-connected inverter, **a** Main circuit, **b** Modulation strategy

S_3 and S_4 are on in negative half period of the grid voltage. It is worth noting that S_1 and S_4 operate with fundamental or high frequency depending on the grid voltage, while S_2 and S_3 operate with line-frequency in SI-NPC.

There are four operating modes according to the polarities of the grid voltage and grid current when SI-NPC works in unity power factor case, as shown in Fig. 4.23. Mode 1 and Mode 2 in Fig. 4.23a, b, respectively, work in positive half period of the grid voltage; symmetrically, Mode 3 and Mode 4 in Fig. 4.23c, d, respectively work in negative half period of the grid voltage.

Mode 1: Refer to Fig. 4.23a; the energy is delivered from the PV side into the grid through S_1, S_2, and the filter L_1 at this stage. According to Fig. 4.22b, the switches S_3 and S_4 are off, and share the input voltage U_{pv} in series, so their turn-off voltage stress is half of the input voltage.

Mode 2: Refer to Fig. 4.23b; only S_2 is on and rest of the switches are off according to Fig. 4.22b. At this stage, the freewheeling current flows through D_5, S_2, and the filter L_1, and keeps feeding the grid. It is worth noting that the potential of the freewheeling path is clamped at the ground.

Mode 3: Refer to Fig. 4.23c; the energy is delivered from the PV side into the grid through S_4, S_3, and the filter L_2 at this stage. According to Fig. 4.22b, the switches

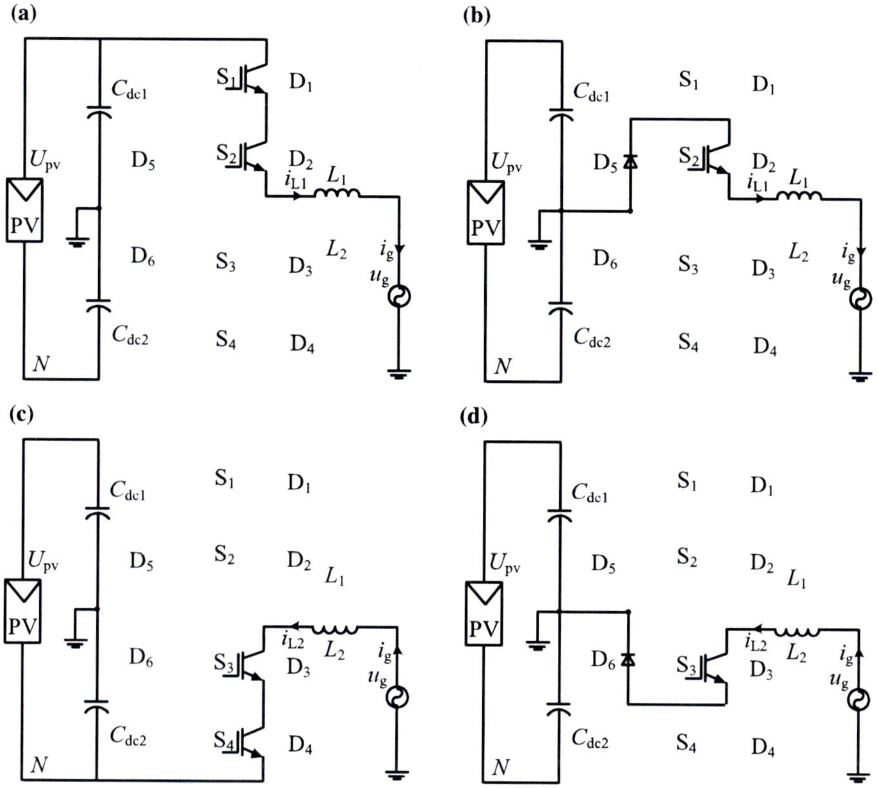

Fig. 4.23 Equivalent circuits of operating modes of SI-NPC, **a** Mode 1, **b** Mode 2, **c** Mode 3, **d** Mode 4

S_1 and S_2 are off, and share the input voltage U_{pv} in series, so their turn-off voltage stress is half of the input voltage.

Mode 4: Refer to Fig. 4.23d; only S_3 is on and rest of the switches are off according to Fig. 4.22b. At this stage, the freewheeling current flows through D_6, S_3, and the filter L_2, and keeps feeding the grid. It is worth noting that the potential of the freewheeling path is clamped at the ground.

Based on the operating modes, the potentials of points 1 and 2 against N, as well as DMV and CMV are shown in Table 4.9. Note that circuit paths of the grid current vary in positive and negative half period of the grid voltage, so the calculation methods of DMV and CMV of Modes 1 and 2 differ from that of Modes 3 and 4. It can be observed that the total CMV of SI-NPC is also kept in a constant value without switching frequency LC. In addition, the risk of shoot through in bridge is eliminated after the modified structure with two output inductors is adopted.

Table 4.9 DMV and CMV levels of different operating modes of SI-NPC

Modes	u_{1N}	u_{2N}	u_{3N}	u_{DM}	u_{CM}	$u_{CM\text{-}DM}$	u_{CM_tot}
(1)	U_{pv}	$u_g + U_{pv}/2$	$U_{pv}/2$	$U_{pv}/2$	$3U_{pv}/4$	$-U_{pv}/4$	$U_{pv}/2$
(2)	$U_{pv}/2$	$u_g + U_{pv}/2$	$U_{pv}/2$	0	$U_{pv}/2$	0	$U_{pv}/2$
(3)	$u_g + U_{pv}/2$	0	$U_{pv}/2$	$-U_{pv}/2$	$U_{pv}/4$	$U_{pv}/4$	$U_{pv}/2$
(4)	$u_g + U_{pv}/2$	$U_{pv}/2$	$U_{pv}/2$	0	$U_{pv}/2$	0	$U_{pv}/2$

2. Simulation results

In order to verify the aforementioned operating principle and property analysis, a simulation model of SI-NPC grid-connected inverter is established in Matlab/Simulink, and the key parameters are listed in Table 4.3.

When the SI-NPC operates with unity power factor, the steady-state waveforms of the grid voltage u_g, reference signal of grid-in current i_{ref}, and actual grid-in current i_g are shown in Fig. 4.24. Furthermore, the simulation waveforms of DMV, CMV, as well as LC are shown in Figs. 4.25, 4.26, and 4.27, respectively. It can be seen that the total CMV is kept in a constant value, and then LC is mitigated in the SI-NPC.

In case of shoot through in the SI-NPC and NPC, the modulation strategy of switches is shown in Fig. 4.28. When shoot through occurs at peak of the grid voltage, the modulation strategy at line-frequency scale and switching frequency scale is shown in Fig. 4.28a, b, respectively. Furthermore, the current waveforms of each switch in NPC and SI-NPC during shoot through stage are shown in Fig. 4.29. It can be noticed that switches of NPC will be damaged due to the sharp rising current. However, the switch currents of SI-NPC are limited by inductances with small rising speed and low amplitude during shoot through stage, which effectively avoids the short circuit of DC side.

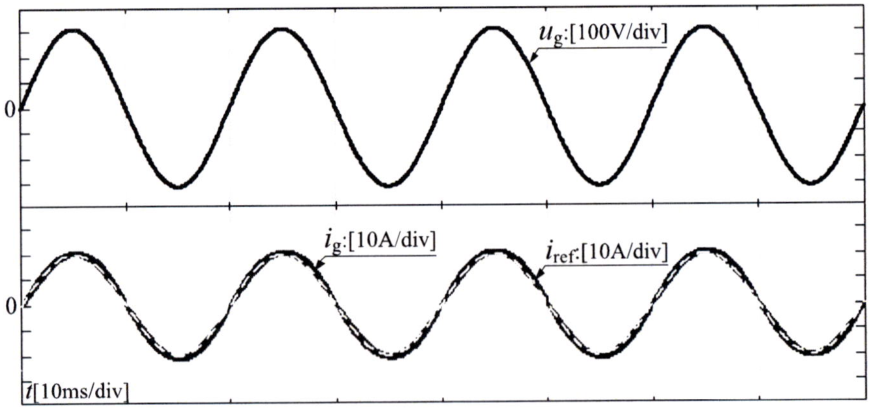

Fig. 4.24 Simulation waveforms of u_g, i_{ref}, and i_g of SI-NPC under unity power factor

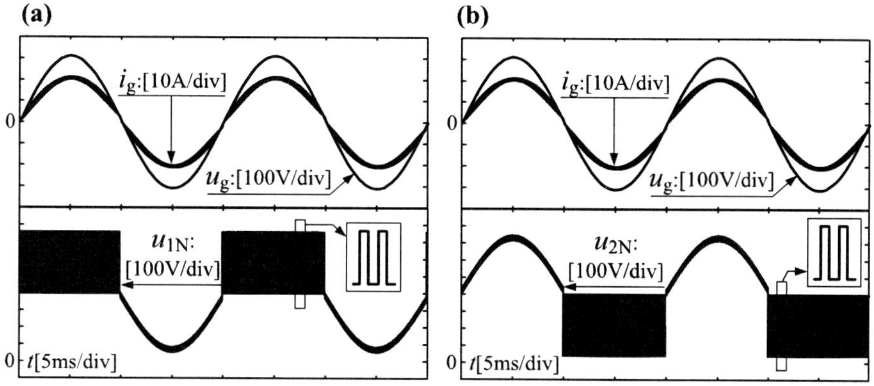

Fig. 4.25 DMV waveforms of SI-NPC, **a** Output voltage 1, **b** Output voltage 2

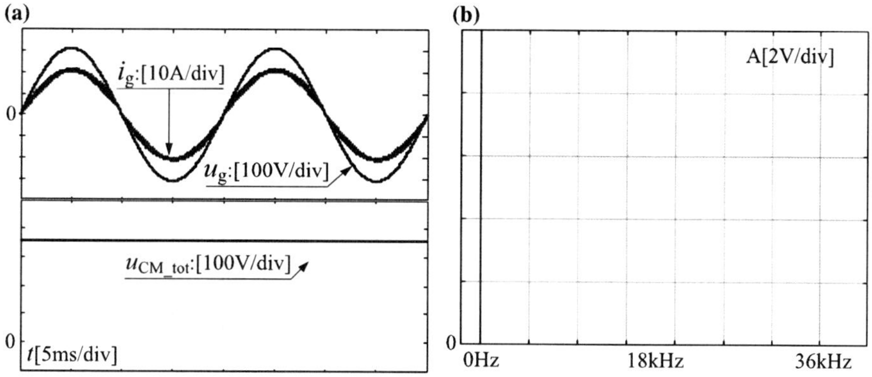

Fig. 4.26 CMV waveforms of SI-NPC, **a** Time domain, **b** Frequency domain

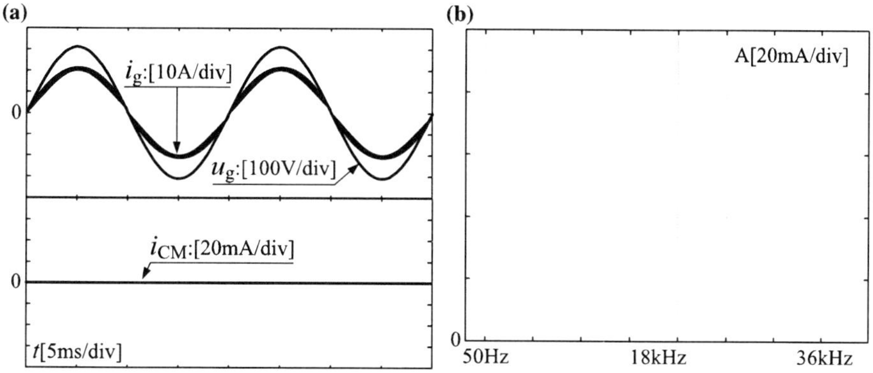

Fig. 4.27 LC waveforms of SI-NPC, **a** Time domain, **b** Frequency domain

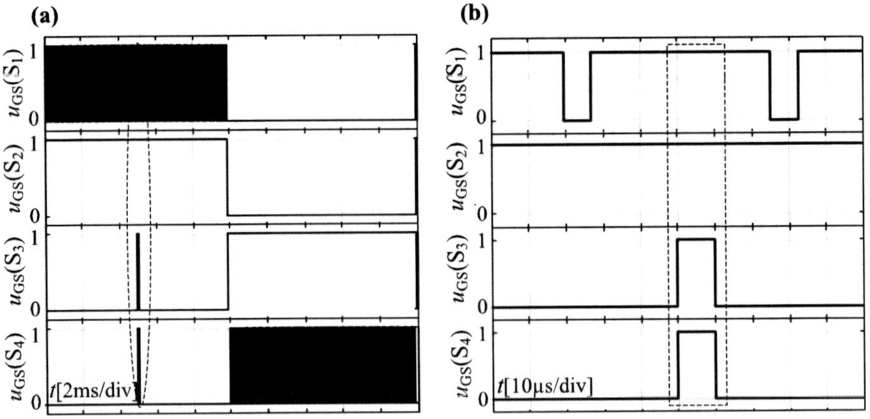

Fig. 4.28 Modulation strategy with shoot through scenario, **a** At line-frequency scale, **b** At switching frequency scale

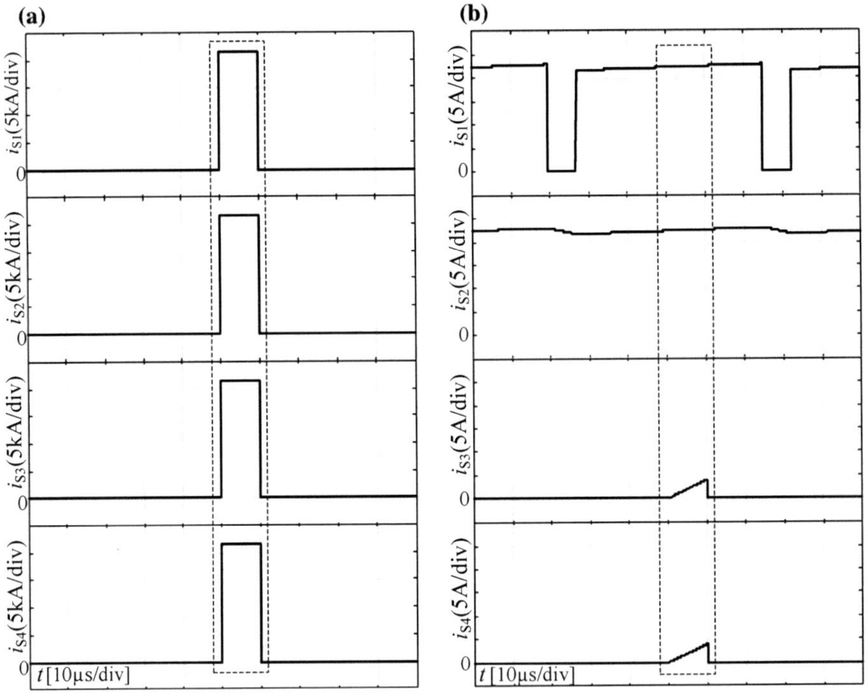

Fig. 4.29 Shoot through currents waveforms from different topologies, **a** NPC, **b** SI-NPC

3. Experimental results

In order to further verify the effectiveness of operating principle and simulation waveforms, a grid-connected experimental platform with SI-NPC circuit is established in lab. The key component parameters are shown in Table 4.4, where values of the filter inductances L_1 and L_2 in the circuit are both 4 mH.

The experimental waveforms of the grid voltage u_g, grid-in current i_g, bridge voltage u_{1N}, u_{2N}, and u_{3N} are shown in Fig. 4.30a, while the experimental current waveforms of the filter inductors L_1 and L_2 are shown in Fig. 4.30b.

The waveforms of the LC i_{CM} are shown in Fig. 4.31. It is necessary to point out that the PV output was simulated by using a DC source with solar array simulation in this experiment. This is the reason that some LC components appear in Fig. 4.31a

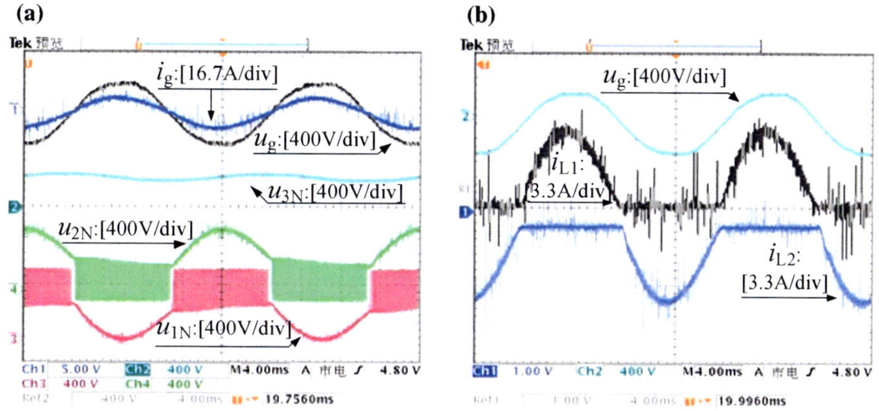

Fig. 4.30 Differential-mode waveforms of SI-NPC, **a** Bridge output voltages, **b** Filter inductor currents

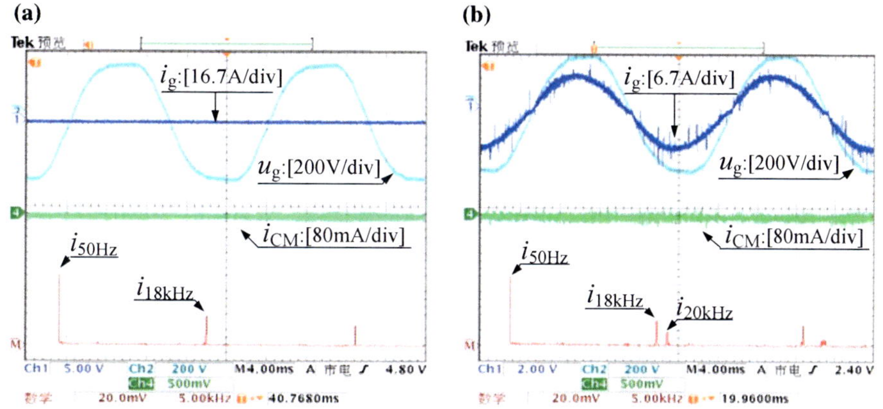

Fig. 4.31 LC waveforms of SI-NPC, **a** Inverter not working, **b** Inverter working

when the circuit breaker is closed and the SI-NPC inverter has not started yet. After the SI-NPC inverter starts working to deliver the energy into the grid, a new LC component appears in Fig. 4.31b, and its frequency is the switching frequency of the SI-NPC inverter (i.e., 20 kHz). By equivalent calculation, the value of LC is about 1 mA, which is equivalent to the common-mode performance of the NPC circuit.

4.2.5 Diode Split-Inductor NPC Inverter

1. Operating principle

The main circuit of diode split-inductor neutral point clamped three-level inverter (D-SI-NPC) is proposed by Michael Frisch and Ernö Ternesi in a US patent, as shown in Fig. 4.32a [9, 10]. Compared with SI-NPC, two diodes are introduced in the D-SI-NPC circuit, and point 1 is connected to the negative pole of the DC bus through D_8. Symmetrically, point 2 is connected to the positive pole of the DC bus through D_7. As a result, the reactive power capacity of SI-NPC is greatly improved. Furthermore, the modulation strategy of different switches is consistent with NPC and SI-NPC in unity power factor case, as shown in Fig. 4.32b. Therefore, the operating modes of D-SI-NPC under unity power factor can also refer to equivalent circuits of Fig. 4.23.

When D-SI-NPC works in non-unity power factor case, the modulation strategy is shown in Fig. 4.32c. It can be noticed that two new operating modes are introduced by the high-frequency operation of S_2 and S_3, as shown in Fig. 4.33.

Mode 1 in Fig. 4.33a is characterized by positive grid current and negative grid voltage while the reactive power is delivered from the PV side into the grid through D_8. Symmetrically, Mode 2 in Fig. 4.33b is characterized by negative grid current and positive grid voltage, and the reactive power is delivering through D_7.

Based on the operating modes, the potentials of points 1 and 2 against N, as well as DMV and CMV are shown in Table 4.10. The calculation methods of DMV and CMV of Modes 1 and 2 differ from that of Modes 3 and 4 due to different circuit paths of grid current in positive and negative half periods of the grid voltage. The operating principle, DMV and CMV characteristics are consistent in unity power factor case of D-SI-NPC and SI-NPC. At the same time, it is worth mentioning that D-SI-NPC has reactive capability compared with SI-NPC.

2. Simulation results

According to the NPC-TLI parameters listed in Table 4.3, the D-SI-NPC simulation model was established in Matlab/Simulink to verify the validity of the theoretical analysis above. When D-SI-NPC operates under unity power factor, the simulation waveforms can refer to Figs. 4.24, 4.26, and 4.27. Figure 4.34 shows the steady-state waveforms of grid voltage u_g, grid-in current reference i_{ref}, and actual grid-in current i_g when D-SI-NPC operates under non-unity power factor. Figure 4.35 shows the

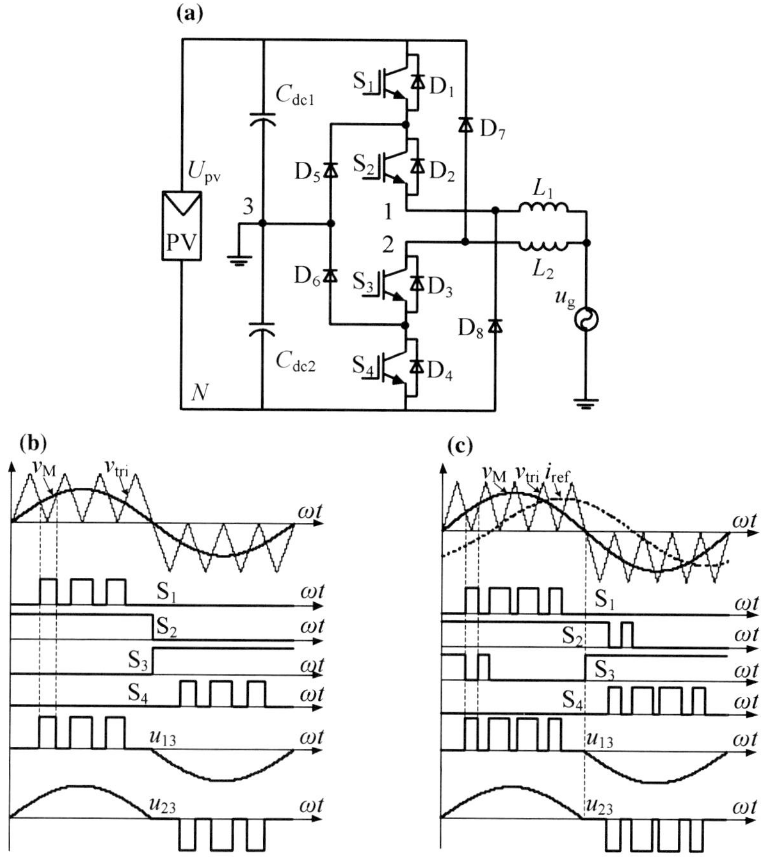

Fig. 4.32 Diode split-inductor neutral point clamping inverter, **a** Main circuit, **b** Modulation strategy I, **c** Modulation strategy II

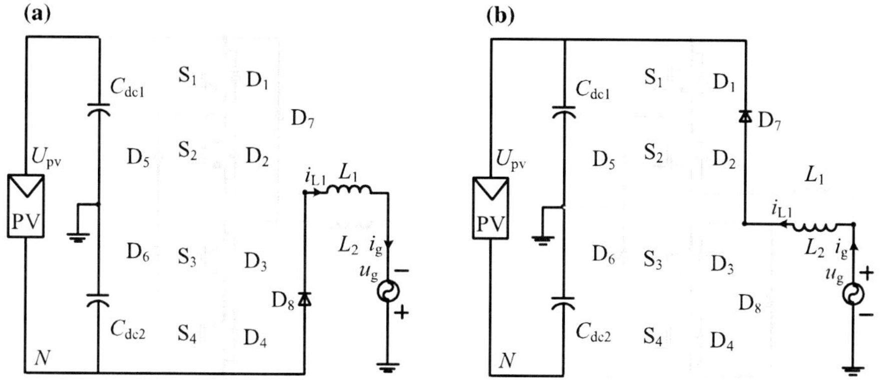

Fig. 4.33 Equivalent circuits of new operating modes of D-SI-NPC with non-unity power factor, **a** Mode 1, **b** Mode 2

Table 4.10 DMV and CMV of different operating modes for D-SI-NPC

	Modes	u_{1N}	u_{2N}	u_{DM}	u_{CM}	$u_{CM\text{-}DM}$	u_{CM_tot}
Unity power factor	(1)	U_{pv}	$U_{pv}/2$	$U_{pv}/2$	$3U_{pv}/4$	$-U_{pv}/4$	$U_{pv}/2$
	(2)	$U_{pv}/2$	$U_{pv}/2$	0	$U_{pv}/2$	0	$U_{pv}/2$
	(3)	0	$U_{pv}/2$	$-U_{pv}/2$	$U_{pv}/4$	$U_{pv}/4$	$U_{pv}/2$
	(4)	$U_{pv}/2$	$U_{pv}/2$	0	$U_{pv}/2$	0	$U_{pv}/2$
Non-unity power factor	(1)	0	$U_{pv}/2$	$-U_{pv}/2$	$U_{pv}/4$	$U_{pv}/4$	$U_{pv}/2$
	(2)	U_{pv}	$U_{pv}/2$	$U_{pv}/2$	$3U_{pv}/4$	$-U_{pv}/4$	$U_{pv}/2$

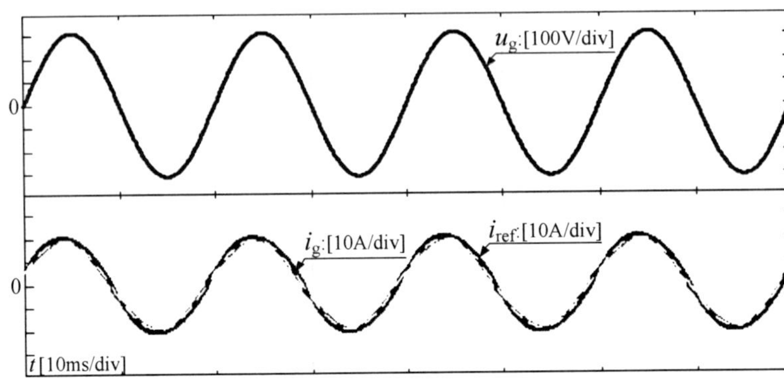

Fig. 4.34 Simulation waveforms of u_g, i_{ref}, and i_g of D-SI-NPC under non-unity power factor

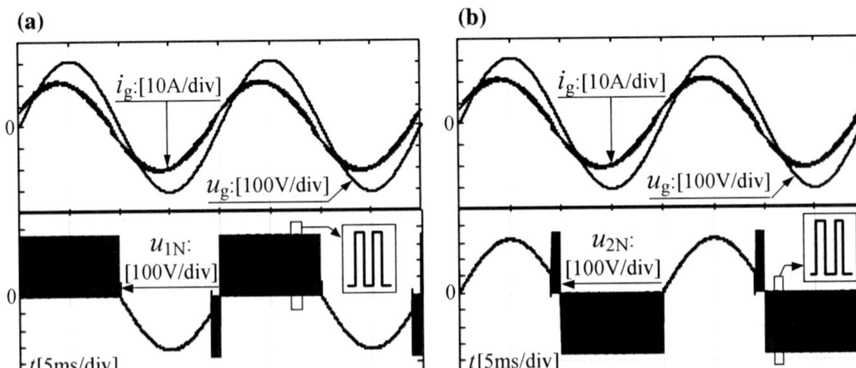

Fig. 4.35 DMV waveforms of D-SI-NPC, **a** Output voltage 1, **b** Output voltage 2

waveforms of D-SI-NPC bridge voltage u_{1N} and u_{2N}. As conclusion, compared with SI-NPC, two diodes added to provide a reactive power channel, so D-SI-NPC can operate under non-unity power factor.

4.2.6 Six-Switch Five-Level Active Neutral Point Clamped Inverter

Different from traditional two-level and three-level inverters, the number of output levels in multi-level TLIs is increased [11–13], including five-level, seven-level, and so on [14]. Furthermore, waveforms of multi-level inverters are more sinusoidal and shaped with less harmonic components; then, smaller filters can be used in the circuit. In this section, the six-switch five-level midpoint clamp TLI will be discussed in detail [15].

1. Operating principle

The main circuit of a six-switch five-level active neutral point clamped inverter (6S-5L-ANPC) is shown in Fig. 4.36a [15, 16], where the midpoint of the input voltage is connected to the neutral point of the grid. Furthermore, refer to the driving logic shown in Fig. 4.36b, and it can be noticed that the waveform of DMV is shaped as five-level, and the voltage levels of different switching stages are listed in Table 4.11.

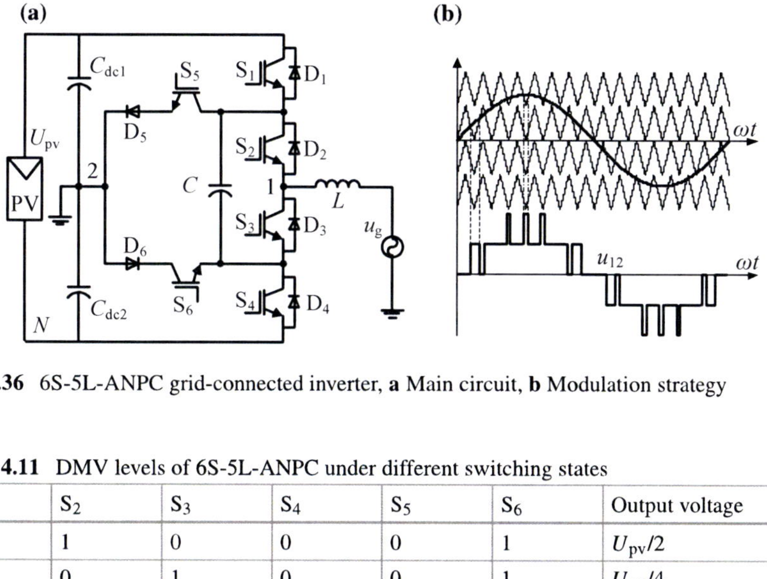

Fig. 4.36 6S-5L-ANPC grid-connected inverter, a Main circuit, b Modulation strategy

Table 4.11 DMV levels of 6S-5L-ANPC under different switching states

S_1	S_2	S_3	S_4	S_5	S_6	Output voltage
1	1	0	0	0	1	$U_{pv}/2$
1	0	1	0	0	1	$U_{pv}/4$
0	1	0	0	0	1	$U_{pv}/4$
0	0	1	0	0	1	0+
0	1	0	0	1	0	0–
0	0	1	0	1	0	$-U_{pv}/4$
0	1	0	1	1	0	$-U_{pv}/4$
0	0	1	1	1	0	$-U_{pv}/2$

There are eight operating modes according to the polarities of the grid voltage and grid current when 6S-5L-ANPC works in unity power factor case, as shown in Fig. 4.37. In positive half period of the grid voltage, the associated modes 1–4 are illustrated in Fig. 4.37a, b, c, and d; symmetrically, the rest of the modes 5–8 work in negative half period of the grid voltage.

Mode 1: Refer to Fig. 4.37a; the energy is delivered from the PV side into the grid through S_1, S_2, and the AC filter at this stage. The switches S_3 and S_4 are off, and their turn-off voltage stress is half of the input voltage.

Mode 2: Refer to Fig. 4.37b; the energy is delivered from the PV side into the grid by charging the capacitor C. Besides, the grid current flows through S_1, D_3, and the AC filter with 1/4 of U_{pv} output in midpoint of the bridge.

Mode 3: Refer to Fig. 4.37c; the energy is delivered from the PV side into the grid by discharging the capacitor C. Besides, the grid current flows through D_6, S_6, S_2, and the AC filter with 1/4 of U_{pv} output in midpoint of the bridge.

Mode 4: Refer to Fig. 4.37d; only S_6 is on and rest of the switches are off. The freewheeling current flows through D_6, S_6, D_3, and the AC filter at this stage. It is worth noting that the potential of the freewheeling path is still clamped at the ground.

Mode 5: Refer to Fig. 4.37e; the energy is delivered from the PV side into the grid through S_4, S_3, and the AC filter at this stage. The switches S_1 and S_2 are off, and their turn-off voltage stress is half of the input voltage.

Mode 6: Refer to Fig. 4.37f; the energy is delivered from the PV side into the grid by the charging capacitor C. Besides, the grid-in current flows through S_4, D_2, and the AC filter with 1/4 of $-U_{pv}$ output in midpoint of the bridge.

Mode 7: Refer to Fig. 4.37g; the energy is delivered from the PV side into the grid by discharging the capacitor C. Besides, the grid current flows through D_5, S_5, S_3, and the AC filter with 1/4 of $-U_{pv}$ output in midpoint of the bridge.

Mode 8: Refer to Fig. 4.37h; only S_4 is on and rest of the switches are off. The freewheeling current flows through D_5, S_5, D_2, and the AC filter at this stage. It is worth noting that the potential of the freewheeling path is still clamped at the ground.

2. Simulation results

In order to verify the aforementioned operating principle and property analysis, a simulation model of 6S-5L-ANPC grid-connected inverter is established in Matlab/Simulink, and the key parameters are listed in Table 4.12.

Figure 4.38 shows the waveforms of the grid voltage u_g, the reference of grid-in current i_{ref}, and the actual grid-in current i_g when the 6S-5L-ANPC operates with unity power factor. The simulation waveforms of DMV u_{DM} in time domain and frequency domain are shown in Fig. 4.39a, b, respectively. Besides the line-frequency component, the amplitude of DMV is prominent around the switching frequency and double switching frequency.

As shown in Fig. 4.40, it can be seen that the capacitor voltage is 1/4 of the input voltage U_{pv} during operation. Since the capacitor is charged and discharged at the switching frequency scale, the balance between charge and discharge stages is

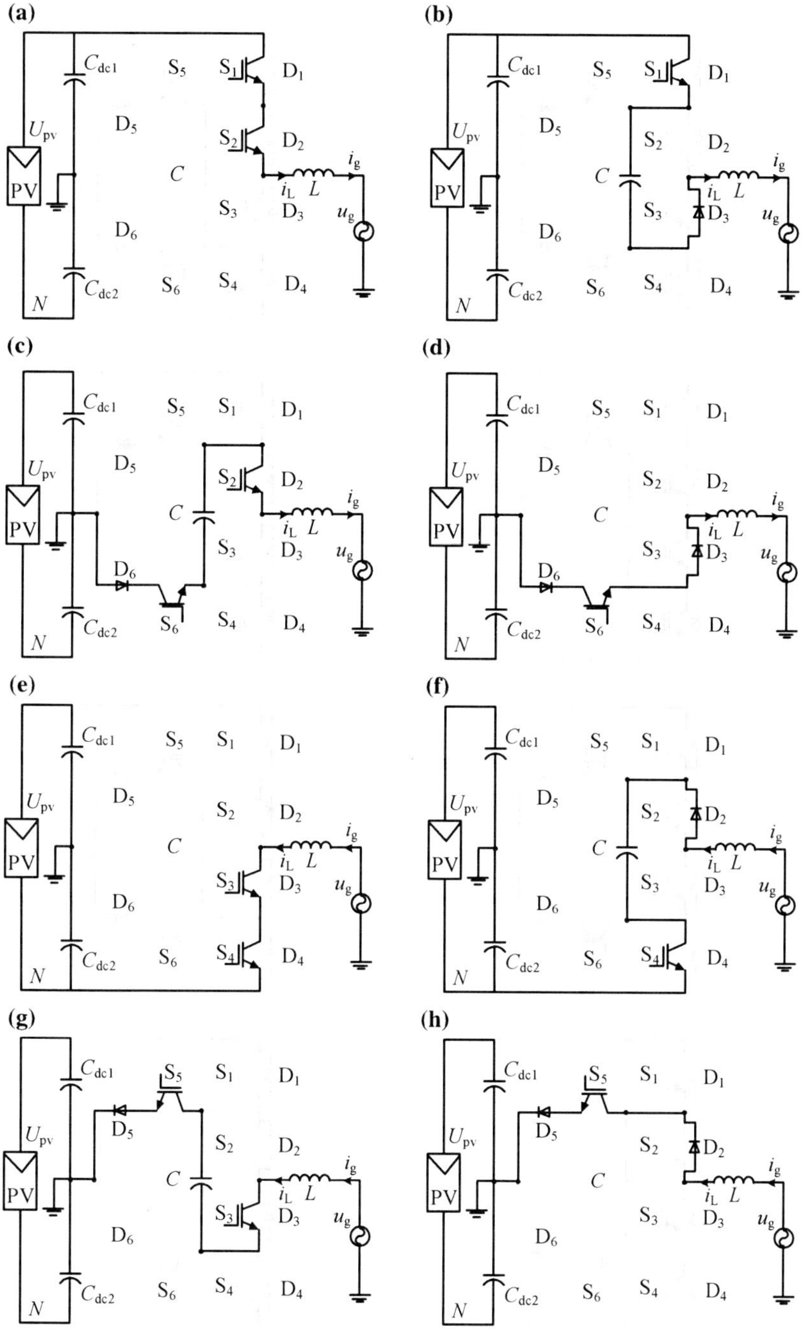

Fig. 4.37 Equivalent circuits of operating modes of 6S-5L-ANPC, **a** Mode 1, **b** Mode 2, **c** Mode 3, **d** Mode 4, **e** Mode 5, **f** Mode 6, **g** Mode 7, **h** Mode 8

Table 4.12 Key parameters of 1 kW 6S-5L-ANPC

Parameters	Value	Parameters	Value
Input voltage U_{pv}/V	400	Switching frequency f_s/kHz	15
Grid voltage u_g/V	110	Capacitance C/μF	310
Line frequency/Hz	50	Filter inductor L_f/mH	1.6
Rated power P/W	1000	Parasitic capacitance of PV panel C_{pv}/nF	50

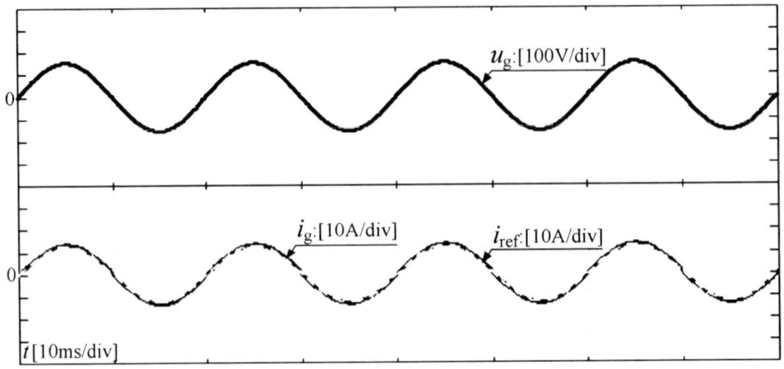

Fig. 4.38 Simulation waveforms of u_g, i_{ref}, and i_g of 6S-5L-ANP under unity power factor

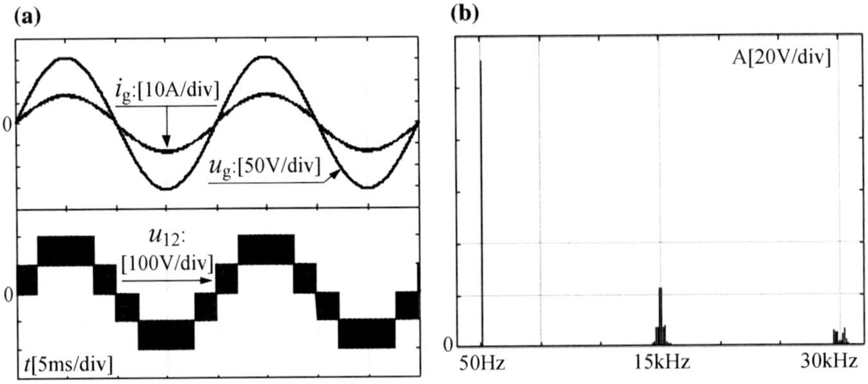

Fig. 4.39 DMV waveforms of 6S-5L-ANPC, **a** Time domain, **b** Frequency domain

effectively realized. As a result, the stability of the voltage can be controlled with a smaller capacitor in the 6S-5L-ANPC circuit.

In the I-type half-bridge circuit with a voltage clamping structure, the operating voltages of switches can be reduced to half of the input voltage, which is suitable for high-voltage applications. When TLIs are used in lower voltage occasions, the conduction loss is larger if the grid-in current flows through two switches during the

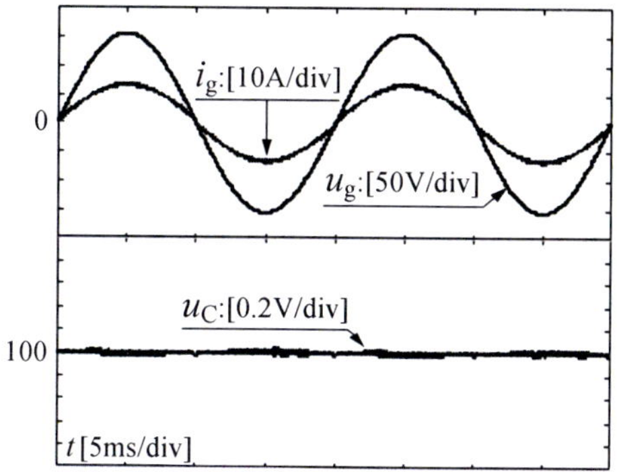

Fig. 4.40 Capacitor voltage waveform in 6S-5L-ANPC

energy-delivering stage. Furthermore, the T-type half-bridge TLI was proposed to reduce the conduction loss of switches.

4.3 T-Type Half-Bridge TLIs

T-type half-bridge TLIs mainly include the T neutral point clamped grid-connected inverter (T-NPC) and its improved topologies. Specifically, the T-NPC was invented by Knaup Peter with an applied international patent [17]. T-NPC, also known as Conergy NPC, has been applied to the IPG series products (2–5 kW) of Conergy Company. According to report of Photo International in July 2007, the European efficiency of T-NPC is 95.1% and the peak efficiency is 96.1%.

4.3.1 Conventional T-NPC Inverter

1. Operating principle

Compared with NPC, only two switches are configured in the bridge of T-NPC, as shown in Fig. 4.41a [17, 18]. Furthermore, the capacitor bridge and switch bridge are connected by two series switches to realize voltage clamping. As the modulation strategy in Fig. 4.41b shows, switches S_1 and S_3 conduct complementarily in positive half period of the grid voltage. Symmetrically, switches S_2 and S_4 conduct complementarily during negative half period of the grid voltage. It is worth mentioning that the waveform of DMV is shaped as three-level voltage by unipolar modulation.

There are four operating modes according to the polarities of the grid voltage and grid current when T-NPC works in unity power factor case, as shown in Fig. 4.42. Mode 1 and Mode 2 in Fig. 4.42a, b, respectively, work in positive half period of the grid voltage; symmetrically, Mode 3 and Mode 4 in Fig. 4.42c, d, respectively, work in negative half period of the grid voltage.

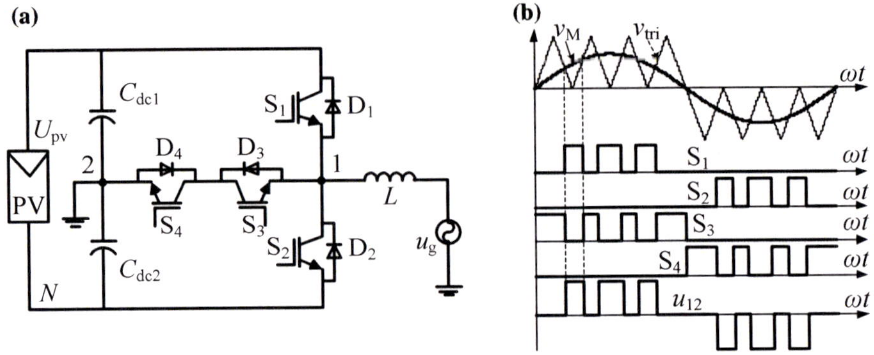

Fig. 4.41 T-NPC grid-connected inverter, **a** Main circuit, **b** Modulation strategy

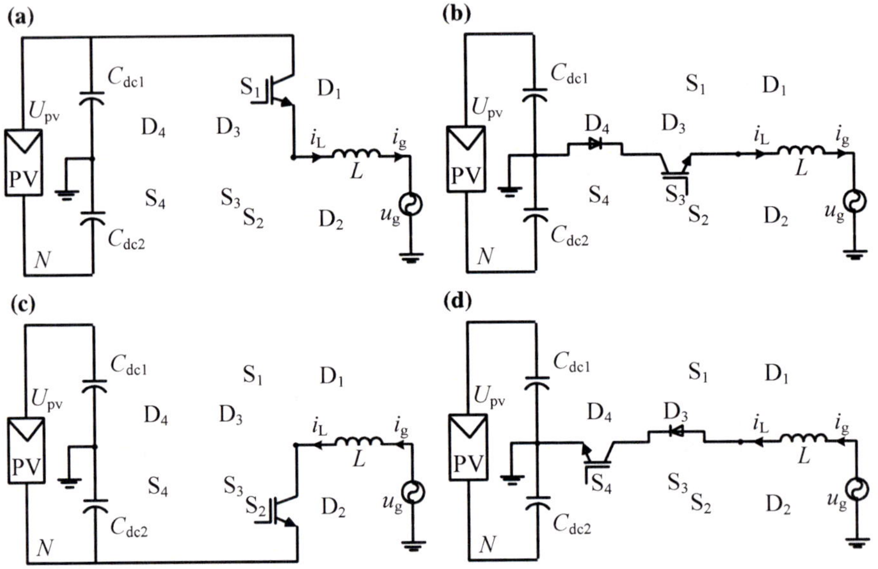

Fig. 4.42 Equivalent circuits of T-NPC in different modes, **a** Mode 1, **b** Mode 2, **c** Mode 3, **d** Mode 4

Mode 1: Refer to Fig. 4.42a; only S_1 is on and rest of the switches are off according to Fig. 4.41b. Similar to the operation mode of conventional half-bridge converters, the energy is delivered from the PV side into the grid through S_1 and AC filter at this stage.

Mode 2: Refer to Fig. 4.42b; only S_3 is on and rest of the switches are off according to Fig. 4.41b. At this stage, the freewheeling current flows through D_4, S_3, and the filter, and keeps feeding the grid. It is worth noting that the potential of the freewheeling path is clamped at the ground.

Table 4.13 DMV and CMV levels of different operating modes for T-NPC

Modes	u_{1N}	u_{2N}	u_{DM}	u_{CM}	$u_{CM\text{-}DM}$	u_{CM_tot}
(1)	U_{pv}	$U_{pv}/2$	$U_{pv}/2$	$3U_{pv}/4$	$-U_{pv}/4$	$U_{pv}/2$
(2)	$U_{pv}/2$	$U_{pv}/2$	0	$U_{pv}/2$	0	$U_{pv}/2$
(3)	0	$U_{pv}/2$	$-U_{pv}/2$	$U_{pv}/4$	$U_{pv}/4$	$U_{pv}/2$
(4)	$U_{pv}/2$	$U_{pv}/2$	0	$U_{pv}/2$	0	$U_{pv}/2$

Mode 3: Refer to Fig. 4.42c; the energy is delivered from the PV side into the grid through S_2 and the AC filter at this stage.

Mode 4: Refer to Fig. 4.42d; only S_4 is on and rest of the switches are off according to Fig. 4.41b. At this stage, the freewheeling current flows through D_3, S_4, and the filter, and keeps feeding the grid. It is worth noting that the potential of the freewheeling path is clamped at the ground.

Based on the operating modes, the potentials of points 1 and 2 against N, as well as DMV and CMV are shown in Table 4.13. Compared with NPC, two diodes are removed in the circuit of T-NPC. The grid current only flows through one switch in the energy-delivering stage, so the switching loss and its distribution are optimized, and bring advantages for design of heat dissipation and improvement of inverter life. However, the voltage stress of switches S_1 and S_2 in T-NPC is U_{pv}, which is twice of that in NPC circuit in the same case. Meanwhile, it is worth noting that the total CMV of T-NPC remains in a constant value in all operating modes.

2. Simulation results

In order to verify the aforementioned operating principle and property analysis, a simulation model of T-NPC grid-connected inverter is established in Matlab/Simulink, and the key parameters are listed in Table 4.3.

When the T-NPC operates with unity power factor, the steady-state waveforms of the grid voltage u_g, reference signal of grid-in current i_{ref}, and actual grid-in current i_g are shown in Fig. 4.43. The simulation waveforms of DMV and CMV in time domain and frequency domain are shown in Figs. 4.44 and 4.45. It can be noticed that the DMV is shaped as quasi-unipolar three-level waveform, while no harmonic appears in the CMV. Furthermore, the simulation waveforms of LC in time domain and frequency domain are shown in Fig. 4.46a, b, respectively. As a conclusion, the LC is effectively eliminated in T-NPC with a constant CMV.

4.3.2 *Reduced Switches T-NPC Inverter*

1. Operating principle

The reduced switches T neutral point clamped three-level inverter (RS-T-NPC) was proposed, as shown in Fig. 4.47a [19]. Compared with conventional T-NPC, a circuit

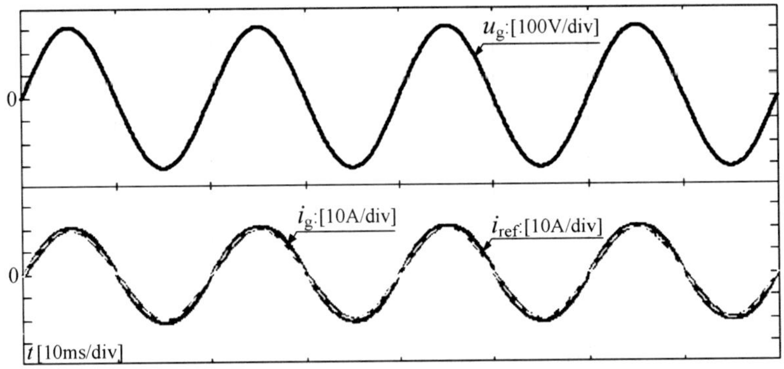

Fig. 4.43 Simulation waveforms of u_g, i_{ref}, and i_g of T-NPC

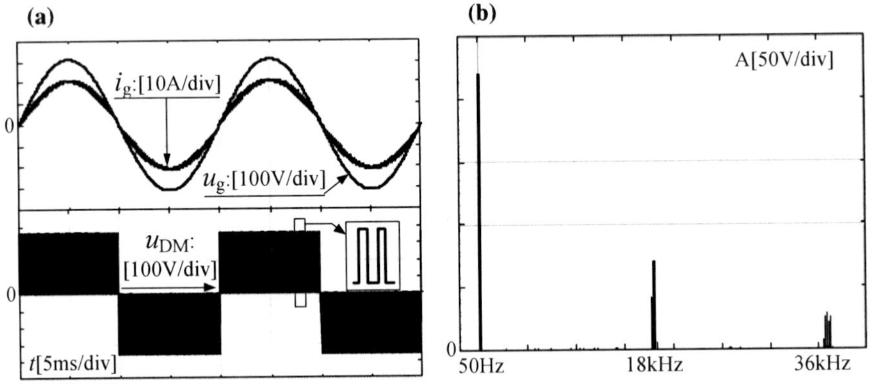

Fig. 4.44 DMV waveforms of T-NPC, **a** Time domain, **b** Frequency domain

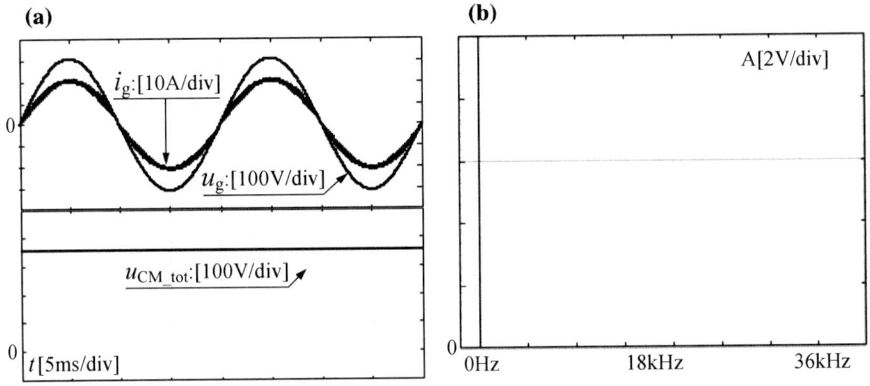

Fig. 4.45 CMV waveforms of T-NPC, **a** Time domain, **b** Frequency domain

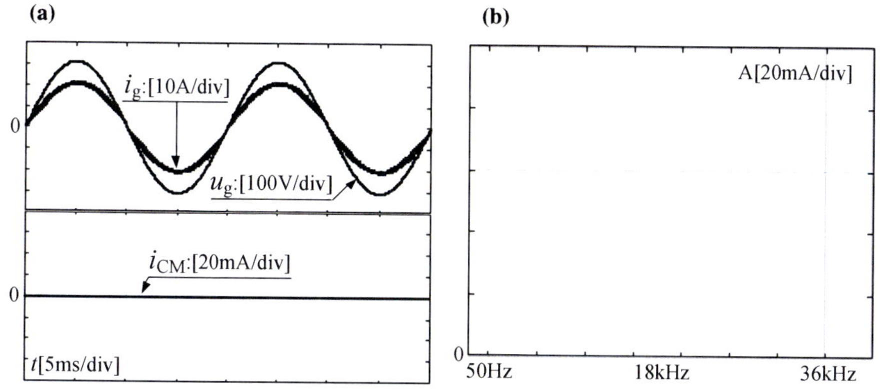

Fig. 4.46 LC waveforms of T-NPC, **a** Time domain, **b** Frequency domain

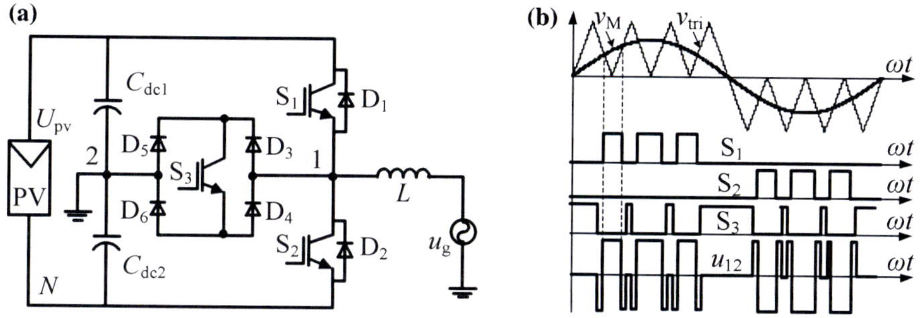

Fig. 4.47 RS-T-NPC grid-connected inverter, **a** Main circuit, **b** Modulation strategy

composed of one switch and four diodes replaces the circuit composed of two switches to realize freewheeling and voltage clamping. It can be noticed that S_3 operates with high frequency throughout whole cycle of the grid voltage, as shown in Fig. 4.47b. In order to prevent short-circuit at DC side, dead time in switching frequency scale is needed between S_3 and the combination of S_1 and S_2. As a result, the waveform of DMV is shaped as three-level voltage by unipolar modulation.

There are six operating modes according to the polarities of the grid voltage and grid current when RS-T-NPC works in unity power factor case, as shown in Fig. 4.48. Mode 1, Mode 2, and Mode 3 in Fig. 4.48a–c, respectively, work in positive half period of the grid voltage; symmetrically, Mode 3, Mode 4, and Mode 5 in Fig. 4.48d–f, respectively, work in negative half period of the grid voltage.

Mode 1: Refer to Fig. 4.48a; only S_1 is on and rest of the switches are off according to Fig. 4.47b. And thus the energy is delivered from the PV side into the grid through S_1 and AC filter at this stage.

Mode 2: Refer to Fig. 4.48b; all of the switches are off according to Fig. 4.47b at this stage. Thus the dead time current flows through D_2 and AC filter with $-U_{pv}/2$ output in midpoint of the bridge.

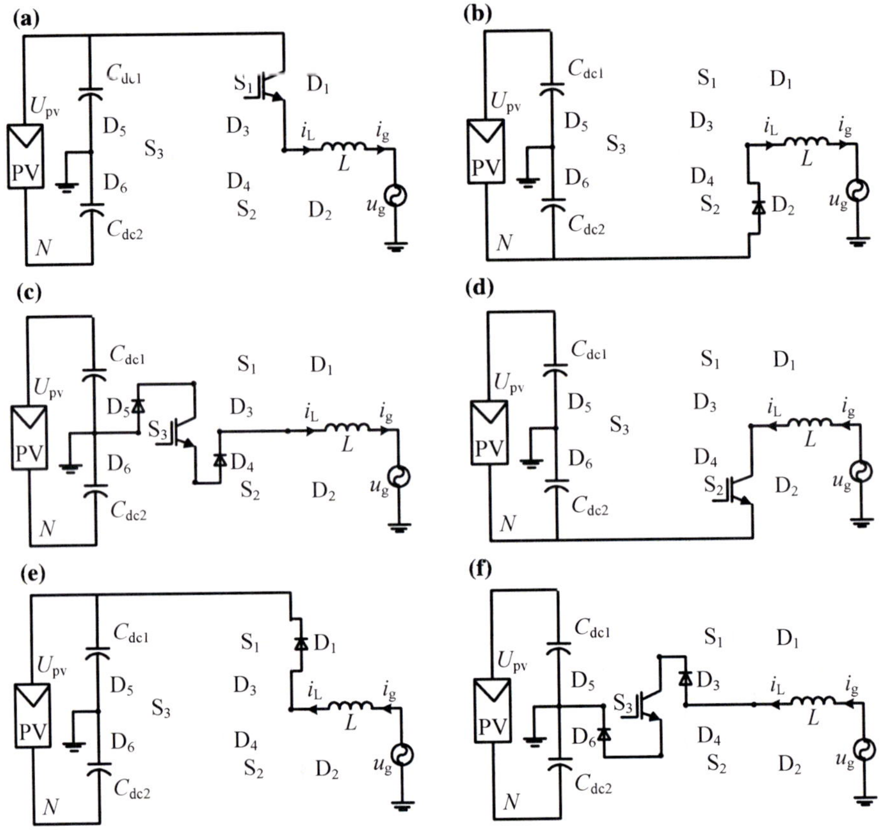

Fig. 4.48 Equivalent circuits of operating modes of RS-T-NPC, **a** Mode 1, **b** Mode 2, **c** Mode 3, **d** Mode 4, **e** Mode 5, **f** Mode 6

Mode 3: Refer to Fig. 4.48c; only S_3 is on and rest of the switches are off according to Fig. 4.47b. Thus the freewheeling current flows through D_5, S_3, D_4, and the AC filter at this stage. It is worth noting that the potential of the freewheeling path is clamped at the ground.

Mode 4: Refer to Fig. 4.48d; only S_2 is on and rest of the switches are off according to Fig. 4.47b. Thus the energy is delivered from the PV side into the grid through S_2 and the AC filter at this stage.

Mode 5: Refer to Fig. 4.48e; all switches are off according to Fig. 4.47b at this stage. Thus the dead time current flows through D_1 and AC filter with $U_{pv}/2$ output in midpoint of the bridge.

Mode 6: Refer to Fig. 4.48f; only S_3 is on and rest of the switches are off according to Fig. 4.47b. Thus the freewheeling current flows through D_6, S_3, D_3, and the AC filter at this stage. Meanwhile, the potential of the freewheeling path is clamped at the ground.

Table 4.14 DMV and CMV of different operating modes for HE-S-NPC

Modes	u_{1N}	u_{2N}	u_{DM}	u_{CM}	$u_{CM\text{-}DM}$	u_{CM_tot}
(1), (5)	U_{pv}	$U_{pv}/2$	$U_{pv}/2$	$3U_{pv}/4$	$-U_{pv}/4$	$U_{pv}/2$
(3)	$U_{pv}/2$	$U_{pv}/2$	0	$U_{pv}/2$	0	$U_{pv}/2$
(2), (4)	0	$U_{pv}/2$	$-U_{pv}/2$	$U_{pv}/4$	$U_{pv}/4$	$U_{pv}/2$
(6)	$U_{pv}/2$	$U_{pv}/2$	0	$U_{pv}/2$	0	$U_{pv}/2$

Based on the operating modes, the potentials of points 1 and 2 against N, as well as DMV and CMV are shown in Table 4.14. It can be noticed that the principle of RS-T-NPC is highly consistent with that of T-NPC. Specifically, the voltage is clamped by one switch and one diode in T-NPC, while the voltage is clamped by one switch and two diodes in RS-T-NPC during freewheeling stage. Considering the rectification function of D_3–D_6, the grid current will flow through S_3 in a certain direction, so an active power device can be saved in RS-T-NPC compared with T-NPC. In the dead-time stage, it is worth mentioning that there are narrow pulses with a reverse polarity in DMV while total CMV remains constant.

2. Simulation results

In order to verify the aforementioned operating principle and property analysis, a simulation model of RS-T-NPC grid-connected inverter is established in Matlab/Simulink, and the key parameters are listed in Table 4.3.

When the RS-T-NPC operates with unity power factor, the steady-state waveforms of the grid voltage u_g, reference signal of grid-in current i_{ref}, and actual grid-in current i_g are shown in Fig. 4.49. The simulation waveforms of DMV u_{DM} in time domain and frequency domain are shown in Fig. 4.50a, b, respectively. It can be seen that the DMV is shaped as a quasi-unipolar three-level waveform. In addition to the characteristics of CMV, RS-T-NPC is consistent with that T-NPC as shown in Figs. 4.45 and 4.46.

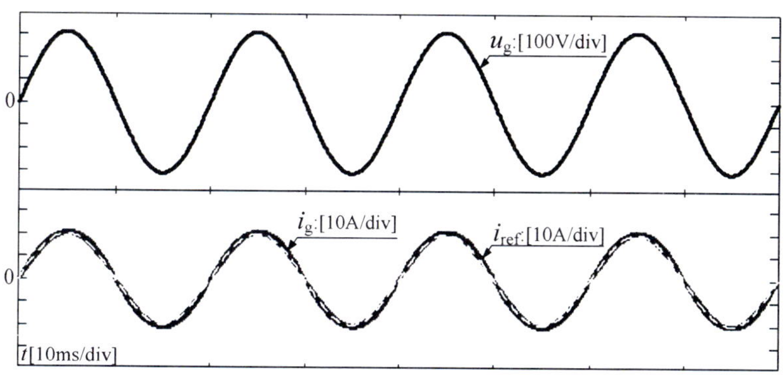

Fig. 4.49 Simulation waveforms of u_g, i_{ref}, and i_g of RS-T-NPC

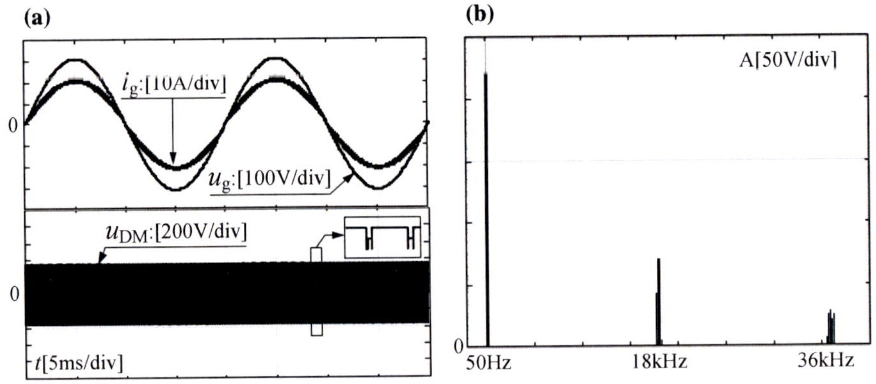

Fig. 4.50 DMV waveforms of RS-T-NPC, **a** Mode 1, **b** Mode 2

4.3.3 Diode-Free T-NPC Inverter

1. Operating principle

To further improve the efficiency, the conduction loss of switches and reverse-recovery loss of diodes in the freewheeling stage need to be reduced, and thus the diode-free T neutral point clamped three-level inverter (DF-T-NPC) was proposed in [20]. Compared with T-NPC, the combination of two IGBT in the freewheeling and clamping branch are replaced by four CoolMosfet S_3-S_6, as shown in Fig. 4.51a. Furthermore, the modulation strategy of unipolar modulation is shown in Fig. 4.51b. In order to avoid short-circuit of DC side, dead time in switching frequency scale is configured between switches S_3-S_6 and S_1, S_2. It can be noticed that the waveform of DMV is finally shaped as three levels.

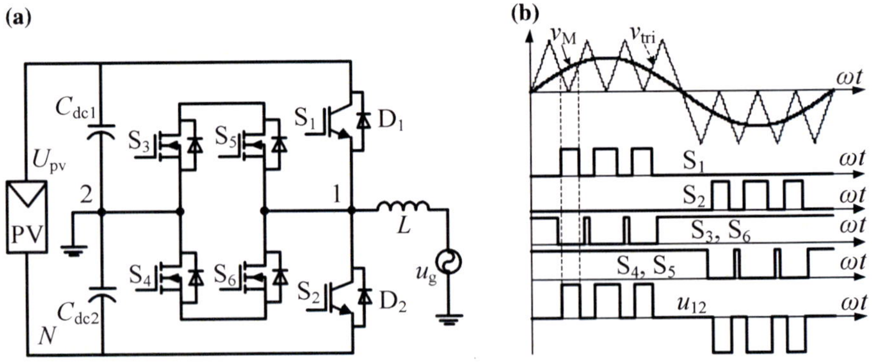

Fig. 4.51 DF-T-NPC grid-connected inverter, **a** Main circuit, **b** Modulation strategy

There are six operating modes according to the polarities of the grid voltage and grid current when DF-T-NPC works in unity power factor case, as shown in Fig. 4.52. Mode 1, Mode 2, and Mode 3 in Fig. 4.52a–c, respectively, work in positive half period of the grid voltage; symmetrically, Mode 3, Mode 4, and Mode 5 (Fig. 4.52d–f), respectively, work in negative half period of the grid voltage. Switches S_3–S_6 are MOSFETs with bidirectionally conductive channels, so the dead time is independent with DMV. In the freewheeling stage, two switches among switches S_3–S_6 are on in the forward direction while the other switches are on in the reverse direction.

Mode 1: Refer to Fig. 4.52a; the energy is delivered from the PV side into the grid through S_1 and AC filter at this stage. Meanwhile, switches S_4 and S_5 are also on according to Fig. 4.51b.

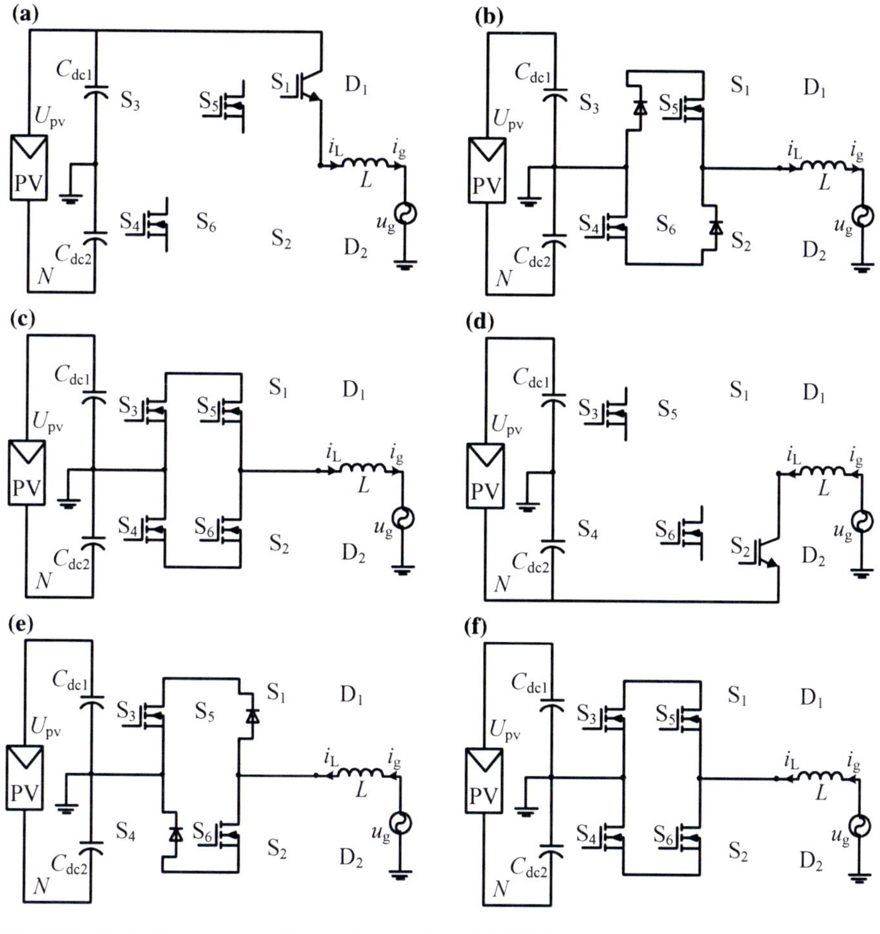

Fig. 4.52 Equivalent circuits of operating modes of DF-T-NPC, **a** Mode 1, **b** Mode 2, **c** Mode 3, **d** Mode 4, **e** Mode 5, **f** Mode 6

Table 4.15 DMV and CMV of different operating modes in DF-T-NPC

Modes	u_{1N}	u_{2N}	u_{DM}	u_{CM}	$u_{CM\text{-}DM}$	$u_{CM\ tot}$
(1)	U_{pv}	$U_{pv}/2$	$U_{pv}/2$	$3U_{pv}/4$	$-U_{pv}/4$	$U_{pv}/2$
(2), (3)	$U_{pv}/2$	$U_{pv}/2$	0	$U_{pv}/2$	0	$U_{pv}/2$
(4)	0	$U_{pv}/2$	$-U_{pv}/2$	$U_{pv}/4$	$U_{pv}/4$	$U_{pv}/2$
(5), (6)	$U_{pv}/2$	$U_{pv}/2$	0	$U_{pv}/2$	0	$U_{pv}/2$

Mode 2: Refer to Fig. 4.52b; only S_4 and S_5 are on at this stage according to Fig. 4.51b, so the dead time current flows through switches S_4, S_5, parallel diodes of S_3 and S_6, and the filter inductors. It is worth noting that the potential of the dead time path is clamped at the ground.

Mode 3: Refer to Fig. 4.52c; only S_1 and S_2 are off and rest of the switches are on at this stage. The freewheeling current flows through switches S_3–S_6 and filter, where switches S_4 and S_5 are forward conducted while switches S_3 and S_6 are reversely conducted. Meanwhile, the potential of the freewheeling path is clamped at the ground.

Mode 4: Refer to Fig. 4.52d; the energy is delivered from the PV side into the grid through S_2 and AC filter at this stage. Meanwhile, switches S_3 and S_6 are also on according to Fig. 4.51b.

Mode 5: Refer to Fig. 4.52e; only S_3 and S_6 are on at this stage according to Fig. 4.51b, so the dead time current flows through switches S_3, S_6, parallel diodes of S_4 and S_5, and the filter inductors. It is worth noting that the potential of the dead time path is clamped at the ground.

Mode 6: Refer to Fig. 4.52f; only S_1 and S_2 are off and rest of the switches are on at this stage. The freewheeling current flows through switches S_3-S_6 and filter, where switches S_4 and S_5 are reversely conducted while switches S_3 and S_6 are forward conducted. Meanwhile, the potential of the freewheeling path is clamped at the ground.

Based on the operating modes, the potentials of points 1 and 2 against N, as well as DMV and CMV are shown in Table 4.15. It can be seen that the grid current of DF-T-NPC at the freewheeling stage is divided into two branches and flows through S_3–S_6, respectively. Furthermore, the conduction loss of switches is optimized due to the reduced equivalent resistor of clamping branch. Additionally, the DMV of DF-T-NPC is clamped at the ground while the total CMV remained constant during the dead time stage.

2. Simulation results

The DMV and CMV characteristics of DF-T-NPC are consistent with that of T-NPC, as illustrated in Figs. 4.43, 4.44, 4.45, and 4.46.

4.4 Stacked Neutral Point Clamped TLIs

Stacked neutral point clamped three-level inverter (S-NPC) was proposed to realize even distribution of the power loss, combining bridge structure and midpoint clamping circuit of NPC and T-NPC, respectively. Meanwhile, the operating principle of S-NPC is similar to that of NPC and T-NPC, and thus the waveform of DMV in S-NPC is also shaped as three-level [21–23].

4.4.1 Conventional S-NPC Inverter

1. Operating principle

The main circuit of S-NPC is shown in Fig. 4.53a [24] with modulation strategies I and II shown in Fig. 4.53b, c.

There are four operating modes according to the polarities of the grid voltage and grid current when S-NPC works in unity power factor case with modulation strategy I, as shown in Fig. 4.54. Mode 1 and Mode 2 in Fig. 4.54a, b, respectively, work in positive half period of the grid voltage; symmetrically, Mode 3 and Mode 4 in Fig. 4.54c, d, respectively, work in negative half period of the grid voltage.

Mode 1: Refer to Fig. 4.54a; the energy is delivered from the PV side into the grid through S_1, S_2, and the AC filter at this stage. The switches S_3 and S_4 are off, and share the input voltage U_{pv} in series, so their turn-off voltage stress is half of the input voltage.

Mode 2: Refer to Fig. 4.54b; only S_2 and S_6 are on at this stage according to Fig. 4.53b. The freewheeling current is divided into two branches, in which branch one consists of S_2 and D_7, while branch two consists of S_6 and D_5. Finally, the currents converge in the AC filter with the potential of the freewheeling path clamped at the ground.

Mode 3: Refer to Fig. 4.54c; the energy is delivered from the PV side into the grid through S_4, S_3, and AC filter at this stage. The switches S_1 and S_2 are off, and evenly share the input voltage.

Mode 4: Refer to Fig. 4.54d; S_3, S_5, and S_6 are on at this stage according to Fig. 4.53b. The freewheeling current is divided into two branches, in which branch one consists of S_3 and D_8, while branch two consists of S_5 and D_6. Finally, the currents converge in the AC filter with the potential of the freewheeling path clamped at the ground.

When S-NPC operates under non-unity power factor with modulation strategy I, there are two new modes as shown in Fig. 4.55. The positive grid current flows through D_3, D_4, and the filter in negative half period of the grid voltage, as shown in Fig. 4.55a. Symmetrically, the negative grid current flows through D_1, D_2, and the filter in positive half period of the grid voltage, as shown in Fig. 4.55b.

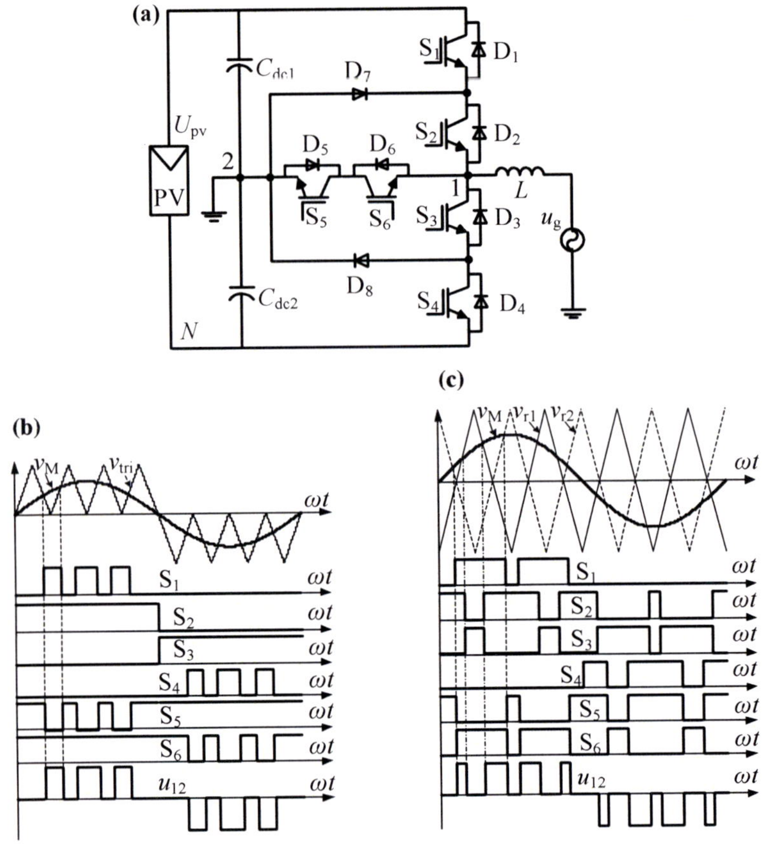

Fig. 4.53 S-NPC, a Main circuit, b Modulation strategy I, c Modulation strategy II

Based on the operating modes, the potentials of points 1 and 2 against N, as well as DMV and CMV are shown in Table 4.16. It can be noticed that the grid current flows through two parallel freewheeling branches, which include not only the diodes but also the switches S_5 and S_6. As a result, the total CMV remains at constant, so the common-mode characteristic is optimized in S-NPC.

The switching loss distribution of S-NPC under unity power factor is shown in Fig. 4.56 [18]. It is worth noting that the switching frequency was configured as 1 kHz, while the value of load-current is configured as 130 A. Since the switch combination S_1 and S_4, S_2 and S_3, S_5 and S_6, the diode combination D_7 and D_8 operate alternately throughout whole cycle of the grid voltage, the switching losses of each combination are an equivalent value, and thus Fig. 4.56 only shows the power loss of S_1, S_2, S_5, and D_7. It can be noticed that switching loss distribution is significantly uneven while the disparity between S_1 and S_2 is the most outstanding. Therefore, modulation strategy II was proposed to address the issue of switching loss distribution.

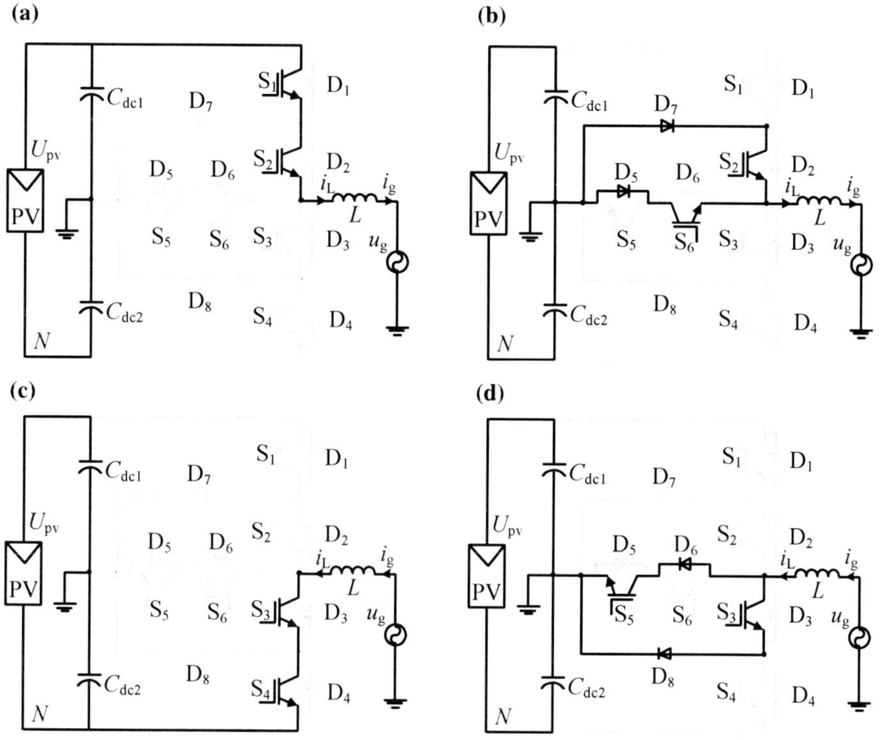

Fig. 4.54 Equivalent circuits of operating modes of S-NPC with modulation strategy I, **a** Mode 1, **b** Mode 2, **c** Mode 3, **d** Mode 4

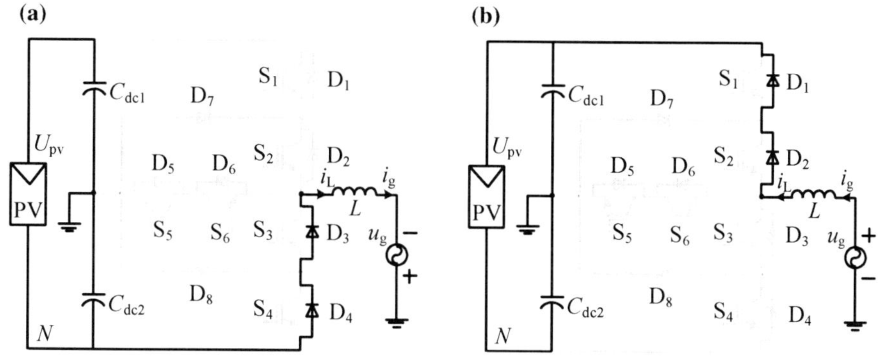

Fig. 4.55 Equivalent circuits of new operating modes of S-NPC with modulation strategy I under non-unity power factor, **a** Mode 1, **b** Mode 2

Table 4.16 DMV and CMV levels of different working stages with modulation strategy I

	Modes	u_{1N}	u_{2N}	u_{DM}	u_{CM}	$u_{CM\text{-}DM}$	u_{CM_tot}
Unity power factor	(1)	U_{pv}	$U_{pv}/2$	$U_{pv}/2$	$3U_{pv}/4$	$-U_{pv}/4$	$U_{pv}/2$
	(2)	$U_{pv}/2$	$U_{pv}/2$	0	$U_{pv}/2$	0	$U_{pv}/2$
	(3)	0	$U_{pv}/2$	$-U_{pv}/2$	$U_{pv}/4$	$U_{pv}/4$	$U_{pv}/2$
	(4)	$U_{pv}/2$	$U_{pv}/2$	0	$U_{pv}/2$	0	$U_{pv}/2$
Non-unity power factor	(1)	0	$U_{pv}/2$	$-U_{pv}/2$	$U_{pv}/4$	$U_{pv}/4$	$U_{pv}/2$
	(2)	U_{pv}	$U_{pv}/2$	$U_{pv}/2$	$3U_{pv}/4$	$-U_{pv}/4$	$U_{pv}/2$

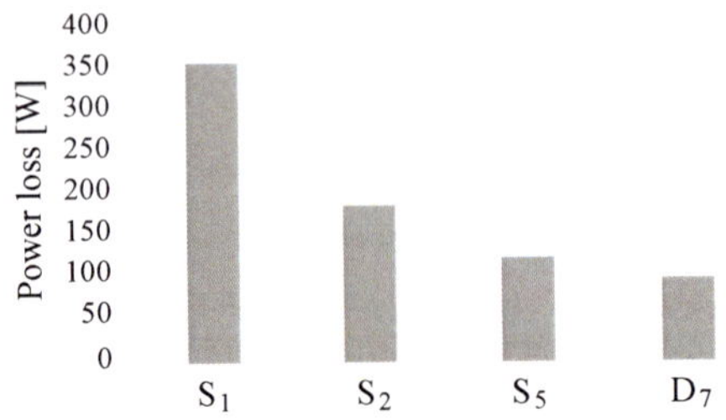

Fig. 4.56 Loss distribution among power devices with modulation strategy I

There are six operating modes according to the polarities of the grid voltage and grid current when S-NPC works in unity power factor with modulation strategy II, as shown in Fig. 4.57. Modes 1, 2, and 3 belong to positive half period of the grid voltage in Fig. 4.57a–c, respectively; symmetrically, Modes 3, 4, and 5 belong to negative half period of the grid voltage in Fig. 4.57d, f.

Mode 1: Refer to Fig. 4.57a; the energy is delivered from the PV side into the grid through S_1, S_2, and the AC filter at this stage. The switches S_3, S_4, and S_5 are off, and S_3 and S_4 share the input voltage U_{pv} in series, so their turn-off voltage stress is half of the input voltage.

Mode 2: Refer to Fig. 4.57b; S_2 and S_5 are on and rest of the switches are off according to Fig. 4.53c. At the first freewheeling stage, the current flows through D_7, S_2, and the filter, and keeps feeding the grid. It is worth noting that the potential of the freewheeling path is clamped at the ground.

Mode 3: Refer to Fig. 4.57c; S_1, S_3, and S_6 are on at the second freewheeling stage according to Fig. 4.53c. The freewheeling current flows through D_5, S_6, and the filter, and keeps feeding the grid. Meanwhile, the potential of the freewheeling path is also clamped at the ground.

Mode 4: Refer to Fig. 4.57d, the energy is delivered from the PV side into the grid through S_4, S_3, and the AC filter at this stage. The switches S_1 and S_2 are off and evenly share the input voltage U_{pv}.

Mode 5: Refer to Fig. 4.57e, only S_3 is on and rest of the switches are off according to Fig. 4.53c. At the third freewheeling stage, the current flows through D_8, S_3, and the filter, and keeps feeding the grid. It can be noticed that the potential of the freewheeling path is clamped at the ground.

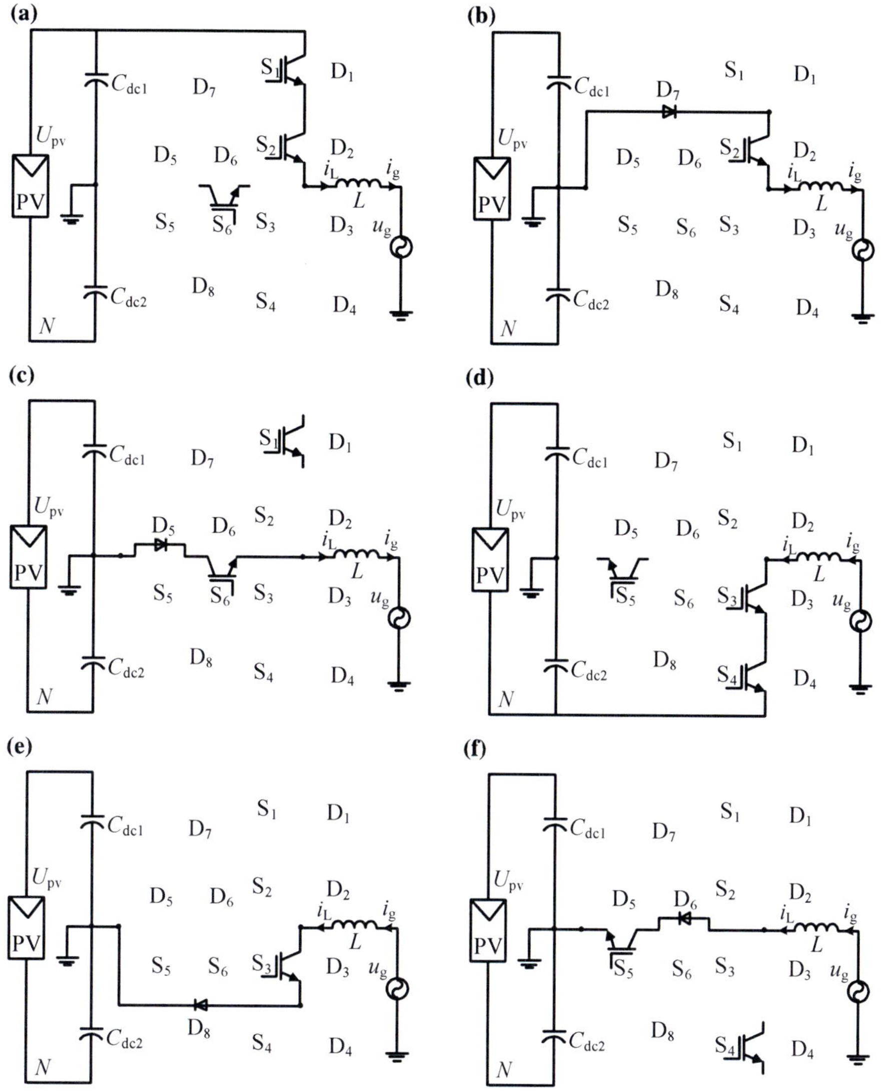

Fig. 4.57 Equivalent circuits of operating modes of S-NPC with modulation strategy II, **a** Mode 1, **b** Mode 2, **c** Mode 3, **d** Mode 4, **e** Mode 5, **f** Mode 6

Mode 6: Refer to Fig. 4.57f; S_2, S_4, and S_5 are on at the fourth freewheeling stage according to Fig. 4.53c. The freewheeling current flows through S_5, D_6, and the filter, and keeps feeding the grid. Meanwhile, the potential of the freewheeling path is still clamped at the ground.

Based on the operating modes, the potentials of points 1 and 2 against N, as well as DMV and CMV are shown in Table 4.17, and the losses of S_1, S_2, S_5, and D_7

Table 4.17 DMV and CMV levels of different working stages with modulation strategy II

Modes	u_{1N}	u_{2N}	u_{DM}	u_{CM}	$u_{CM\text{-}DM}$	$u_{CM\ tot}$
(1)	U_{pv}	$U_{pv}/2$	$U_{pv}/2$	$3U_{pv}/4$	$-U_{pv}/4$	$U_{pv}/2$
(2)	$U_{pv}/2$	$U_{pv}/2$	0	$U_{pv}/2$	0	$U_{pv}/2$
(3)	$U_{pv}/2$	$U_{pv}/2$	0	$U_{pv}/2$	0	$U_{pv}/2$
(4)	0	$U_{pv}/2$	$-U_{pv}/2$	$U_{pv}/4$	$U_{pv}/4$	$U_{pv}/2$
(5)	$U_{pv}/2$	$U_{pv}/2$	0	$U_{pv}/2$	0	$U_{pv}/2$
(6)	$U_{pv}/2$	$U_{pv}/2$	0	$U_{pv}/2$	0	$U_{pv}/2$

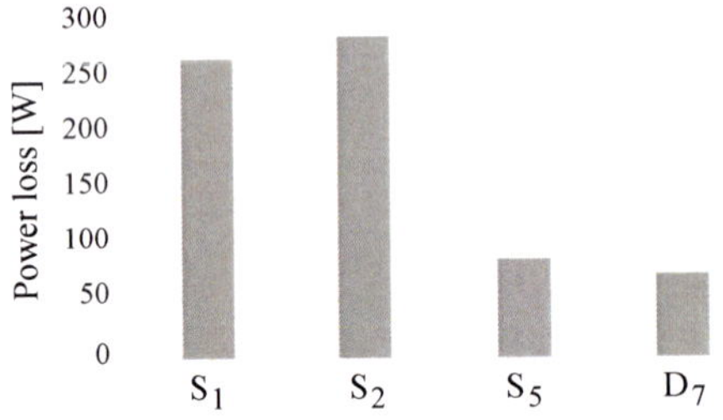

Fig. 4.58 Loss distribution among different devices with modulation strategy II

in S-NPC with modulation strategy II [18] is shown in Fig. 4.58. Compared with modulation strategy I, it can be noticed that the loss distribution of S_1 and S_2 is more balanced due to the existence of two freewheeling stages in every switching cycle. Additionally, there are two voltage pulses in each switching cycle, which can yield an equivalent effect of twice switching frequency.

2. Simulation results

In order to verify the aforementioned operating principle and property analysis, a simulation model of S-NPC grid-connected inverter is established in Matlab/Simulink, and the key parameters are listed in Table 4.3. When the S-NPC operates with unity power factor and modulation strategy I, the steady-state waveforms of the grid voltage u_g, reference signal of grid-in current i_{ref}, and actual grid-in current i_g are shown in Fig. 4.59.

The simulation waveforms of DMV u_{DM} and CMV u_{CM_tat} in time domain and frequency domain are shown in Figs. 4.60 and 4.61. It can be noticed that the amplitude of DMV is prominent around the switching frequency while no harmonic appears in the waveform of CMV. Furthermore, the simulation waveforms of the LC in time domain and frequency domain are shown in Fig. 4.62a, b, respectively. It is worth mentioning that the LC is eliminated in S-NPC due to the constant total CMV.

Figure 4.63 shows the steady-state waveforms of the grid voltage u_g, the grid-in current i_g, and its reference i_{ref} when T-NPC operates with non-unity power factor and modulation strategy I, which indicates that there is reactive power capacity in T-NPC.

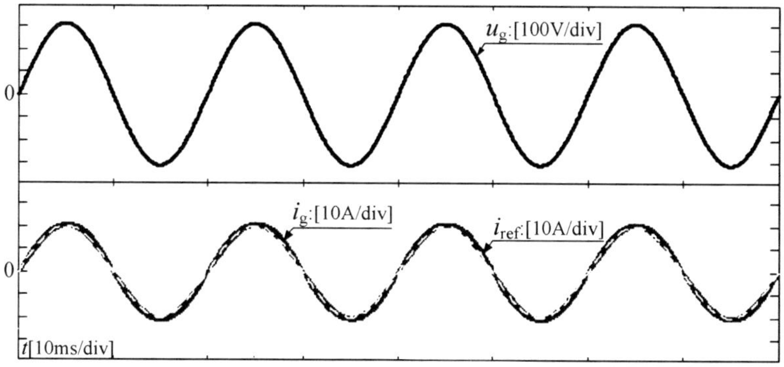

Fig. 4.59 Simulation waveforms of u_g, i_{ref}, and i_g of S-NPC under unity power factor

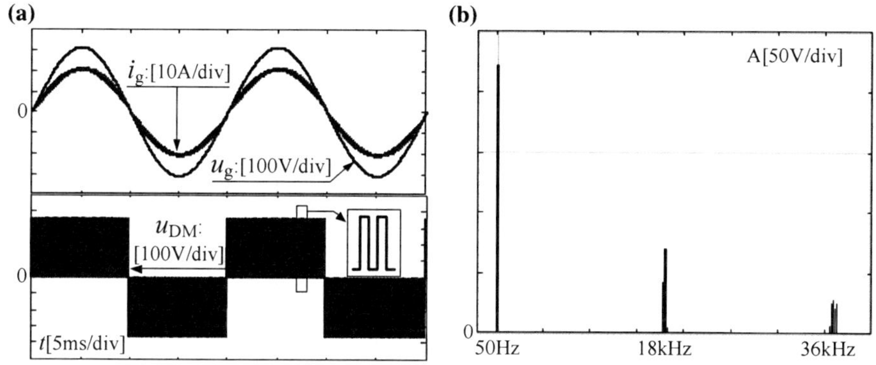

Fig. 4.60 DMV waveforms of S-NPC with modulation strategy I, **a** Time domain, **b** Frequency domain

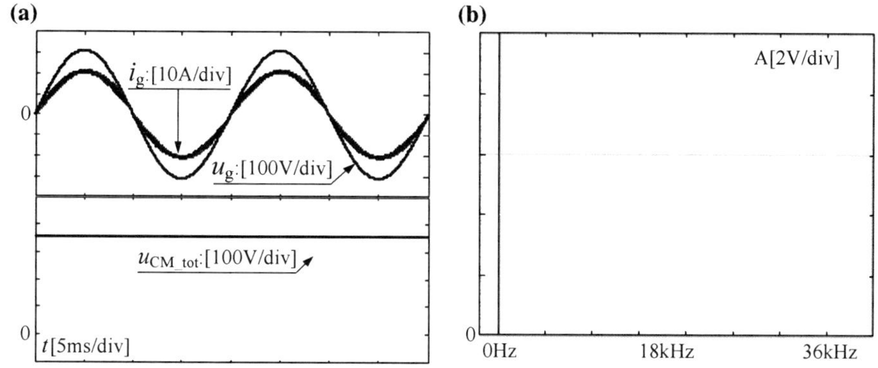

Fig. 4.61 CMV waveforms of S-NPC with modulation strategy I, **a** Time domain, **b** Frequency domain

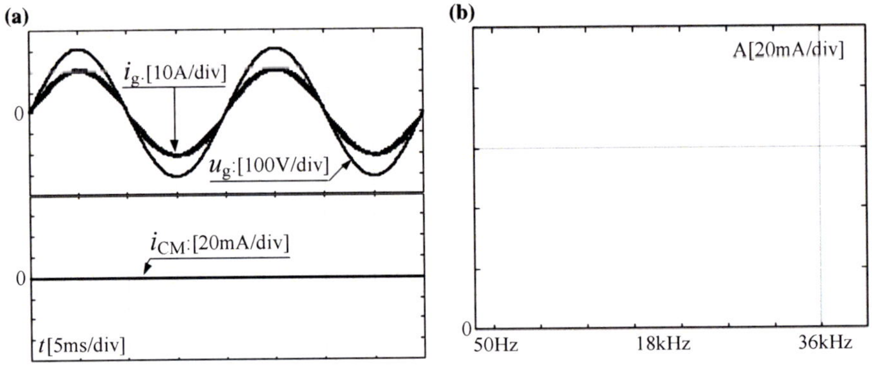

Fig. 4.62 LC waveforms of S-NPC with modulation strategy I, **a** Time domain, **b** Frequency domain

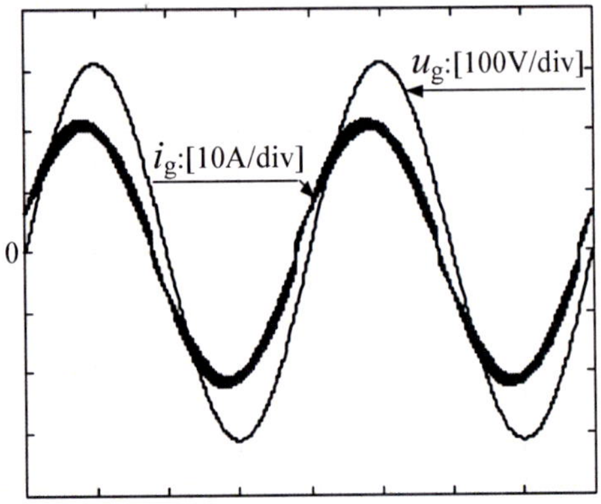

Fig. 4.63 Simulation waveforms of u_g, and i_g in S-NPC with non-unity power factor

When the S-NPC operates with non-unity power factor, the steady-state waveforms of the grid voltage u_g and grid-in current i_g are shown in Fig. 4.63, which indicates that S-NPC can be equipped with reactive power capacity.

When the S-NPC operates with non-unity power factor and modulation strategy II, the simulation waveforms of DMV u_{DM} in time domain and frequency domain are shown in Fig. 4.64a, b, respectively. It can be seen that the amplitude of DMV is zero at switching frequency, but double switching frequency component is significant. As a conclusion, modulation strategy II in S-NPC can breed an equivalent effect of twice switching frequency, and bring advantages for the design of filter.

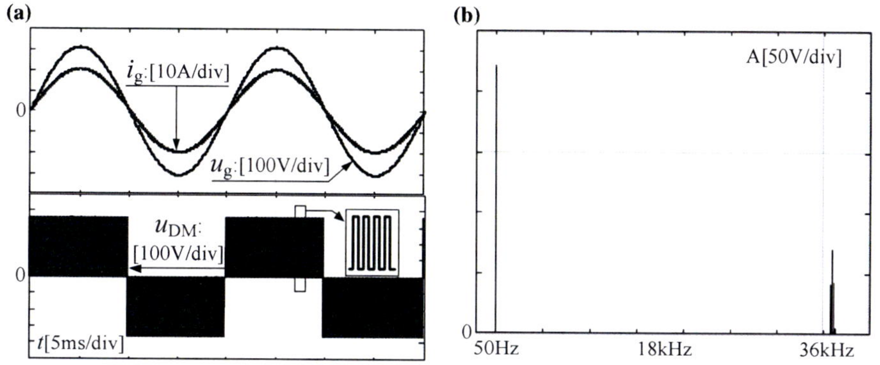

Fig. 4.64 DMV waveforms of S-NPC with modulation strategy II, **a** Time domain, **b** Frequency domain

4.4.2 Active Stacked NPC Inverter

1. Operating principle

To further realize the even distribution of switching loss, the active stacked neutral point clamped three-level inverter (AS-NPC) was proposed, as shown in Fig. 4.65a [25]. It can be noticed that the active switch is configured to replace the clamping diodes of S-NPC. Modulated by the unipolar double frequency strategy, the waveform of DMV is shaped as unipolar and three-level voltage, as shown in Fig. 4.65b.

There are six operating modes according to the polarities of the grid voltage and grid current when AS-NPC works in unity power factor case, as shown in Fig. 4.66.

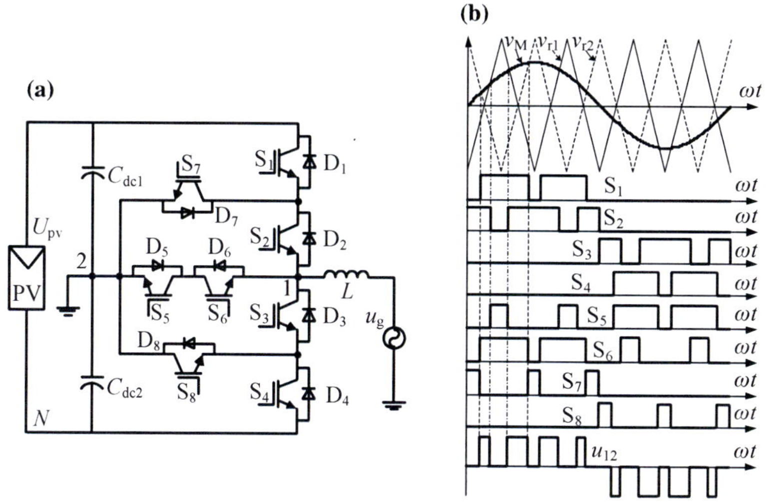

Fig. 4.65 AS-NPC grid-connected inverter, **a** Main circuit, **b** Modulation strategy

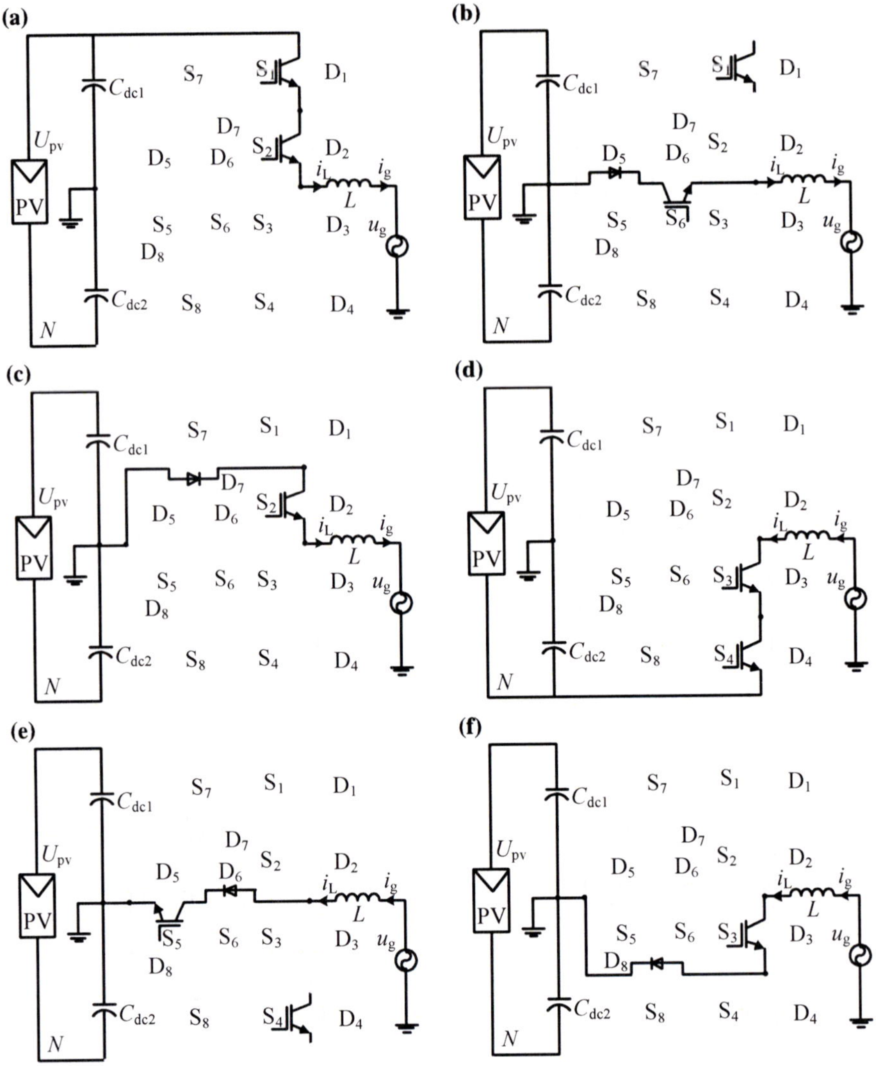

Fig. 4.66 Equivalent circuits of operating modes of AS-NPC, **a** Mode 1, **b** Mode 2, **c** Mode 3, **d** Mode 4, **e** Mode 5, **f** Mode 6

Modes 1, 2, and 3 in Fig. 4.66a–c, respectively, work in positive half period of the grid voltage; symmetrically, Modes 3, 4, and 5 in Fig. 4.66d–f, respectively, work in negative half period of the grid voltage.

Mode 1: Refer to Fig. 4.66a; the energy is delivered from the PV side into the grid through S_1, S_2, and the AC filter at this stage. The switches S_3 and S_4 are off, and share the input voltage U_{pv} in series, so their turn-off voltage stress is half of the input voltage.

Mode 2: Refer to Fig. 4.66b; S_1 and S_6 are on at the first freewheeling stage according to Fig. 4.65b. The freewheeling current flows through D_5, S_6, and the filter, and keeps feeding the grid. Meanwhile, the potential of the freewheeling path is clamped at the ground.

Mode 3: Refer to Fig. 4.66c; S_2 and S_7 are on and rest of the switches are off according to Fig. 4.65b. At the second freewheeling stage, the current flows through D_7, S_2, and the filter, and keeps feeding the grid. It is worth noting that the potential of the freewheeling path is clamped at the ground.

Mode 4: Refer to Fig. 4.66d; the energy is delivered from the PV side into the grid through S_4, S_3, and the AC filter at this stage. The switches S_1 and S_2 are off and evenly share the input voltage U_{pv}.

Mode 5: Refer to Fig. 4.66e; S_4 and S_5 are on at the third freewheeling stage according to Fig. 4.65b. The freewheeling current flows through S_5, D_6, and the filter, and keeps feeding the grid. Meanwhile, the potential of the freewheeling path is still clamped at the ground.

Mode 6: Refer to Fig. 4.66f; S_3, S_8 are on and rest of the switches are off according to Fig. 4.65b. At the fourth freewheeling stage, the current flows through D_8, S_3 and filter, and keeps feeding the grid. It can be noticed that the potential of the freewheeling path is still clamped at the ground.

Based on the operating modes, the potentials of point 1 and 2 against N, as well as DMV and CMV are shown in Table 4.18. At the freewheeling stage, two branches of S-NPC work simultaneously while branches of AS-NPC conduct separately. Furthermore, the pulse frequency of the output voltage in AS-NPC is twice the switching frequency.

2. Simulation results

In order to verify the aforementioned operating principle and property analysis, a simulation model of AS-NPC grid-connected inverter is established in Matlab/Simulink, and the key parameters are listed in Table 4.3. The simulation waveforms of DMV u_{DM} in time domain and frequency domain are shown in Fig. 4.67a, b, respectively. It can be seen that the amplitude of DMV is zero at switching frequency, but double switching frequency component is significant. In accordance with the theoretical analysis, the AS-NPC can yield an equivalent effect of double switching

Table 4.18 DMV and CMV of different operating modes for AS-NPC

Modes	u_{1N}	u_{2N}	u_{DM}	u_{CM}	$u_{CM\text{-}DM}$	u_{CM_tot}
(1)	U_{pv}	$U_{pv}/2$	$U_{pv}/2$	$3U_{pv}/4$	$-U_{pv}/4$	$U_{pv}/2$
(2)	$U_{pv}/2$	$U_{pv}/2$	0	$U_{pv}/2$	0	$U_{pv}/2$
(3)	$U_{pv}/2$	$U_{pv}/2$	0	$U_{pv}/2$	0	$U_{pv}/2$
(4)	0	$U_{pv}/2$	$-U_{pv}/2$	$U_{pv}/4$	$U_{pv}/4$	$U_{pv}/2$
(5)	$U_{pv}/2$	$U_{pv}/2$	0	$U_{pv}/2$	0	$U_{pv}/2$
(6)	$U_{pv}/2$	$U_{pv}/2$	0	$U_{pv}/2$	0	$U_{pv}/2$

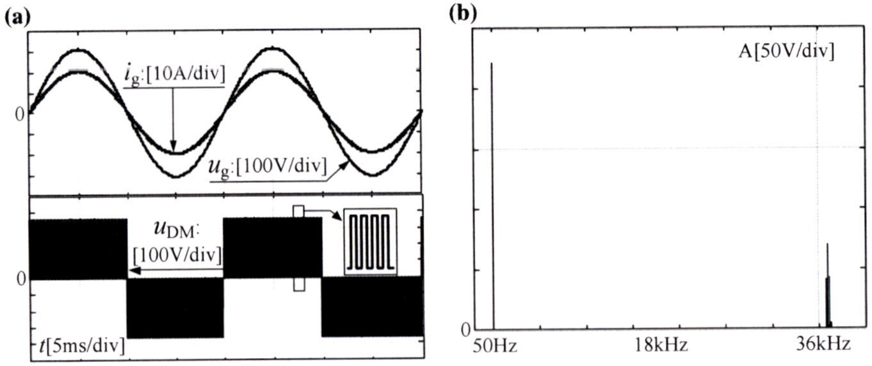

Fig. 4.67 DMV waveforms of AS-NPC, **a** Time domain, **b** Frequency domain

frequency in DMV. Furthermore, the CMV characteristics of AS-NPC are consistent with that of S-NPC, which can refer to Figs. 4.60, 4.61, and 4.62.

4.4.3 High-Efficiency Stacked NPC Inverter

1. Operating principle

To further reduce the switching loss of S-NPC, high-efficiency stacked neutral point clamped three-level inverter (HE-S-NPC) was proposed, as shown in Fig. 4.68a [26]. It can be noticed that CoolMosfet is configured to replace switches in the middle and outside of S-NPC, and the midpoint clamping branch is verified from series switches to parallel switches. Furthermore, the modulation strategy of HE-S-NPC is shown in

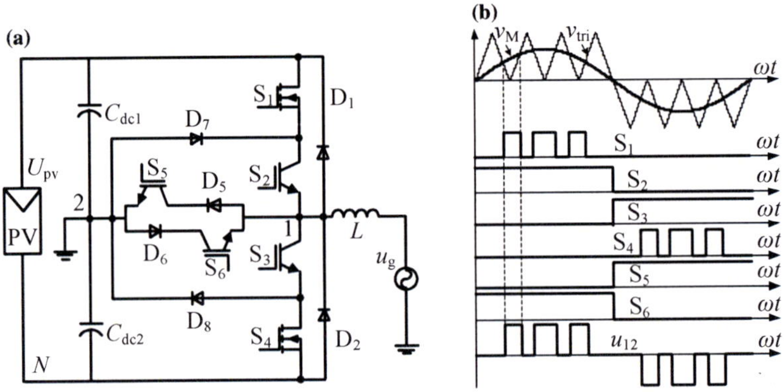

Fig. 4.68 HE-S-NPC grid-connected inverter, **a** Main circuit, **b** Modulation strategy

Fig. 4.68b. It is worth noting that the waveform of DMV is shaped as unipolar and three-level voltage.

There are four operating modes according to the polarities of the grid voltage and grid current when HE-S-NPC works in unity power factor case, as shown in Fig. 4.69. Modes 1 and 2 in Fig. 4.69a, b, respectively, work in positive half period of the grid voltage; symmetrically, Modes 3 and 4 in Fig. 4.69c, d, respectively, work in negative half period of the grid voltage.

Mode 1: Refer to Fig. 4.69a; the energy is delivered from the PV side into the grid through S_1, S_2, and the AC filter at this stage. The switches S_3 and S_4 are off, and share the input voltage U_{pv} in series, so their turn-off voltage stress is half of the input voltage.

Mode 2: Refer to Fig. 4.69b; S_2 and S_6 are on and rest of the switches are off according to Fig. 4.68b. The freewheeling current is divided into two branches, in which branch one consists of S_2 and D_7 while branch two consists of S_6 and D_6. Finally, the currents converge in the AC filter with the potential of the freewheeling path clamped at the ground.

Mode 3: Refer to Fig. 4.69c; the energy is delivered from the PV side into the grid through S_4, S_3, and the AC filter at this stage. Meanwhile, the switches S_1 and S_2 are off and evenly share the input voltage U_{pv}.

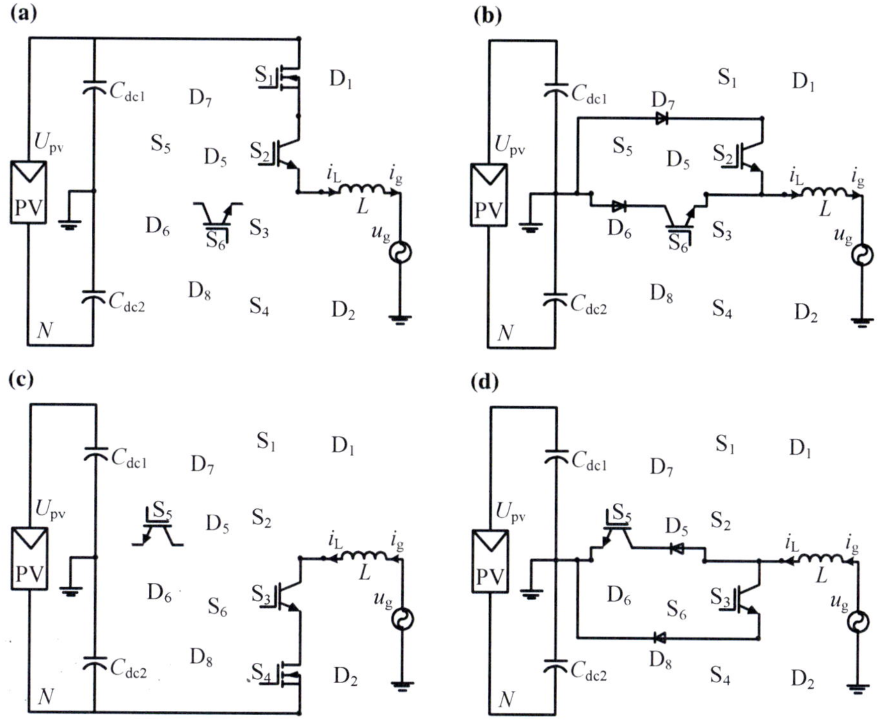

Fig. 4.69 Equivalent circuits of operating modes of HE-S-NPC with unity power factor, **a** Mode 1, **b** Mode 2, **c** Mode 3, **d** Mode 4

Mode 4: Refer to Fig. 4.69d; S_3 and S_5 are on and rest of the switches are off according to Fig. 4.68b. The freewheeling current is divided into two branches, in which branch one consists of S_5 and D_5 while branch two consists of S_3 and D_8. Finally, the currents converge in the AC filter with the potential of the freewheeling path clamped at the ground.

When HE-S-NPC operates with non-unity power factor, there are two new modes as shown in Fig. 4.70. As shown in Fig. 4.70a, the current flows to the DC side through diode D_2 during the mode with negative grid voltage and positive grid-in current. As shown in Fig. 4.70b, the current flows to the DC side through diode D_1 during the mode with positive grid voltage and negative grid-in current. It is revealed that the current does not flow through the parallel diodes of the switches during the whole cycle of the grid voltage.

Based on the operating modes, the potentials of points 1 and 2 against N, as well as DMV and CMV are shown in Table 4.19. Considering that only two switches (CoolMosfet) of HE-S-NPC operate at high frequency with better performance, the switching loss of HE-S-NPC will be obviously less than S-NPC.

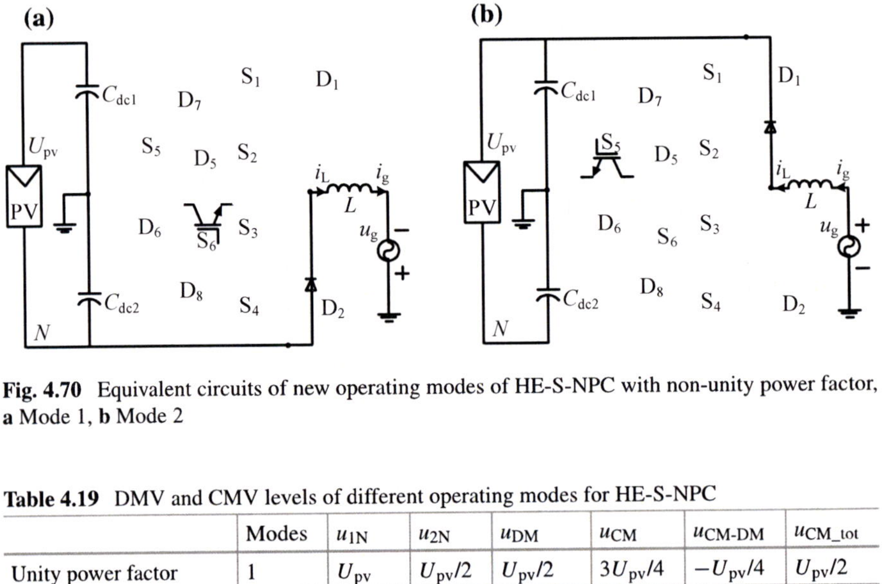

Fig. 4.70 Equivalent circuits of new operating modes of HE-S-NPC with non-unity power factor, **a** Mode 1, **b** Mode 2

Table 4.19 DMV and CMV levels of different operating modes for HE-S-NPC

	Modes	u_{1N}	u_{2N}	u_{DM}	u_{CM}	$u_{CM\text{-}DM}$	u_{CM_tot}
Unity power factor	1	U_{pv}	$U_{pv}/2$	$U_{pv}/2$	$3U_{pv}/4$	$-U_{pv}/4$	$U_{pv}/2$
	2	$U_{pv}/2$	$U_{pv}/2$	0	$U_{pv}/2$	0	$U_{pv}/2$
	3	0	$U_{pv}/2$	$-U_{pv}/2$	$U_{pv}/4$	$U_{pv}/4$	$U_{pv}/2$
	4	$U_{pv}/2$	$U_{pv}/2$	0	$U_{pv}/2$	0	$U_{pv}/2$
Non-unity power factor	1	0	$U_{pv}/2$	$-U_{pv}/2$	$U_{pv}/4$	$U_{pv}/4$	$U_{pv}/2$
	2	U_{pv}	$U_{pv}/2$	$U_{pv}/2$	$3U_{pv}/4$	$-U_{pv}/4$	$U_{pv}/2$

SW, D, L and C denote the number of switche, diode, inductor and capacitor, respectively. EDS is the energy-delivering stage while FWS is the freewheeling stage

Table 4.20 Switch numbers of S-NPC and HE-S-NPC in conduction loops

	SW	D	Unity power factor		Non-unity power factor	
			SW(EDS)	SW(FWS)	SW3(EDS)	SW4(FWS)
S-NPC	6*IGBT	8	2*IGBT	2//2	2	2
HE-S-NPC	4*IGBT+ 2*CoolMosfet	6	1*IGBT+ 1*CoolMosfet	2//2	1	2

Specifically, the numbers of conducting components in S-NPC and HE-S-NPC during different operating modes are listed in Table 4.20. It can be noticed that the switching loss of HE-S-NPC is less than that of S-NPC during the energy-delivering stage. In addition, the conduction loss in HE-S-NPC can be reduced by using the diodes D_5/D_6, whose loss is less than IGBT's reverse parallel diode of S-NPC. In addition to the freewheeling stage, the switching loss of HE-S-NPC is less than S-NPC even though the two circuits have equivalent current in both branches of S-NPC. It is worth noting that the switching loss during positive and negative half period is a consistent value. As a conclusion, HE-S-NPC has more promising efficiency compared with S-NPC.

As indicated by aforementioned discussion, the grid current of HE-S-NPC flows through two parallel branches in the freewheeling stage to reduce the conduction resistor and improve operation efficiency. Compared with S-NPC, it is worth noting that the internal parallel diodes of switches will not conduct in HE-S-NPC. As S_5 and S_6 are separately connected with an external diode, the conduction loss of switches is lower than that of switches' internal parallel diodes in the freewheeling stage. In addition, the switching frequency can be further increased with CoolMosfet used in the main circuit, which brings advantages for the filter design of HE-S-NPC.

2. Simulation results

The operating principle and simulation waveforms of HE-S-NPC are highly consistent with that of S-NPC, which can refer to Figs. 4.59, 4.60, 4.61, 4.62 and 4.63.

4.5 Summary

The half-bridge NPC-TLI is discussed in this chapter with I-type, and T-type and stacked-type are also included. The total CMV of half-bridge NPC-TLIs can be kept in a constant value at high-frequency range which is equal to or larger than the switching frequency. In addition to low-frequency range, the CMV is maintained as half of the input voltage, and thus there is no LC as expected. Besides, distributions of the switching loss, the ability to eliminate DC components, the reliability of preventing short-circuit, and the reactive power are discussed. Comparison of characteristics among different half-bridge NPC TLIs discussed in this chapter is shown in Table 4.21, which can be a reference for engineers and technicians.

Table 4.21 Comparison of DMV performance among different NPC-TLIs

Name		Circuit structure				Power losses			Remark
		SW	D	L	C	SW(HF)	SW(EDS)	SW(FWS)	HF: high-frequency
I-type	NPC [1]	4	4 + 2	1	2	2 + 1	2	2	–
	A-NPC [2]	6	6	1	2	2	2	2	Uniform distribution of switching losses
	NPC-C [7]	4	6	1	4	2	2	2	Less DC component
	SI-NPC [8]	4	6	2	2	2	2	2	Prevention of short-circuit
	D-SI-NPC [9]	4	8	2	2	2	2	2	Large reactive power capability
	6S-5L-ANPC [15]	6	6	4	6	2	2	3	Five-level output voltage
T-type	T-NPC [17]	4	4	1	2	4	1	2	–
	RS-T-NPC [19]	3	6	1	2	2 + 1	1	3	–
	DF-T-NPC [20]	6	6	1	2	6	1	2//2	–
Stacked -type	S-NPC [24]	6	8	1	2	4	2	2//2	Uniform distribution of switching losses
	AS-NPC [25]	8	8	1	2	6 + 2	2	2	Uniform distribution of switching losses
	HE-S-NPC [26]	6	6	1	2	2	2	2//2	–

HF denotes high-frequency while other Abbreviations are consistent with those of Table 4.20

References

1. Nabae A, Takahashi I, Akagi H (1981) A new neutral-point-clamped PWM inverter. IEEE Trans Ind Appl 17(5):518–523
2. Brückner T, Bernet S, Güldner H (2005) The active NPC converter and its loss-balancing control. IEEE Trans Ind Electron 52(3):855–868
3. Floricau D, Floricau E, Dumitrescu M, Natural doubling of the apparent switching frequency using three-level ANPC converter. In: 2008 international school on Nonsinusoidal currents and compensation, Lagow, pp 1–6
4. Ma L, Kerekes T, Rodriguez P et al (2015) A new PWM strategy for grid-connected half-bridge active NPC converters with losses distribution balancing mechanism. IEEE Trans Power Electron 30(9):5331–40
5. Brückner T, Bernet S, Loss balancing in three-level voltage source inverters applying active NPC switches In 2001 IEEE Annual Power Electronics Specialists Conference, Vancouver, pp 1135–1140
6. Li J, Liu JJ, Boroyevich D et al (2011) Three-level active neutral-point clamped zero-current-transition converter for sustainable energy systems. IEEE Trans Power Electron 26(12):3680–3693
7. Gonzalez R, Gubia E, Lopez J et al (2008) Transformerless single-phase multilevel photovoltaic inverter. IEEE Trans Ind Electron 55(7):2694–2702
8. Xiao HF, Xie SJ (2012) Transformerless split-inductor neutral point clamped three-level PV grid-connected inverter. IEEE Trans Power Electron 27(4):1799–1808
9. Frisch M, Temesi E (2011)Inverter topologies usable with reactive power. America, US Patent: US20 110 013 438 A1
10. Poliseno MC, Mastromauro RA, Liserre M et al (2012) High efficiency transformerless PV power converters. In: 2012 international symposium on power electronics power electronics, electrical drives, automation and motion, Sorrento, pp 93–98
11. Grigoletto FB (2019) Five-level transformerless inverter for single-phase solar photovoltaic applications. IEEE J Emerg Sel Topics Power Electron, Early Access
12. Vazquez G, Martinez-Rodriguez PR, Sosa JM et al, H5-HERIC based transformerless multi-level inverter for single-phase grid connected PV systems. In: 2015 Annual conference of the IEEE industrial electronics society, pp 1026–1031
13. Rahim NA, Selvaraj J (2010) Multistring five-level inverter with novel PWM control dcheme for PV application. IEEE Trans Industr Electron 57(6):2111–2123
14. Buticchi G, Barater D, Lorenzani E et al (2014) A nine-level grid-connected converter topology for single-phase transformerless PV systems. IEEE Trans Industr Electron 61(8):3951–3960
15. Wang HL, Kou L, Liu YF et al (2017) A new six-switch five-level active neutral point clamped inverter for PV applications. IEEE Trans Power Electron 32(9):6700–6715
16. Teymour HR, Sutanto D, Muttaqi KM et al (2015) A novel modulation technique and a new balancing control strategy for a single-phase five-level ANPC converter. IEEE Trans Ind Appl 51(2):1215–1227
17. Knaup P (2007) International Patent Application. WO2007048420A1
18. Schweizer M, Kolar JW (2013) Design and implementation of a highly efficient three-level T-type converter for low-voltage applications. IEEE Trans Power Electron 28(2):899–907
19. Hasanzadeh A, Edrington CS, Leonard J, Reduced switch NPC-based transformerless PV inverter by developed switching pattern. In: 2012 annual IEEE applied power electronics conference and exposition, Orlando, pp 359–360
20. Wang Y, Shi WW, Xie N et al (2014) Diode-free T-type three level neutral-point- clamped inverter for low voltage renewable energy system. IEEE Trans Ind Electron 61(11):6168–6174
21. Bhagwat P, Stefanovic VR (1980) Generalized structure of a multilevel PWM inverter. IEEE Trans Ind Appl 19(6):761–766
22. Brumsickle WE, Divan DM, Lipo TA, Reduced switching stress in high-voltage IGBT inverters via a three-level structure. In: 1998 annual applied power electronics conference and exposition, Anaheim, pp 544–550

23. Meynard TA, Foch H, Forest F et al (2002) Multicell converters: derived topologies. IEEE Trans Industr Electron 49(5):978–986
24. Floricau D, Gateau G, Dumitrescu M et al, A new stacked NPC converter: 3L-topology and control. In: 2007 IEEE European conference on power electronics and applications, Aalborg, pp 1–10
25. Floricau D, Gateau G, Leredde A (2010) New active stacked npc multilevel converter: operation and features. IEEE Trans Industr Electron 57(7):2272–2278
26. Wang Y, Li R (2013) Novel high-efficiency three-level stacked-neutral-point-clamped grid-tied inverter. IEEE Transactions on Industiral Electronics 60(9):3766–3774

Chapter 5
Common-Ground Transformerless Grid-Connected Inverters

Abstract *Rule* 1 for eliminating leakage current of full-bridge TLIs and *Rule* 2 for eliminating leakage current of half-bridge TLIs have been discussed in Chaps. 3 and 4, respectively. This chapter focuses on *Rule* 3 used in common-ground TLIs. Common-ground TLIs mean that the neutral line of the grid is directly connected to the negative terminal of the PV output terminal. According to *Rule* 3, the total CMV u_{CM_tot} of common-ground TLIs is zero, so common-ground TLIs have the potential of the best CMV performance.

Keywords Transformerless PV grid-connected inverter · Common-ground · Capacitor-based · Inductor-based

5.1 Capacitor-Based Common-Ground TLIs

The challenge of the common ground TLI topology is how to provide negative voltage or negative current during the grid negative half period. The implementation measures of common-ground structure can be classified into two categories, that is, capacitor-based [1–6] and inductor-based circuits [7–11]. This section discusses the circuit structures, operating principle, and characteristics of capacitor-based common-ground TLIs.

5.1.1 Capacitor-Based Common-Ground Inverter I

1. Operating principle

The circuit structure of the first common-ground transformerless grid-connected inverter (CGT-I) is shown in Fig. 5.1a [1], the PV negative terminal is directly connected to the neutral line of the grid. There are four power switches and a capacitor C in CGT-I, and the modulation strategy is illustrated in Fig. 5.1b. It is worth mentioning that switches S_1 and S_2 are working complementarily with dead time,

H. Xiao and X. Wang, *Transformerless Photovoltaic Grid-Connected Inverters*, CPSS Power Electronics Series, https://doi.org/10.1007/978-981-15-8525-8_5

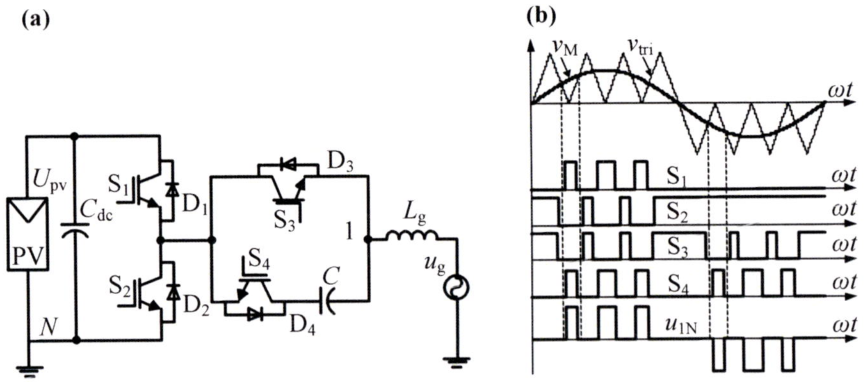

Fig. 5.1 CGT-I grid-connected inverter, **a** Main circuit, **b** Modulation strategy

and switches S_3 and S_4 are complementary with each other as well. As a result, the output voltage u_{1N} is a three-level voltage with unipolar SPWM characteristic.

There are four operating modes according to the polarities of the grid voltage and grid-in current when the power factor of CGT-I is equal to 1, as shown in Fig. 5.2. It is necessary to point out that the voltage on capacitor C is half of the input DC voltage U_{pv}. Mode 1 and Mode 2 in Fig. 5.2a, b are modes in positive half period of the grid, respectively; symmetrically, Mode 3 and Mode 4 in Fig. 5.2c, d belong to negative half period of the grid.

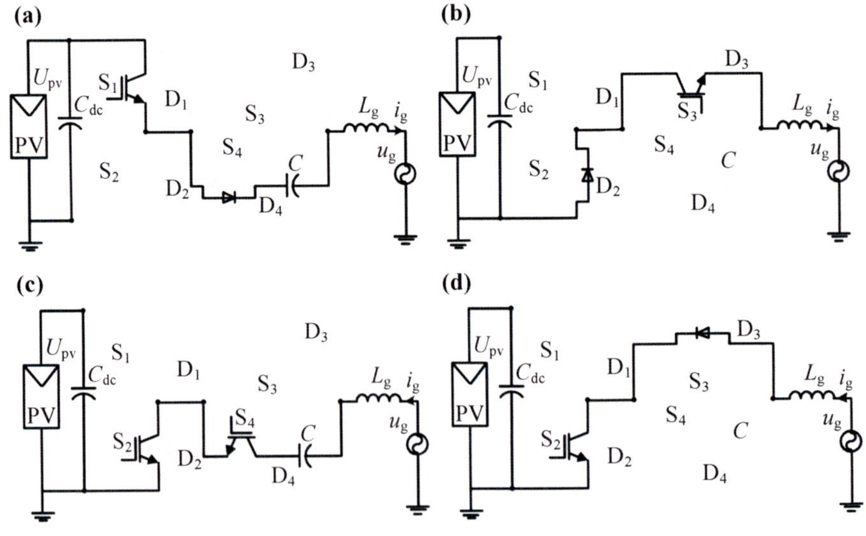

Fig. 5.2 Equivalent circuits of operating modes of CGT-I, **a** Mode 1, **b** Mode 2, **c** Mode 3, **d** Mode 4

Mode 1: Refer to Fig. 5.2a; the energy is delivered from the PV side into the grid through S_1, D_4, *C*, and the AC filter at this stage. Due to the amplitude and polarity of the voltage across capacitor *C*, the amplitude of the output DMV is half the DC input voltage.

Mode 2: Refer to Fig. 5.2b; at this stage, the freewheeling current flows through D_2, S_3, and the AC filter, and keeps feeding the grid. It is worth noting that the potential of the freewheeling path is clamped at the ground.

Mode 3: Refer to Fig. 5.2c; the energy is delivered from capacitor *C* into the grid through S_2, S_4, and the AC filter at this stage.

Mode 4: Refer to Fig. 5.2d; the freewheeling current flows through D_3, S_2, and the AC filter, and keeps feeding the grid. Obviously, it is the freewheeling mode in the negative half grid period.

According to the analysis of operating modes, capacitor *C* is charged during the positive half grid period and discharged during the negative half cycle, so the capacitance of *C* should be large enough to keep its voltage constant during the whole grid cycle. In addition, the voltage across capacitor *C* has to be maintained at half the input DC voltage.

2. Simulation Research

In order to further verify the presented theoretical analysis of CGT-I, a 1 kW CGT-I simulation model has been built in Matlab/Simulink. The simulation parameters are listed in Table 5.1.

The simulation waveforms of CGT-I with unity power factor are presented in Figs. 5.3 and 5.4. While Fig. 5.3 shows the simulation waveforms of the grid voltage u_g, grid-in current reference i_{ref}, and the grid-in current i_g when the power factor is 1, the simulation waveforms of the output voltage u_{1N} with unipolar SPWM characteristic is illustrated in Fig. 5.4a, verifying the amplitude of u_{1N} is indeed maintained at half the input DC voltage. Obviously, the amplitude of u_{1N} decreases with the capacitor *C* charged and discharged during each half grid period, but it only drops a little because of the large capacitance of *C*. The switches operate at 15 kHz, so the harmonic component of u_{1N} concentrates at 15 kHz and its multiples, as is shown in Fig. 5.4b. The simulation waveforms of the current through S_1, S_4, and D_4 are shown in Fig. 5.5a, b, respectively. It is clear that only the output current flows through S_1, S_4, and D_4, proving that the current stress on the switches is small. The operating period of *C* is the same as the grid period, but the phase is delayed as shown in Fig. 5.6. Additionally, CGT-I can operate perfectly when the power factor is not 1 as shown in Fig. 5.7.

Table 5.1 Simulation parameters of 1 kW CGT-I

Parameter	Value	Parameter	Value
Input voltage U_{pv}/V	680	Switching frequency f_s/kHz	15
Grid voltage u_g/V	220	Capacitor C/μF	1100
Grid frequency/Hz	50	Filter inductor L_f/mH	8
Rated power P/W	1000	PV panel parasitic capacitor C_{pv}/nF	50

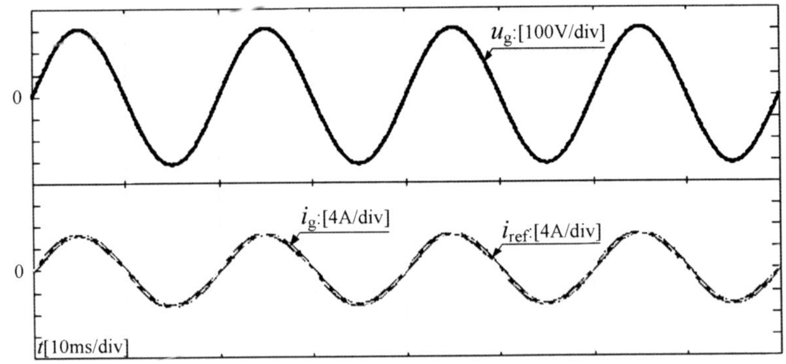

Fig. 5.3 Simulation waveforms of grid voltage and grid-in current of CGT-I with unity power factor

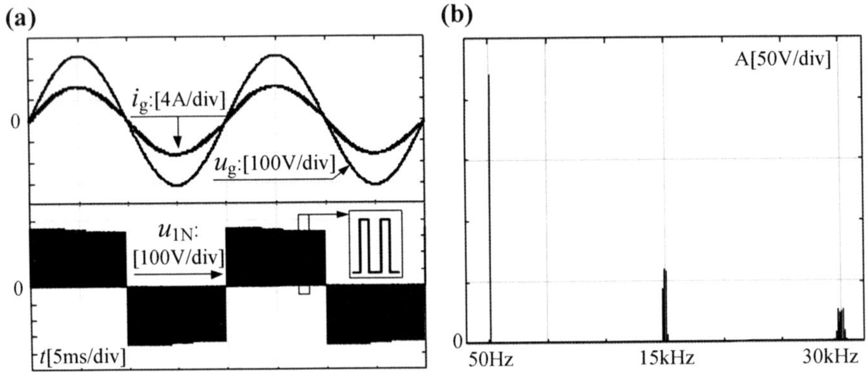

Fig. 5.4 Simulation waveforms of output voltage U_{1N}, **a** Time domain, **b** Frequency domain

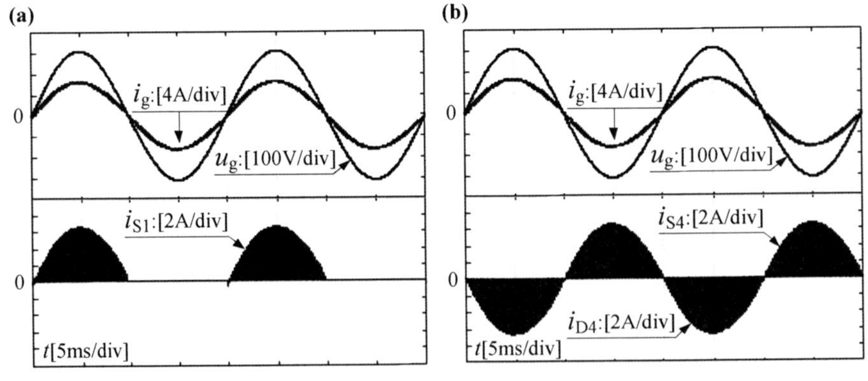

Fig. 5.5 Simulation waveforms of switches current, **a** Current waveform of S_1, **b** Current waveforms of S_4 and D_4

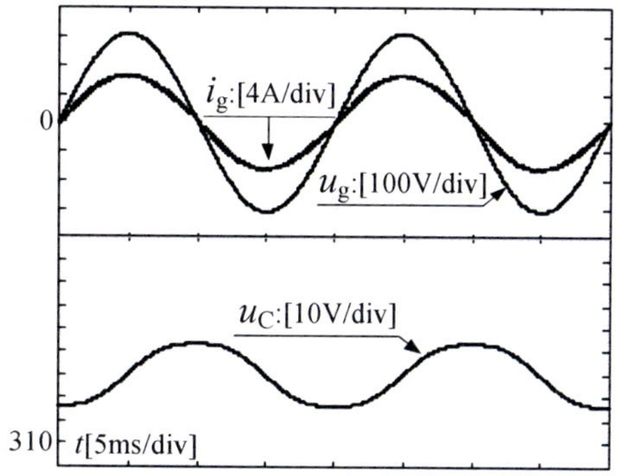

Fig. 5.6 Simulation waveforms of voltage across C

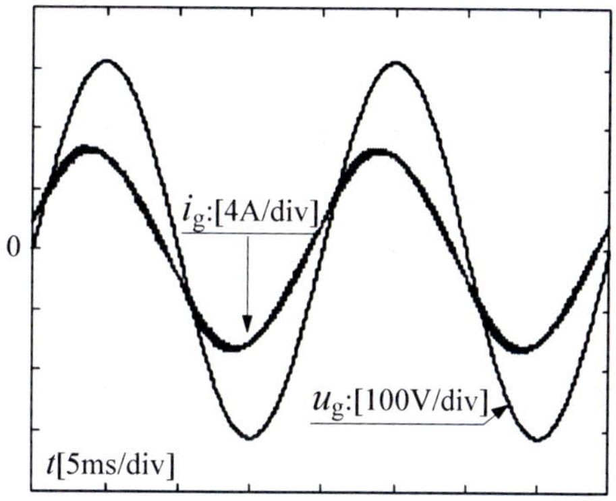

Fig. 5.7 Simulation waveforms of grid voltage and grid-in current of CGT-I with non-unity power factor

5.1.2 Capacitor-Based Common-Ground Inverter II

As analyzed in Sect. 5.1, the capacitor C in CGT-I can only be charged during the positive half grid period and discharged during the negative half grid period. Therefore, it needs the large capacitor C and long time to be charged, which leads to bigger capacitor volume and shorter lifetime. In order to reduce capacitor volume, the virtual DC bus inverter (VDCBI) which capacitor can be charged and discharged at high frequency was proposed in [2].

1. Operating principle

The circuit structure of the second common-ground transformerless grid-connected inverter (CGT-II) is shown in Fig. 5.8a, which is also called VDCBI [2]. Different from CGT-I, except for the capacitor C_1 connected to the PV panels forming the real positive DC bus, another capacitor C_2 connected to the PV panels through S_1 and S_3 forming the virtual negative DC bus is added into CGT-II. It is worth mentioning that while switches S_1 and S_3 are working simultaneously, switches S_1/S_3 and S_2 are working complementarily with dead time and switches S_4 and S_5 are complementary

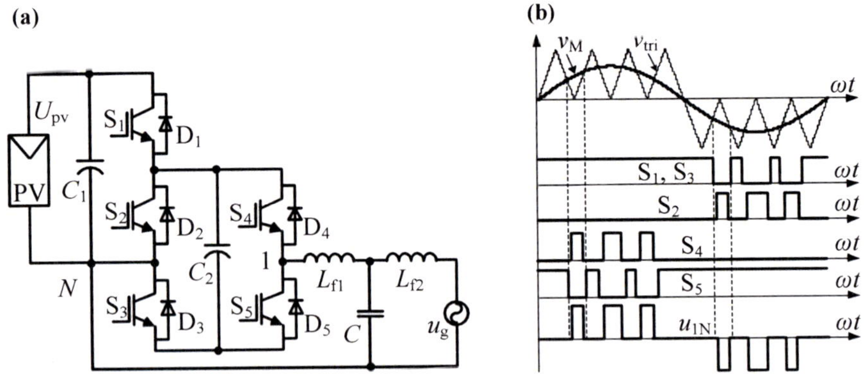

Fig. 5.8 VDCBI grid-connected inverter, **a** Main circuit, **b** Modulation strategy

with each other as well. As a result, the output voltage u_{1N} is a three-level voltage with unipolar SPWM characteristic.

There are four operating modes according to the polarities of the grid voltage and grid-in current when the power factor of VDCBI is 1, as shown in Fig. 5.9. Mode 1 and Mode 2 in Fig. 5.9a, b belong to positive half period of the grid voltage,

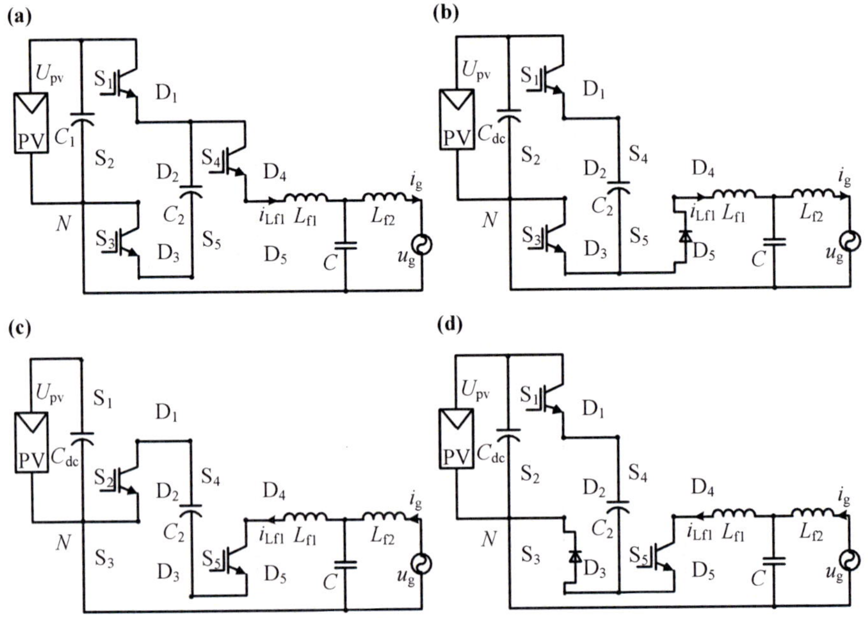

Fig. 5.9 Equivalent circuits of operating modes of VDCBI, **a** Mode 1, **b** Mode 2, **c** Mode 3, **d** Mode 4

respectively; symmetrically, Mode 3 and Mode 4 in Fig. 5.9c, d work in negative half period of the grid voltage.

Mode 1: Refer to Fig. 5.9a; the energy is delivered from the PV side into the grid through S_1, S_4, and the AC filter at this stage. Simultaneously, C_2 is charged by the input DC voltage U_{pv} through S_1 and D_3.

Mode 2: Refer to Fig. 5.9b; at this stage, the freewheeling current flows through S_3, D_5, and the AC filter, and keeps feeding the grid. C_2 continues to be charged by the input DC voltage U_{pv} through S_1 and S_3.

Mode 3: Refer to Fig. 5.9c; the energy is delivered from capacitor C_2 into the grid through S_2, S_5, and the AC filter at this stage.

Mode 4: Refer to Fig. 5.9d; the freewheeling current flows through S_5, D_3, and the filter, and keeps feeding the grid. C_2 restarts to be charged by the input DC voltage U_{pv} through S_1 and D_3.

According to the analysis of operating modes, the DMV and CMV in each mode of VDCBI are listed in Table 5.2. Note that the voltage u_{2N} is clamped at the ground.

It is revealed from the analysis of operating modes that a small capacitor C_2 has the capability to keep the grid voltage constant because C_2 operates at switching frequency. In Mode 4, the charging current and the freewheeling current both flow through D_3, which leads to large current stress on it. More specifically, the pre-charge of C_2 can be accomplished by keeping S_1 ON in the standby mode, so the voltage of C_1 and C_2 are synchronized with U_{pv}, limiting the inrush current.

2. Simulation Research

In order to further verify the presented theoretical analysis of VDCBI, a 500 W VDCBI simulation model has been built in Matlab/Simulink. The simulation parameters are listed in Table 5.3.

Table 5.2 DMV and CMV in each mode of VDCBI

Modes	u_{1N}	u_{2N}	u_{DM}	u_{CM}	u_{CM_DM}	u_{CM_tot}
(1)	U_{pv}	0	U_{pv}	$U_{pv}/2$	$-U_{pv}/2$	0
(2)	0	0	0	0	0	0
(3)	$-U_{pv}$	0	$-U_{pv}$	$-U_{pv}/2$	$U_{pv}/2$	0
(4)	0	0	0	0	0	0

Table 5.3 Simulation parameters of 500 W VDCBI

Parameter	Value	Parameter	Value
Input voltage U_{pv}/V	400	Capacitor C_2/μF	940
Grid voltage u_g/V, Frequency/Hz	220, 50	Filter inductance L_{f1}/mH	8
Rated power P/W	500	Filter capacitor C/μF	0.34
Switching frequency f_s/kHz	20	Filter inductance L_{f2}/mH	0.8
Capacitor C_1/μF	470	PV panel parasitic Capacitor C_{pv}/nF	75

The simulation waveforms of VDCBI with unity power factor are presented in Figs. 5.10 and 5.11. While Fig. 5.10 shows the simulation waveforms of the grid voltage u_g, grid-in current reference i_{ref}, and the grid-in current i_g when the power factor is 1, the simulation waveforms of the output voltage u_{1N} with unipolar SPWM characteristic is illustrated in Fig. 5.11a. The simulation waveforms of the current through the S_1 and S_3 are shown in Fig. 5.12a, b, respectively. It is clear that the current stress on S_1 and S_3 is bigger during the negative half grid period.

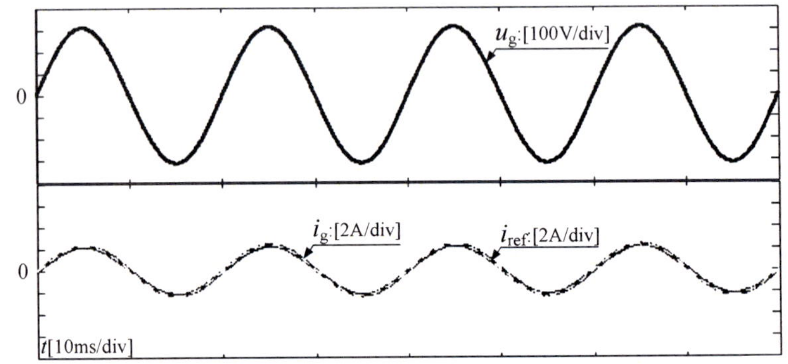

Fig. 5.10 Simulation waveforms of grid voltage, grid-in current reference, and grid-in current of VDCBI with unity power factor

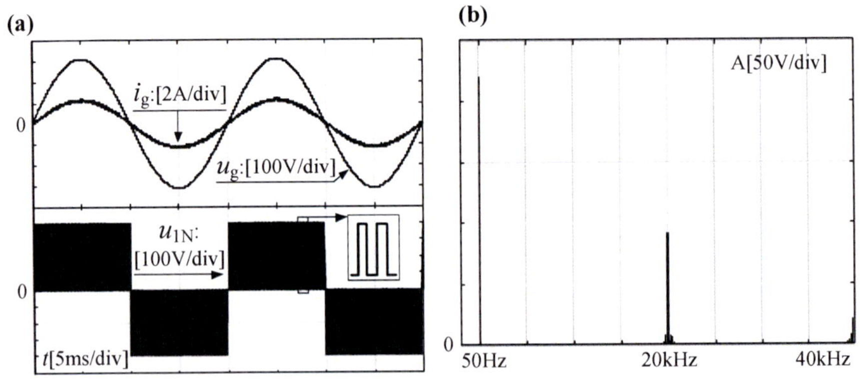

Fig. 5.11 Simulation waveforms of output voltage u1N, **a** Time domain, **b** Frequency domain

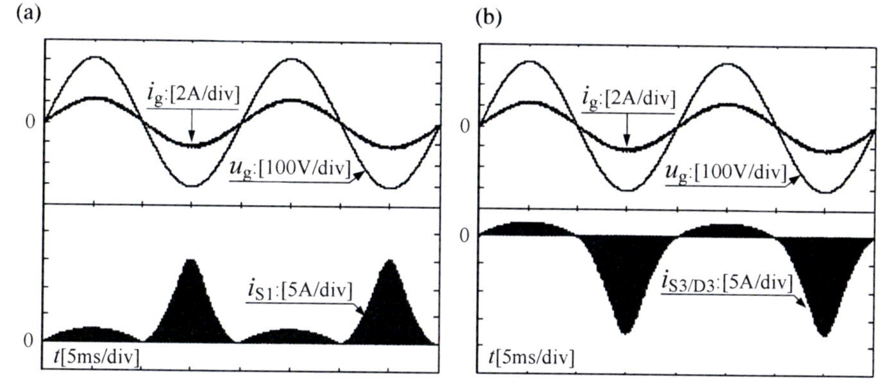

Fig. 5.12 Simulation waveforms of switches current, **a** Current waveform of S_1, **b** Current waveforms of S_4/D_4

5.1.3 Capacitor-Based Common-Ground Inverter III

1. Operating principle

The circuit structure of the third common-ground transformerless grid-connected inverter (CGT-III) is shown in Fig. 5.13a [3, 4], and the PV negative terminal is directly connected to the neutral line of the grid. There are four power switches, two diodes, and two capacitors C_1 and C_2 in CGT-III, and the modulation strategy is illustrated in Fig. 5.13b. It is worth mentioning that switches S_1 and S_3 are working complementarily with dead time, and switches S_2 and S_4 are complementary with each other as well. As a result, the output voltage u_{1N} is a three-level voltage with unipolar SPWM characteristic. Compared with CGT-I and VDCBI, CGT-III has the operation capability when the power factor is not 1.

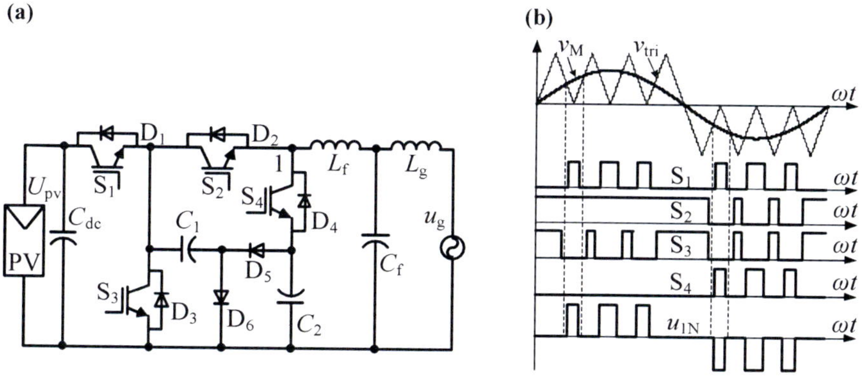

Fig. 5.13 CGT-III grid-connected inverter, **a** Main circuit, **b** Modulation strategy

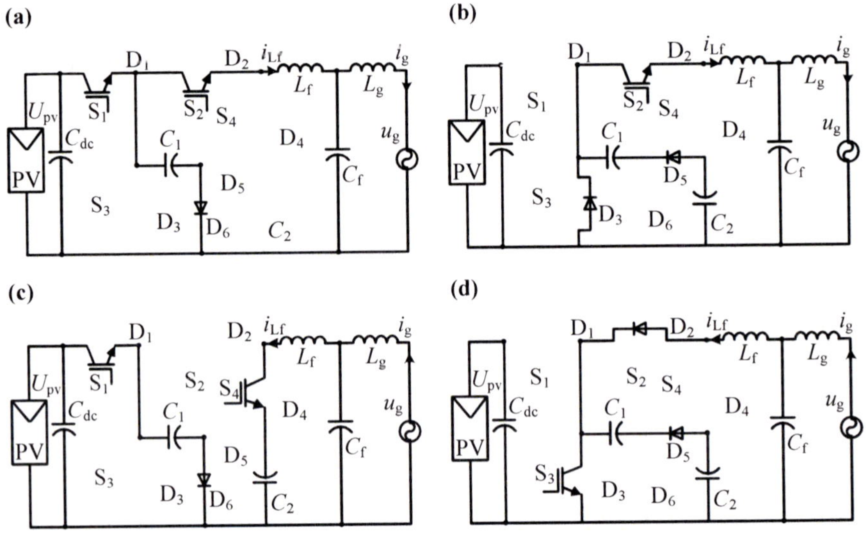

Fig. 5.14 Equivalent circuits of operating modes of CGT-III with unity power factor, **a** Mode 1, **b** Mode 2, **c** Mode 3, **d** Mode 4

There are four operating modes according to the polarities of the grid voltage and grid-in current when the power factor of CGT-III is 1, as shown in Fig. 5.14. Mode 1 and Mode 2 shown in Fig. 5.14a, b belong to positive half period of the grid voltage, respectively; symmetrically, Mode 3 and Mode 4 shown in Fig. 5.14c, d belong to negative half period of the grid voltage.

Mode 1: Refer to Fig. 5.14a; the energy is delivered from the PV side into the grid through S_1, S_2, and the AC filter at this stage. Simultaneously, C_1 is charged by the input DC voltage U_{PV} through S_1 and D_6.

Mode 2: Refer to Fig. 5.14b; at this stage, the freewheeling current flows through S_2, D_3, and the filter, and keeps feeding the grid. With S_1 OFF, C_1 begins to charge C_2 through D_3 and D_5.

Mode 3: Refer to Fig. 5.14c; the energy is delivered from capacitor C_2 into the grid through S_4 and AC filter at this stage. Same as Mode 1, C_1 is re-charged by U_{PV} through S_1 and D_6.

Mode 4: Refer to Fig. 5.14d; the freewheeling current flows through S_3, D_2, and filter, and keeps feeding the grid. With S_1 OFF and S_3 ON, C_1 begins to charge C_2 through S_3 and D_5.

There are four new operating modes when the power factor of CGT-III is not 1, as shown in Fig. 5.15. Compared with Fig. 5.14, Mode 2 and Mode 4 with non-unity power factor is the same as that with unity power factor. Differences mainly reflected in Mode 1and Mode 3, and the reactive energy is exchanged between the grid and C_2 with negative grid voltage and positive grid-in current, as is shown in Fig. 5.15a;

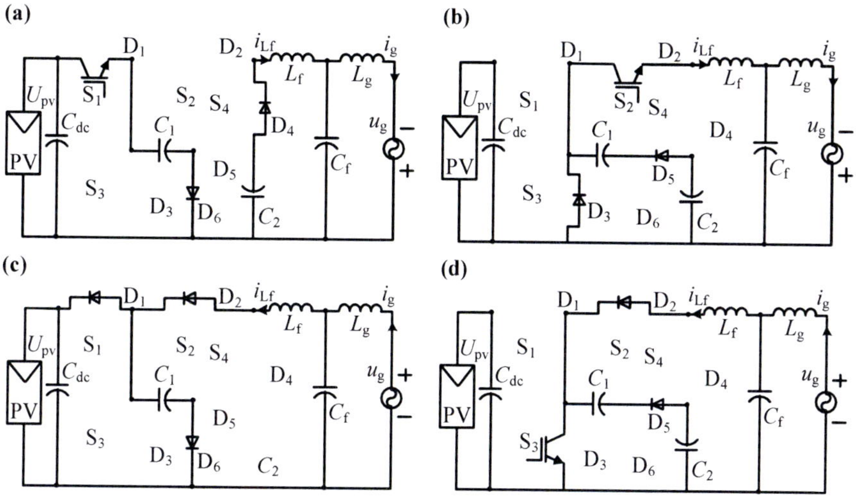

Fig. 5.15 Equivalent circuits of new operating modes of CGT-III with non-unity power factor, **a** Mode 1, **b** Mode 2, **c** Mode 3, **d** Mode 4

conversely, the reactive energy is exchanged between the grid, C_2, and C_{dc}, as is shown in Fig. 5.15c.

According to the analysis of operating modes with unity power factor, C_1 is charged by U_{pv} in Mode 3, and then charges C_2 to keep the output voltage constant in Mode 4. The two transmissions of energy incur a certain amount of loss, and the charging and freewheeling current also results in large current stress on S_1 and S_3 during the negative grid period.

2. Simulation Research

In order to further verify the presented theoretical analysis of CGT-III, a 500 W CGT-III simulation model has been built in Matlab/Simulink. The simulation parameters are listed in Table 5.4.

The simulation waveforms of CGT-III with unity power factor are presented in Figs. 5.16 and 5.17. While Fig. 5.16 shows the simulation waveforms of the grid voltage u_g and the grid-in current i_g when the power factor is 1, the simulation

Table 5.4 Circuit parameters of the 500 W CGT-III topology

Parameter	Value	Parameter	Value
Input voltage U_{pv}/V	400	Capacitor C_2/μF	330
Grid voltage u_g/V, Frequency/Hz	220, 50	Filter inductance L_f/mH	4
Rated power P/W	500	Filter inductance L_g/mH	2
Switching frequency f_s/kHz	24	Filter capacitor C_f/μF	4.7
Capacitor C_1/μF	220	PV panel parasitic capacitor C_{pv}/nF	75

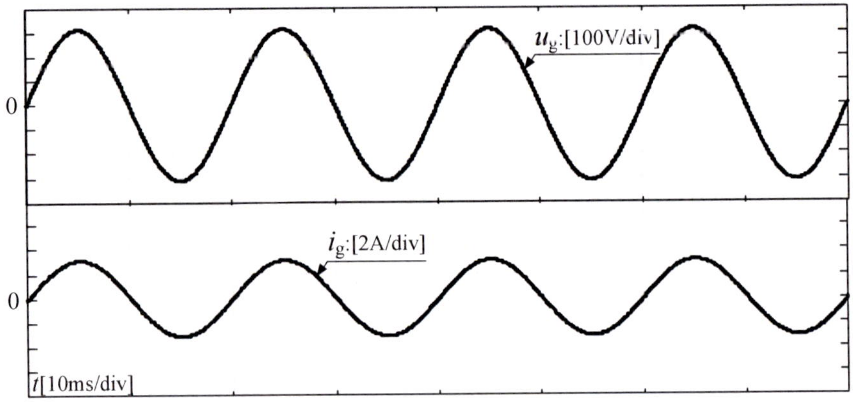

Fig. 5.16 Simulation waveforms of grid voltage and grid-in current of CGT-III with unity power factor

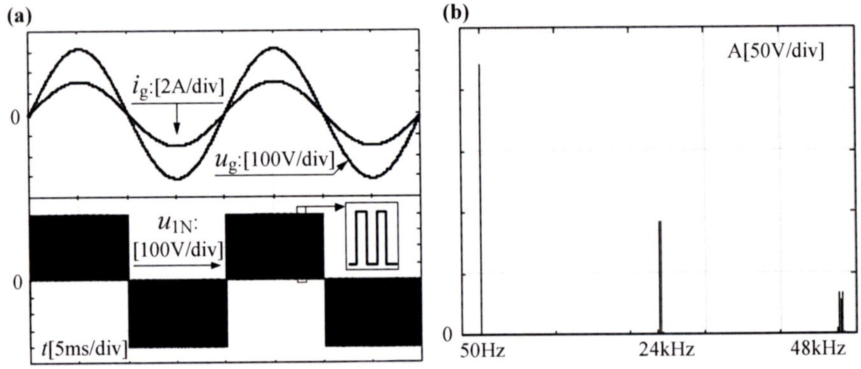

Fig. 5.17 Simulation waveforms of output voltage u1N, **a** Time domain, **b** Frequency domain

waveforms of the output voltage u_{1N} with unipolar SPWM characteristic is illustrated in Fig. 5.17a. The switches operate at 24 kHz, so the harmonic component of u_{1N} concentrates at 24 kHz and its multiples, as is shown in Fig. 5.17b. The simulation waveforms of the current through S_1 and S_3 are shown in Fig. 5.18a, b, respectively. The charging current into C_1 flows through S_1, and both the freewheeling current and the charging current into C_2 flows through S_3. As a result, the current stress on S_1 and S_3 is relatively high.

The simulation waveforms of the grid voltage and grid-in current waveforms of CGT-III with non-unity power factor are presented in Fig. 5.19, which proves that CGT-III can operate with good output characteristics when the power factor is not 1.

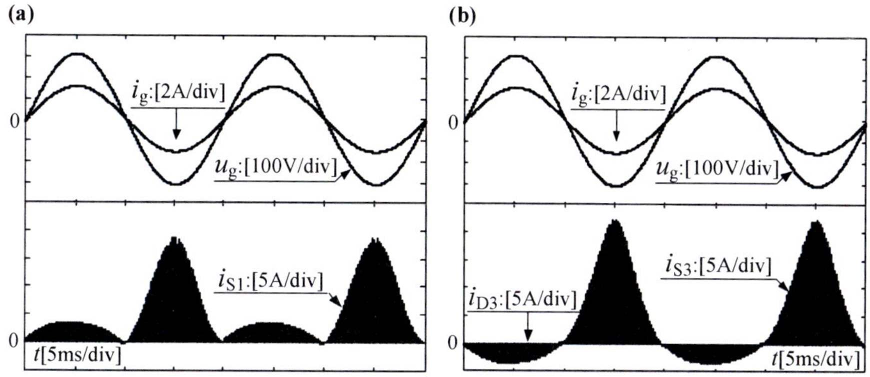

Fig. 5.18 Simulation waveforms of switches current, **a** Current waveform of S_1, **b** Current waveforms of S_3/D_3

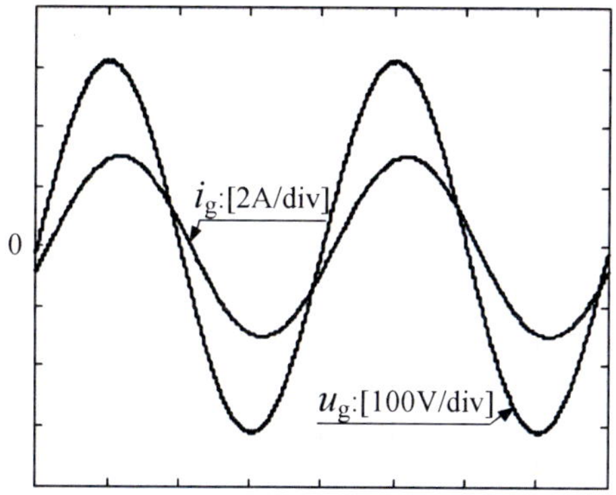

Fig. 5.19 Simulation waveforms of grid voltage and grid-in current of CGT-III with non-unity power factor

5.1.4 Capacitor-Based Common-Ground Inverter IV

1. Operating principle

The circuit structure of the fourth common-ground transformerless grid-connected inverter (CGT-IV) is shown in Fig. 5.20a [5], and the PV negative terminal is directly connected to the neutral line of the grid. There are four power switches, a diode, and a capacitor C_1 in CGT-IV, and the modulation strategy is illustrated in Fig. 5.20b. It is worth mentioning that switches S_1 and S_2 are working complementarily with dead time, and switches S_3 and S_4 are complementary working as well. As a result, the output voltage u_{1N} is a three-level voltage with unipolar SPWM characteristic. Same as CGT-III, CGT-IV has the operation capability when the power factor is not 1 as well.

There are four operating modes according to the polarities of the grid voltage and grid-in current when the power factor of CGT-IV is 1, as shown in Fig. 5.21. Mode 1

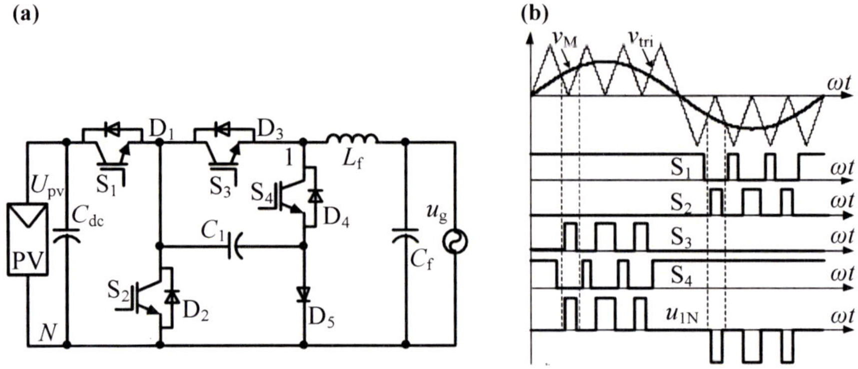

Fig. 5.20 CGT-IV grid-connected inverter, **a** Main circuit, **b** Modulation strategy

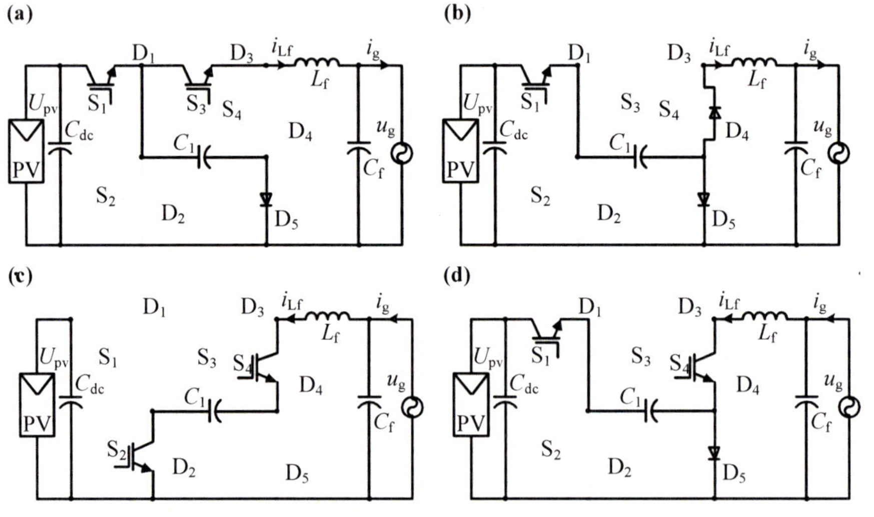

Fig. 5.21 Equivalent circuits of operating modes of CGT-IV with unity power factor, **a** Mode 1, **b** Mode 2, **c** Mode 3, **d** Mode 4

and Mode 2 in Fig. 5.21a, b are the modes in positive half period of the grid voltage; symmetrically, Mode 3 and Mode 4 in Fig. 5.21c, d belong to negative half period of the grid voltage.

Mode 1: Refer to Fig. 5.21a; the energy is delivered from the PV side into the grid through S_1, S_3, and the AC filter at this stage. Simultaneously, C_1 is charged by the input DC voltage U_{PV} through S_1 and D_5.

Mode 2: Refer to Fig. 5.21b; at this stage, the freewheeling current flows through C_{dc}, S_1, C_1, D_4, and AC filter, and keeps feeding the grid with the output voltage u_{1N} clamped to zero. Different from other topologies, except for switches and diodes, the capacitors are included in the freewheeling loop.

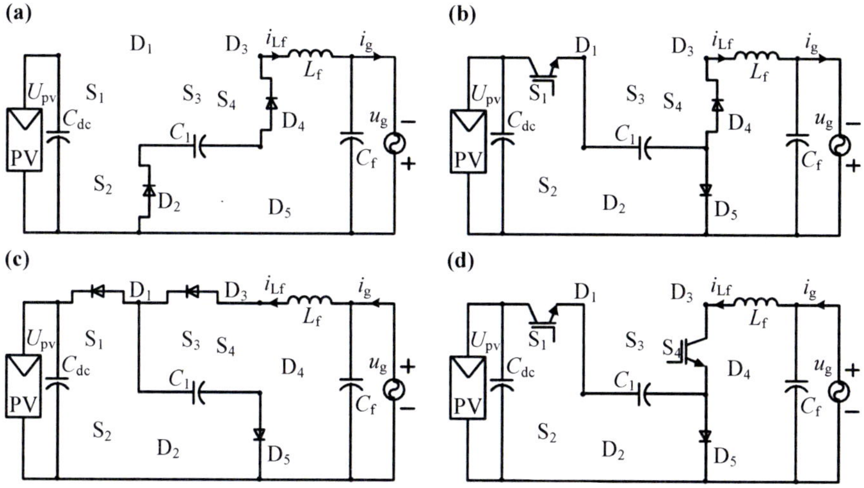

Fig. 5.22 Equivalent circuits of operating modes of CGT-IV with non-unity power factor, **a** Mode 1, **b** Mode 2, **c** Mode 3, **d** Mode 4

Mode 3: Refer to Fig. 5.21c; the energy is delivered from capacitor C_1 into the grid through S_2, S_4, and the AC filter at this stage.

Mode 4: Refer to Fig. 5.21d; the freewheeling current flows through S_4, D_5, and the AC filter, and keeps feeding the grid. Same as Mode 1, C_1 is re-charged by U_{PV} through S_1 and D_5.

There are four new operating modes when the power factor of CGT-IV is not 1, as shown in Fig. 5.22. Compared with Fig. 5.21, Mode 2 and Mode 4 with non-unity power factor are the same as that with unity power factor. Differences mainly reflected in Mode 1 and Mode 3, and the reactive energy is exchanged between the grid and C_1 with negative grid voltage and positive grid-in current, as is shown in Fig. 5.22a; conversely, the reactive energy is exchanged between the grid, C_1, and C_{dc}, as is shown in Fig. 5.22c.

According to the analysis of operating modes with unity power factor, the freewheeling current flows through C_1 in Mode 2. Therefore, C_1 is charged during the positive half grid period, leading to the voltage across C_1 higher than the input voltage U_{pv}. Consequently, in this situation, the modes in the negative half grid period are no more consistent with that shown in Fig. 5.21c, d, and two new modes will appear between Mode 2 and Mode 3 in Fig. 5.21, as is shown in Fig. 5.23a, b. The energy is delivered from capacitor C_1 into the grid and the DC side, as is shown in Fig. 5.23a. With S_2 OFF, the freewheeling current flows through S_4, C_1, and D_1, and keeps feeding the DC side, as is shown in Fig. 5.23b. Once the voltage across C_1 is equal to U_{pv}, the circuit continues to operate in Mode 3 and Mode 4, as is shown in Fig. 5.21c, d.

It is also revealed from the analysis of operating modes that the energy is delivered from capacitor C_1 into the grid in Mode 3, and then C_1 is charged by U_{pv} in Mode

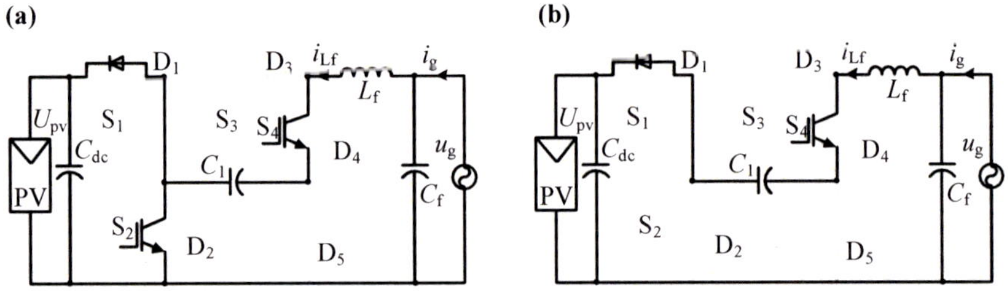

Fig. 5.23 Additional equivalent circuits of operating modes of CGT-IV with unity power factor, **a** Mode 1, **b** Mode 2

4 during the negative grid period. As the charging current flows through S_1 and D_5, the current stress of them is large.

2. Simulation Research

In order to further verify the presented theoretical analysis of CGT-IV, a 1 kW CGT-IV simulation model has been built in Matlab/Simulink. The simulation parameters are listed in Table 5.5.

The simulation waveforms of CGT-IV with unity power factor are presented in Figs. 5.24 and 5.25. While Fig. 5.24 shows the simulation waveforms of the grid voltage u_g and the grid-in current i_g, the simulation waveforms of the output voltage u_{1N} with unipolar SPWM characteristic are illustrated in Fig. 5.25a. For the switches to operate at 50 kHz, the harmonic component of u_{1N} concentrates at 50 kHz and its multiples, as is shown in Fig. 5.25b.

As can be observed in Fig. 5.26, the voltage across C_1 is higher than the DC bus voltage with the energy delivering into it during the positive grid period; once the grid voltage u_g is negative, C_1 discharged until the input voltage. With the voltage across C_1 continuing to decrease, C_1 is conversely charged by U_{pv}, which also accounts for the asymmetry of the output voltage u_{1N}. The simulation waveforms of the current through S_1/D_1 and D_5 are shown in Fig. 5.27a, b, respectively. It is clear that the charging current flows through S_1, and both the charging current and the freewheeling current flow through D_5, which result in large current stress on them. Additionally, CGT-IV has the operation capability when the power factor is not 1 as well, and the simulation waveforms of the grid voltage u_g and the grid-in current i_g under non-unity power factor are shown in Fig. 5.28.

Table 5.5 Simulation parameters of 1 kW CGT-IV

Parameters	Value	Parameters	Value
Input voltage U_{pv}/V	400	Capacitor C_1/μF	470
Grid voltage u_g/V, Frequency/Hz	230, 50	Filter inductance L_f/mH	0.35
Rated power P/W	1000	Filter capacitor C_f/μF	4.7
Switching frequency f_s/kHz	50	PV panel parasitic capacitor C_{pv}/nF	75

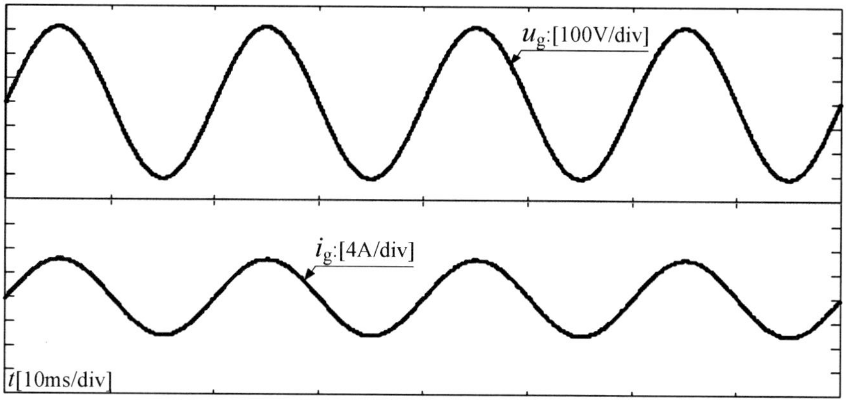

Fig. 5.24 Simulation waveforms of grid voltage and grid-in current of CGT-IV with unity power factor

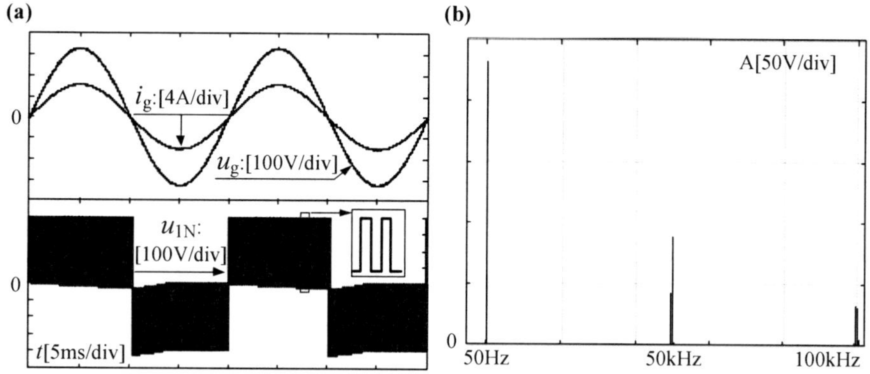

Fig. 5.25 Simulation waveforms of output voltage u1N, **a** Time domain, **b** Frequency domain

Fig. 5.26 Simulation waveform of the voltage across C_1

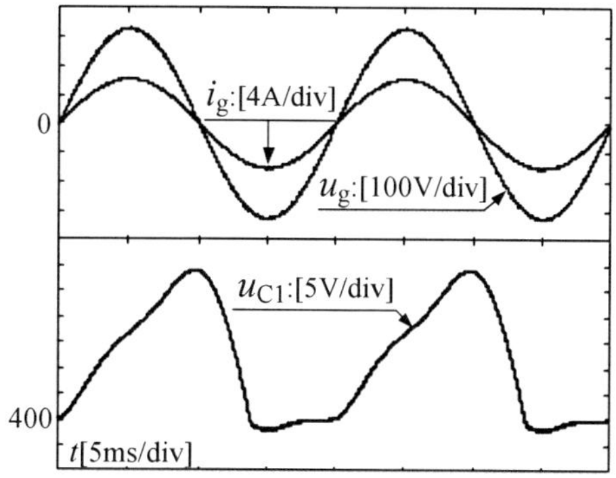

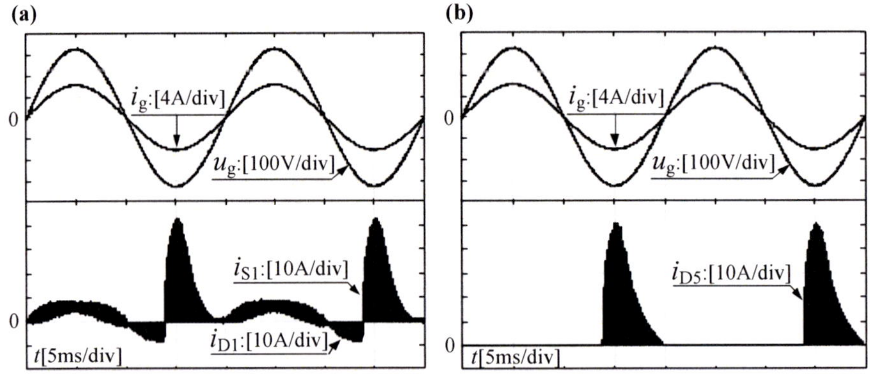

Fig. 5.27 Simulation waveforms of switches current, **a** Current waveform of S_1/D_1, **b** Current waveforms of D5

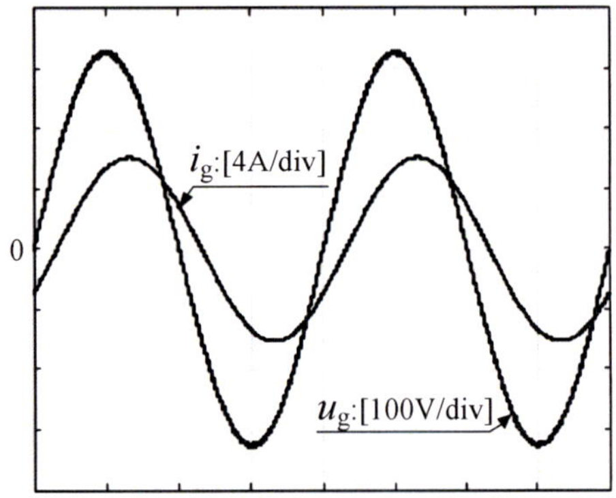

Fig. 5.28 Simulation waveforms of grid voltage and grid-in current of CGT-IV with non-unity power factor

The intermediate capacitors of the above common-ground TLIs have the capability of operating at switching frequency during the negative grid period, which brings the high reduction of the capacitor volume. Despite that, the high current stress on some switches is a crucial challenge required to be studied and solved in capacitor-based common-ground TLIs.

5.2 Inductor-Based Common-Ground TLIs

Capacitor-based common-ground TLIs with capacitors delivering energy have been discussed in Sect. 5.1. The above topologies overcome the difficulty of providing negative voltage or negative current during the grid negative half period. Nevertheless, they can only work in buck mode with high input DC voltage. This is the major limitation in low-power distributed generation applications. Inductors are used as the middle stage and own the ability of buck/boost in inductive-based common-ground TLIs. Early inductive-based common-ground TLIs are mostly two-stage inverters [7–9], which have negative influence on the improvement of inverter efficiency. In order to solve this problem, single-stage inductive-based common-ground TLIs that integrate buck/boost structures have been proposed [10–13].

5.2.1 Karschny Common-Ground Inverter

Karschny common-ground inverter was first patented by Karschny Dietrich in 1998 [11]. It is worth mentioning that Karschny common-ground inverter can work in buck/boost mode with only an inductor storing the energy.

The circuit structure of the conventional Karschny common-ground transformerless grid-connected inverter, which is also called the flying inductor-based common-ground TLI, is shown in Fig. 5.29a. There are five power switches, two diodes, and one inductor in Karschny inverter. The modulation strategy is illustrated in Fig. 5.29b. It is worth noting that the switches S_3–S_5 operate at grid frequency. During the positive grid period, the Karschny inverter works in two-switch forward buck/boost mode with S_3 OFF, S_4 and S_5 ON, and S_1 and S_2 operating at high frequency; symmetrically, the inverter works in two-switch reverse buck/boost mode with S_2 and S_3 ON, S_4 and S_5 OFF, and only S_1 operating at high frequency.

There are four operating modes when the power factor of Karschny inverter is 1, as shown in Fig. 5.30. Mode 1 and Mode 2 in Fig. 5.30a, b are the modes of positive half period of the grid voltage; symmetrically, Mode 3 and Mode 4 in Fig. 5.30c, d are the modes of negative half period of the grid voltage.

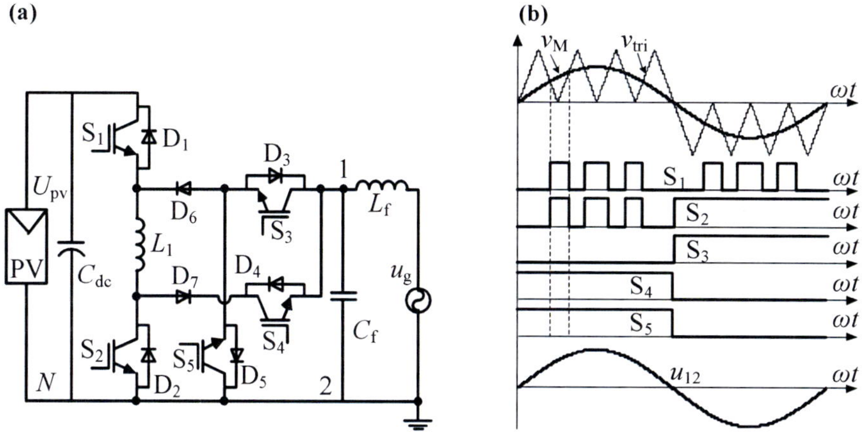

Fig. 5.29 Karschny common-ground inverter, **a** Main circuit, **b** Modulation strategy

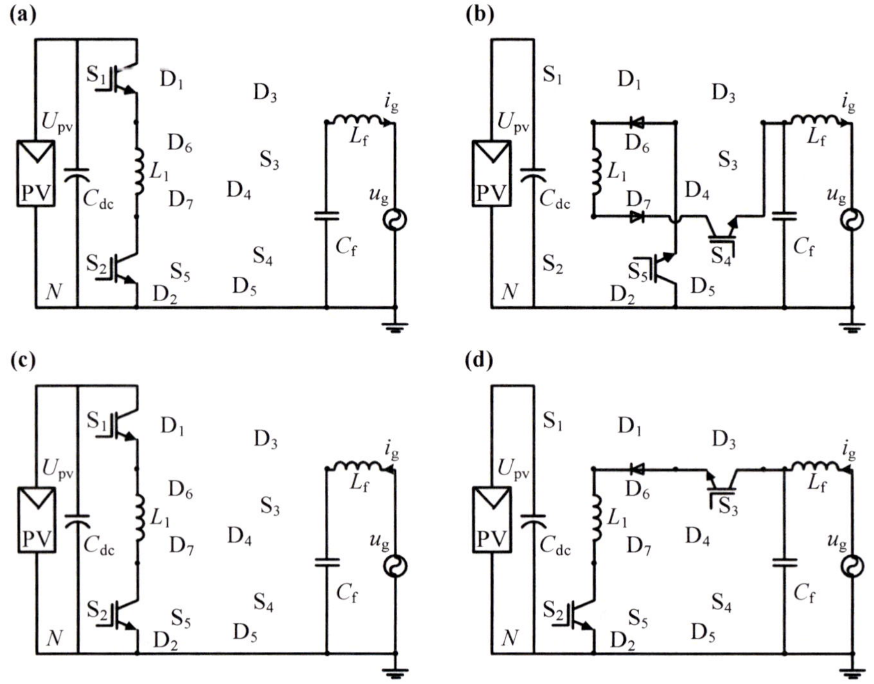

Fig. 5.30 Equivalent circuits of operating modes of Karschny common-ground inverter, **a** Mode 1, **b** Mode 2, **c** Mode 3, **d** Mode 4

Mode 1: Refer to Fig. 5.30a; at this stage, the charging current flows through S_1 and S_2 into the inductor L_1. The energy stored in L_f and C_f keeps feeding the grid.

Mode 2: Refer to Fig. 5.30b; the energy is delivered from inductor L_1 into the grid through S_4, S_5, D_6, and D_7 at this stage. Simultaneously, Karschny inverter works in two-switch forward buck/boost mode with L_f and C_f charged.

Mode 3: Refer to Fig. 5.30c; same as in Mode 1, the charging current flows through S_1 and S_2 into the inductor L_1. The energy stored in L_f and C_f keeps feeding the grid.

Mode 4: Refer to Fig. 5.30d; the energy is delivered from inductor L_1 into the grid through S_2, S_3, and D_6 at this stage. Simultaneously, Karschny inverter works in two-switch backward buck/boost mode with L_f and C_f charged.

According to the analysis of operating modes, the input voltage of Karschny inverter is lower than the peak grid voltage in buck/boost mode; however, the full-power processing has an influence in improving efficiency.

5.2.2 Flying Inductor Transformerless Inverter

Based on one inductor, same as Karschny inverter, the second flying inductor transformerless inverter (FITLI) was proposed in [12].

The structure of the flying inductor transformerless inverter (FITLI) is shown in Fig. 5.31a. There are five power switches and one inductor in FITLI, and the modulation strategy is illustrated in Fig. 5.31b. Note that the switches S_2, S_4, and S_5 operate at grid frequency. During the positive grid period, FITLI works in two-switch forward buck/boost mode with S_2 and S_5 ON, S_4 OFF, while S_1 and S_3 operating at high frequency. Symmetrically, FITLI works in two-switch backward buck/boost mode with S_3 and S_4 ON, S_2 and S_5 OFF, and only S_1 operating at high frequency.

There are four operating modes according to the polarities of the grid voltage and grid-in current of FITLI, as shown in Fig. 5.32. Mode 1 and Mode 2 in Fig. 5.32a, b are the modes of positive half period of the grid voltage; symmetrically, Mode 3 and Mode 4 in Fig. 5.32c, d are the modes of negative half period of the grid voltage.

Mode 1: Refer to Fig. 5.32a; at this stage, the charging current flows through S_1 and S_3 into the inductor L_1. The energy stored in L_f and C_f keeps feeding the grid.

Mode 2: Refer to Fig. 5.32b; the energy is delivered from inductor L_1 into the grid through S_2 and S_5 at this stage. Simultaneously, FITLI works in two-switch forward buck/boost mode with L_f and C_f charged.

Mode 3: Refer to Fig. 5.32c; same as in mode 1, the charging current flows through S_1 and S_3 into the inductor L_1. The energy stored in L_f and C_f keeps feeding the grid.

Mode 4: Refer to Fig. 5.32d; the energy is delivered from inductor L_1 into the grid through S_3 and S_4 at this stage. Simultaneously, FITLI works in two-switch backward buck/boost mode with L_f and C_f charged.

According to the analysis of operating modes, FITLI works in buck/boost mode, and thus the full-power processing has an influence in improving efficiency as well.

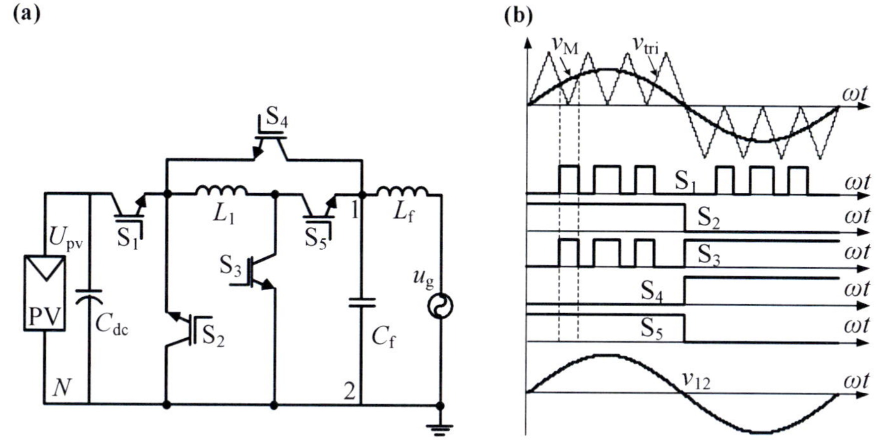

Fig. 5.31 FITLI grid-connected inverter, **a** Main circuit, **b** Modulation strategy

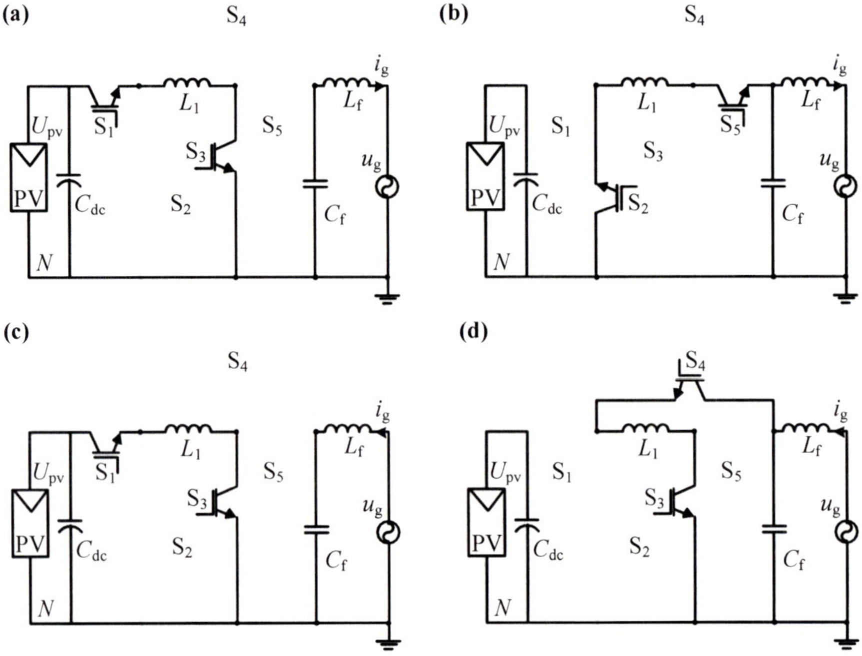

Fig. 5.32 Equivalent circuits of operating modes of FITLI, **a** Mode 1, **b** Mode 2, **c** Mode 3, **d** Mode 4

As is analyzed in Sects. 5.2.1 and 5.2.2, Karschny common-ground inverter and FITLI both have the capability of working in buck/boost mode, but the full-power processing becomes the main limitation in efficiency improvement. In order to achieve high frequency, the Aalborg transformerless inverter (ATLI) with part-power processing in buck/boost mode has been proposed in [13].

5.2.3 Aalborg Transformerless Inverter

The Aalborg transformerless inverter (ATLI) was first proposed by the researchers at the University of Aalborg in Denmark. This section discusses the circuit structures, operation characteristics, and simulation verification.

1. Operating principle

The symmetrical circuit structure of the Aalborg transformerless inverter (ATLI) is shown in Fig. 5.33a [13]. The PV neutral terminal is directly connected to the neutral line of the grid. There are six power switches, four diodes, and two inductors L_p and L_n in ATLI. The modulation strategies operating in buck and boost mode are illustrated in Fig. 5.33b, c, respectively. When the DC capacitor voltage is larger than

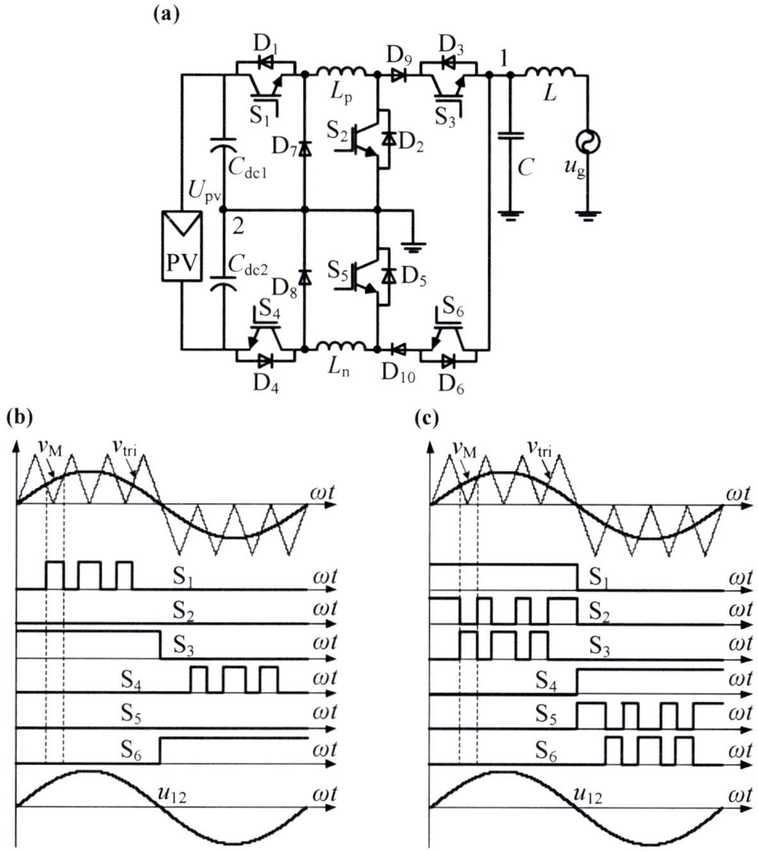

Fig. 5.33 ATLI grid-connected inverter, **a** Main circuit, **b** Buck mode, **c** Boost mode

the amplitude of grid voltage, ATLI works in buck mode; symmetrically, when the DC capacitor voltage is smaller than the amplitude of grid voltage, ATLI works in boost mode.

There are four operating modes according to the polarities of the grid voltage and grid-in current when ATLI works in buck mode, as shown in Fig. 5.34. Mode 1 and Mode 2 in Fig. 5.34a, b belong to positive half period of the grid voltage; symmetrically, Mode 3 and Mode 4 in Fig. 5.34c, d belong to negative half period of the grid voltage.

Mode 1: Refer to Fig. 5.34a; the energy is delivered from the PV side into the grid through S_1, S_3, and the AC filter at this stage. Simultaneously, L_p is charged by C_{dc1}.

Mode 2: Refer to Fig. 5.34b; at this stage, the freewheeling current flows through S_3, D_7, and the AC filter, and keeps feeding the grid.

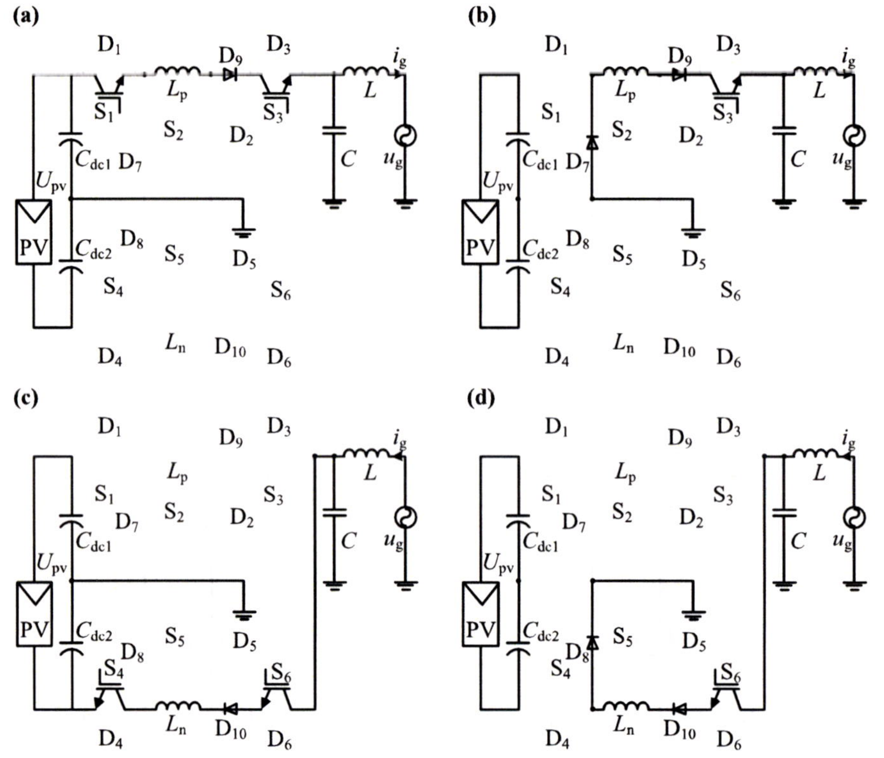

Fig. 5.34 Equivalent circuits of ATLI working in buck mode, **a** Mode 1, **b** Mode 2, **c** Mode 3, **d** Mode 4

Mode 3: Refer to Fig. 5.34c; symmetrical to Mode 1, the energy is delivered from the PV side into the grid through S_4, S_6, and the AC filter at this stage. Simultaneously, L_n is charged by C_{dc2}.

Mode 4: Refer to Fig. 5.34d; symmetrical to Mode 2, the freewheeling current flows through S_6, D_8, and the AC filter, and keeps feeding the grid. Obviously, it is the freewheeling mode in the negative half grid period.

There are four operating modes according to the polarities of the grid voltage and grid-in current when ATLI works in boost mode, as shown in Fig. 5.35. Mode 1 and Mode 2 in Fig. 5.35a, b belong to positive half period of the grid voltage, respectively; symmetrically, Mode 3 and Mode 4 in Fig. 5.35c, d are the modes of negative half period of the grid.

Mode 1: As shown in Fig. 5.35a, it is the positive half grid cycle inductance energy storage mode. Switches S_1 and S_2 are turned ON and the inductor L_p is charged. The filter capacitor C is discharged to maintain the current injected into the grid.

Mode 2: As shown in Fig. 5.35b, it is the positive half grid cycle energy transfer mode. Switch S_2 is turned OFF and switches S_1 and S_3 are turned ON. The inductor L_p is discharged and the current flows through S_1, S_3, and the filter to the power grid.

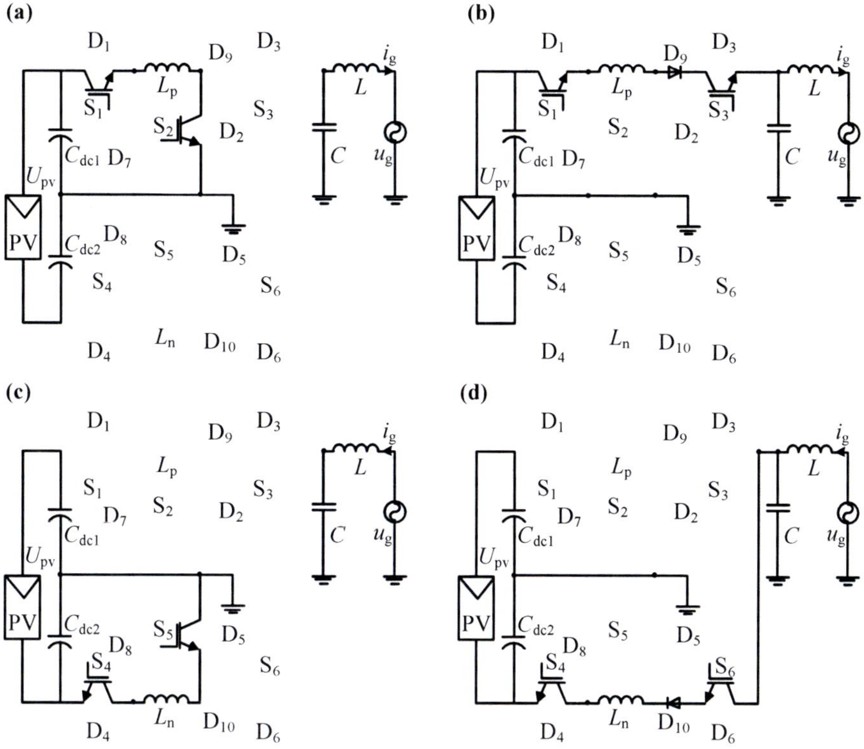

Fig. 5.35 Equivalent circuits of ATLI working in boost mode, **a** Mode 1, **b** Mode 2, **c** Mode 3, **d** Mode 4

Mode 3: As shown in Fig. 5.35c, it is the negative half grid cycle inductance energy storage mode. Switches S_4 and S_5 are turned ON and the inductor L_n is charged. The filter capacitor C is discharged to maintain the current injected into the grid.

Mode 4: As shown in Fig. 5.35d, it is the negative half grid cycle energy transfer mode. Switch S_5 is turned OFF and switches S_4 and S_6 are turned ON. The inductor L_n is discharged and the current flows through S_4, S_6, and the filter to the power grid.

According to the analysis of operating modes, ATLI works in buck/boost mode with only processing a fraction of the total power during the grid period. As a result, the improvement of efficiency of ATLI can be achieved.

2. Simulation Research

In order to further verify the presented theoretical analysis of ATLI, a 2 kW ATLI simulation model has been built in Matlab/Simulink. The simulation parameters are listed in Table 5.6.

The simulation waveforms of ATLI with unity power factor are presented in Figs. 5.36 and 5.37. ATLI works in buck mode in this simulation. While Fig. 5.36 shows the simulation waveforms of the grid voltage u_g and the grid-in current i_g

Table 5.6 Simulation parameters of 2 kW ATLI

Parameter	Value	Parameter	Value
Input voltage U_{pv}/V	700	Inductance L_p, L_n/mH	0.6
Grid voltage u_g/V, Frequency/Hz	220, 50	Filter capacitor C_1/μF	4.7
Rated power P/W	2000	Filter inductance L_f/mH	0.6
Switching frequency f_s/kHz	40	PV panel parasitic capacitor C_{pv}/nF	75

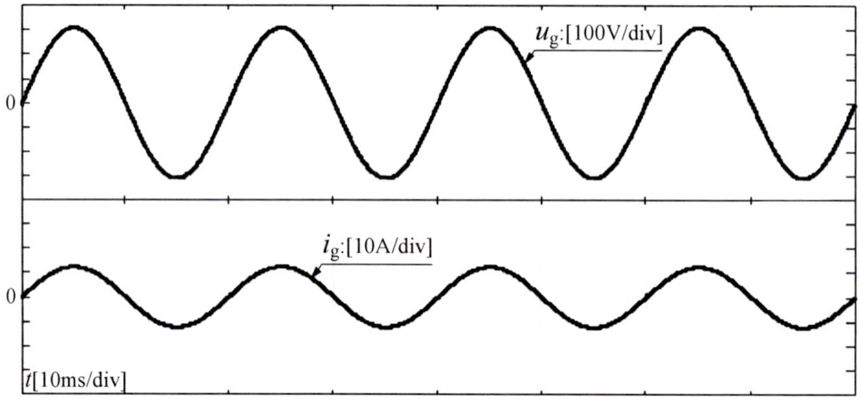

Fig. 5.36 Simulation waveforms of grid voltage ug and grid-in current ig of ATLI

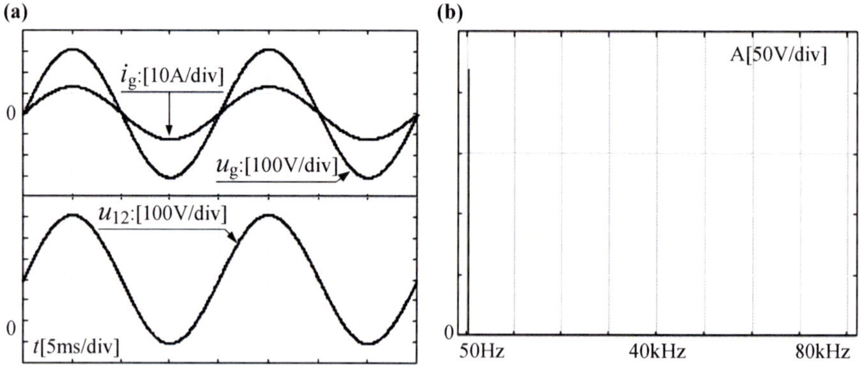

Fig. 5.37 Simulation waveforms of output voltage U_{12}, **a** Time domain, **b** Frequency domain

when the power factor is 1, the simulation waveforms of the output voltage u_{12} is illustrated in Fig. 5.37. The simulation waveforms of the current through L_p and L_n are shown in Fig. 5.38a, b, respectively. The energy is delivered from inductor L_p into the grid during the positive grid period, and the energy is delivered from inductor L_n into the grid during the negative grid period.

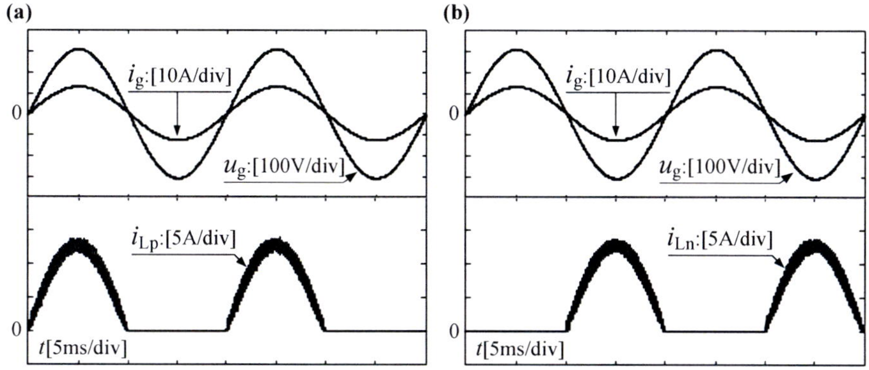

Fig. 5.38 Simulation waveforms of inductors current, **a** Current waveform of Lp, **b** Current waveforms of Ln

5.3 Summary

This chapter discusses the capacitor-based and the inductor-based common-ground TLIs based on *Rule* 3, revealing that the common-ground TLIs has the optimal LC characteristics. Specifically, the inductor-based common-ground TLIs, which can work in buck/boost mode, are more appropriate for low-power distributed generation applications. The comparison results of the capacitor-based and the inductor-based common-ground TLIs are concluded in Tables 5.7 and 5.8, which may provide reference for other designs.

Table 5.7 Comparison results of capacitor-based common-ground TLIs

Topologies	SW/D	SW (High frequency)	Negative bus capacitance
CGT-I [1]	4/0	4/2	1100 μF@1 kW
CGT-II [2]	5/0	2/3	940 μF@500 W
CGT-III [3, 4]	4/2	2/4	220 μF, 330 μF@500 W
CGT-IV [5]	4/1	2/2	470 μF@1 kW

Table 5.8 Comparison results of inductor-based common-ground TLIs

Topologies	Operating mode	SW/D	SW (High frequency)	Negative bus inductance
Karschny [11]	Buck/Boost	5/2	2	Not listed [11]
FITLI [12]	Buck/Boost	5/0	2	0.3mH@200 W
ATLI [13]	Buck/Boost	6/4	1/2	0.6mH@2 kW

References

1. Kadam A, Shukla A (2017) A multilevel transformerless inverter employing ground connection between pv negative terminal and grid neutral point. IEEE Trans Industr Electron 64(11):8897–8907
2. Gu YJ, Li WH, Zhao Y et al (2013) Transformerless inverter with virtual DC bus Concept for cost-effective grid-connected PV power systems. IEEE Trans Power Electron 28(2):793–805
3. Ardashir JF, Sabahi M, Hosseini SH et al (2016) Transformerless inverter with charge pump circuit concept for PV application. IEEE J Emerg Sel Topics Power Electron, Early Access
4. Ardashir JF, Sabahi M, Hosseini SH et al (2017) A single-phase transformerless inverter with charge pump circuit concept for grid-tied PV applications. IEEE Trans Industr Electron 64(7):5403–5415
5. Siwakoti YP, Blaabjerg F (2018) Common-ground-type transformerless inverters for single-phase solar photovoltaic systems. IEEE Trans Industr Electron 65(3):2100–2111
6. Liao ZL, Cao C, Qiu D et al (2019) Single-phase common-ground-type transformerless PV grid-connected inverters 7:63277–63287
7. Saha S, Sundarsingh VP (1996) Novel grid-connected photovoltaic inverter. IEEE Proc Gener Trans Distrib 143(2):219–224
8. Funabiki S, Tanaka T, Nishi T, A new buck-boost-operation-based sinusoidal inverter circuit. In: 2002 power electronics specialists conference, Cairns, pp 1624–1629
9. Prasad BS, Jain S, Agarwal V (2008) Universal single-stage grid-connected inverter. IEEE Trans Energy Convers 23(1):128–137
10. Araujo SV, Zacharias P, Mallwitz R (2010) Highly efficient single-phase transformerless inverters for grid-connected photovoltaic systems. IEEE Trans Industr Electron 57(9):3118–3128
11. Dietrich K. Wechselrichter. DE 19 642 522 C1.2010
12. Azary MT, Sabahi M, Babaei E et al (2018) Modified single-phase single-stage grid-tied flying inductor inverter with MPPT and suppressed leakage current. IEEE Trans Industr Electron 65(1):221–231
13. Wu W, Ji J, Blaabjerg F (2013) Aalborg Inverter—A new type of "buck in buck, boost in boost" grid-tied inverter. IEEE Trans Power Electron 30(9):4784–4793

Chapter 6
DC Current Rejection for Transformerless Grid-Connected Inverters

Abstract Leakage current eliminating techniques for TLIs have been discussed from Chaps. 2, 3, 4 and 5 of this book. In distributed TLI-PVPG systems, DC current in grid-in current is another issue due to the absence of line-frequency transformers at AC side, which could result in safety risks, such as transformer saturation of upstream distribution wires, accelerating degradation of insulation layer of electric cables, and misaction of protection measures. Suppressing methods of DC current injection are discussed in this chapter in detail.

Keywords Transformerless grid-connected inverter · DC current · Detection and feedback · Capacitor blocking · Intelligent algorithm

6.1 DC Component in Grid-In Current

6.1.1 Summarizing Analysis of DC Components

Taking a typical single-phase full-bridge grid-connected TLI system as an example in this section, there are six aspects of causing DC current (DCC), as shown in Fig. 6.1. It is worth noting that the DC voltage component in the grid belongs to the external factor of causing DCC, and the rest five aspects can be classified as internal factors.

1. Different properties among power devices

The reasons of resulting in DCC from power devices can be divided into two groups.

(1) Differences in gate driving circuits, for instance, the asymmetric widths of PWM signals, unequal rising time and falling time of amplifier circuits, different dead time of bridge switches, and so on.
(2) Various characteristics in power devices, such as conduction voltage drop, leakage current, junction capacitors, parasitic parameters, etc.

H. Xiao and X. Wang, *Transformerless Photovoltaic Grid-Connected Inverters*, CPSS Power Electronics Series, https://doi.org/10.1007/978-981-15-8525-8_6

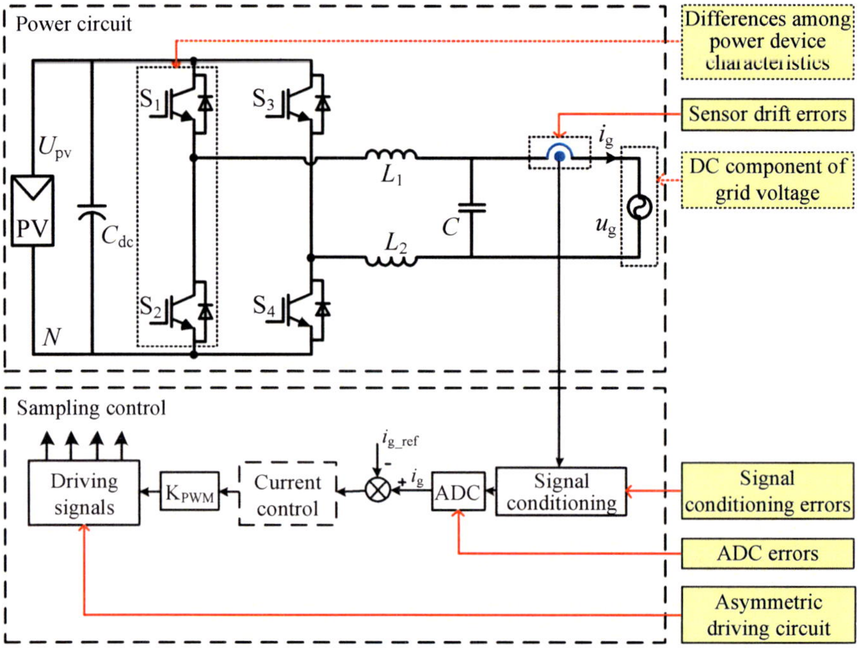

Fig. 6.1 Factors of DC components

2. Sensing errors

Commonly, Hall Effect based current sensors are used in current detection. It has errors in practical use. The sensing errors are from two factors, one is the residual magnetism caused by magnetic hysteresis effect; another one is from potential drift of Hall elements. It can be seen that the sensing errors are unavoidable, it can be reduced by using high precision sensor but expensive.

3. Errors of signal conditioning circuits

The purpose of signal conditioning circuits is to match the output signal of sensors with the input signal of analog-to-digital converters (ADC) in amplitudes. Therefore, the operational amplifier is popularly used to amplify or minify the detection signals. However, some technical limitations of the operational amplifier may cause additional detection errors, such as non-linear region, offset voltage and offset current, etc.

4. Resolution errors of ADCs

The reasons resulting in ADC errors can be classified into two categories, i.e. external factors, and internal factors. For instance, the quality of power supply and ground noise belong to external factors; internal errors result from nonlinearity of differential and integral parts, offset and gain errors, and so on. The digit number of resolutions is usually used to represent the error introduced by the ADC, whose unit is LSB.

5. DC voltage component of the grid

When the grid voltage contains a certain amount of DC voltage component, DCC will be generated in grid-in current through the filter and line impedance by the DC voltage component.

6.1.2 Influence of DCC

Once the amplitude of DCC reaches a limitation value, some damages may be occurred, including but not limited as,

(1) DC magnetic offset in line-frequency transformers, which further causes magnetic saturation of magnetic core in transformer unsymmetrically. As a consequence, the distortion of excitation current will be distorted and harmonic current components increased a lot. It may also generate vibration and noise of transformers. Finally, the transformer may be damaged and loss will be increased.
(2) Power wire loss and accelerating the corrosion of the ground cable and ground electrode.
(3) Eddy current and torque ripples in motor loads, then the motor noise, vibration and overheating will be unacceptable.

6.1.3 Standards Limitations of DCC

In order to avoid aforementioned damages in power system, some relevant technical standards and limitations have been filed as the design criteria and testing guidelines. Several typical Standards are listed in Table 6.1, including the standards and allowed DCC values.

It can be seen that China, the United States, etc., set the limitation at certain ratio of rated current. The United Kingdom, Germany, etc. directly set the limitation at absolute magnitude of DCC. These standards are mandatory complied in associated areas, inverter manufactures have to follow.

Table 6.1 Standards in terms of DCC from different countries

Countries	Standards	Maximum DCC
China	GB/T 33,593–2017	0.5% of rated output current
United States	IEEE 929–2000	0.5% of rated output current
Japan	IEC 61,727–2004	1% of rated output current
Germany	DIN VDE 0126	1000 mA
England	ER G83/1	5 mA
Australia	AS 4777.2	5 mA

6.1.4 Quantitative Analysis of DCC

1. Power circuits of TLIs

There are many factors in power circuits discussed in Sect. 6.1.1 that cause the DCC, but all of them can be represented by the output voltage of the bridge leg. Taking a bipolar modulation single-phase TLI as example, the effective on-state time of the switch action during a grid period is accumulated to T (generally less than grid period), as shown in Fig. 6.2. Where the on-time of positive output voltage is equal to αT, if the turn-on times of positive and negative output voltage are different, i.e., α is not equal to 0.5, DDC will appear in the grid-in current.

Furthermore, the amplitude of DDC can be calculated by FFT (fast Fourier transform),

$$I_{g0} = 2 \cdot I_{ref} \cdot (0.5 - \alpha) \tag{6.1}$$

where I_{ref} is the reference current amplitude, and α is the ratio of the accumulated time of positive output levels and total time of effective output levels. For example, assuming that the inverter power is 3 kW, the grid voltage is 220 V, and $\alpha = 0.4995$, as a result, $I_{\text{ref}} = 3000/220 \times 1.414$ A $= 19.281$ A, and $I_{g0} = 2 \times I_{\text{ref}} \times (0.5-\alpha) =$

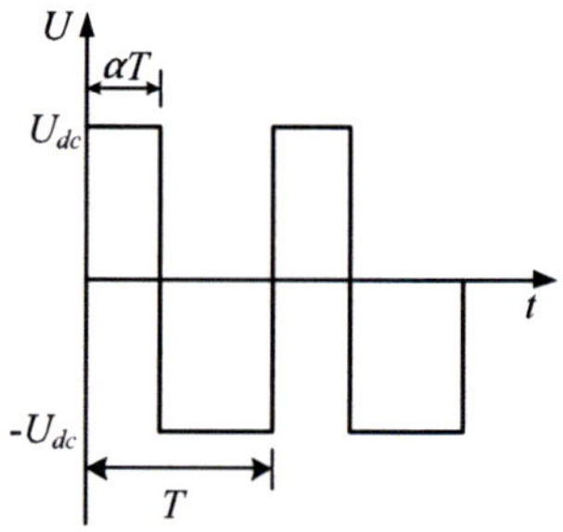

Fig. 6.2 Effective output voltage with accumulation

19 mA, based on (6.1). According to Table 6.1, the asymmetry of power circuits will fail TLIs on markets of the United Kingdom and Australia; some special measures have to be adopted to deal with this problem.

2. Sampling circuits of TLIs

Measuring error of grid-in current is decided by the resolution of ADCs, the accuracy of the current sensor, and the signal conditioning circuit. By summing up, the current error can be expressed as (6.2),

$$\Delta I_{g\_\max} = \left|\Delta I_{g1}\right| + \left|\Delta I_{g2}\right| + \left|\Delta I_{g3}\right| \tag{6.2}$$

Here, ΔI_g represents the total maximum current detection error, and ΔI_{g1}, ΔI_{g2}, and ΔI_{g3} represent the detection errors from the ADC resolution, the current sensor, and the signal conditioning circuit, respectively.

(1) ADC error

Assuming that the sampling error of the ADC is I_{gs}, the inverter overcurrent coefficient is β, the peak-to-peak value of grid-in current is I_{p-p}, and the bit number of ADC is N, respectively, the detection error ΔI_{g1} from the ADC is

$$\Delta I_{g1} = \frac{(1+\beta)\cdot I_{p-p}\cdot \Delta I_{gs}}{2^N} \tag{6.3}$$

Taking a popular DSP TMS320F28335 from TI as example, $\Delta I_{\mathrm{gs}} = \pm$ 1.5LSB is generally acceptable, i.e., the sampling value error is approximately equal to 1.5 ADC resolution bits. For a 3 kW TLI connected with the grid of 220 V and 50 Hz, the grid-in current amplitude is calculated as 3000/220 × 1.414 A = 19.281A, then the calculated detection error ΔI_{g1} is 18 mA if $\beta = 0.3$, and $N = 12$.

Obviously, the higher the resolution of the ADC is, the smaller the DCC caused by the ADC. But, the cost will be high.

(2) Sensor error

The DCC is introduced by the detection accuracy of the current sensor, generally, the allowable error can be expressed as the accuracy level times the capacity range,

$$\Delta I_{g2} = \mathrm{FS}\cdot\mathrm{X} \tag{6.4}$$

Here, X represents the accuracy level of the current sensor, and FS represents the maximum capacity range. Taking above 3 kW TLI as example, a Hall current sensor CSM025NPT5 produced by Nanjing Chahua-Electric is adopted in the TLI, the accuracy X is 0.1%, and the full-scale FS is 25 × 2 A = 50 A by referring datasheet. Therefore, the sensor error ΔI_{g2} is 50 × 0.1% mA = 50 mA.

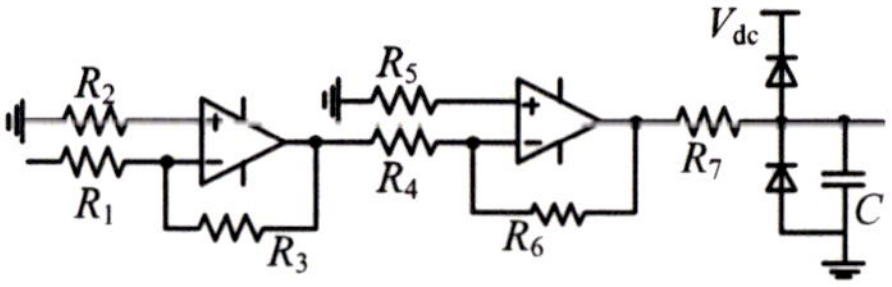

Fig. 6.3 Signal conditioning circuit

(3) Conditioning error

The error of the signal conditioning circuit is mainly dominated by offset voltage U_{io} and offset current I_{io} of operational amplifiers. Taking an inverse proportional amplifier circuit as example, the resulted error [1] is represented as,

$$\Delta I_{g3} = \frac{K_{cs}}{A_{uf}}\left[U_{io}\left(1+\frac{R_F}{R_1}\right)+I_{io}R_F\right] \tag{6.5}$$

Here, ΔI_{g3} represents the current error, R_1 is the resistor of negative terminal of operational amplifiers, R_F is the negative feedback resistor, A_{uf} is the amplification coefficient of the inverse proportional operation circuit, and K_{cs} is the conversion coefficient from the primary side to the secondary side of the current sensor.

Figure 6.3 illustrates a two-stage conditioning circuit, TL074C with low input drift and offset current from TI is selected as the operational amplifier. In this figure, $R_1 = R_4 = R_5 = R_6 = 10\ \text{k}\Omega$, $R_2 = R_3 = 8.2\ \text{k}\Omega$, $R_7 = 300\ \Omega$, $C = 0.1\ \mu\text{F}$, $V_{\text{dc}} = 3.3\ \text{V}$, the input offset voltage $U_{io} = 3\ \text{mV}$, and offset current $I_{io} = 5\ \text{pA}$ can be obtained from the technical datasheet.

Therefore, the conditioning error can be calculated from two steps as bellow,

$$\begin{aligned} err_1 &= \frac{40}{-0.82}\left[3\times10^{-3}\times\left(1+\frac{8.2\times10^3}{10\times10^3}\right)+5\times10^{-12}\times8.2\times10^3\right] \\ &= -266\ \text{mA} \end{aligned} \tag{6.6}$$

$$err_2 = \frac{40}{-1\times(-0.82)}[3\times10^{-3}\times(1+1)+5\times10^{-12}\times10\times10^3] = 293\ \text{mA} \tag{6.7}$$

$$\Delta I_{g3} = err_1 + err_2 = 27\ \text{mA} \tag{6.8}$$

1. DC components of grid voltage

When the grid voltage contains a certain DC voltage component U_{g_dc}, DCC will be generated through the filter and line impedance. As shown in Fig. 6.4, the inverter output voltage is equivalent to an ideal AC voltage source U_{bridge}, which is equal to the frequency components U_g without DC component in the grid voltage; the lumped impedance of the AC filter, power line and other components among the power lines is R_s; and I_g is the grid-in current.

Therefore, DCC produced by the DC voltage component in grid-in current can be expressed as

Fig. 6.4 Equivalent circuit of DC voltage component in the grid voltage

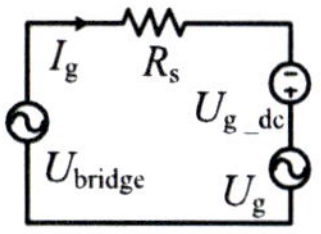

$$I_{g1} = \frac{U_{g\_dc}}{R_s} \tag{6.9}$$

Still taking the 3 kW TLI connected with the grid voltage of 220 V and frequency of 50 Hz as example, if $U_{g\_dc} = 0.1$ V and $R_S = 10\ \Omega$, then $I_{g1} = 0.1/10$ A $= 10$ mA.

In summary, the maximum DCC can be expressed as

$$I_{dc\_\max} = \left|I_{g0}\right| + \left|\Delta I_{g\_\max}\right| + \left|I_{g1}\right| \tag{6.10}$$

Here, $\Delta I_{g\_\max}$ can be obtained based on (6.2), $\Delta I_{g\_\max} = \Delta I_{g1} + \Delta I_{g2} + \Delta I_{g3} = 18 + 50 + 27$ mA $= 95$ mA, finally, $I_{dc\_\max}$ can be obtained: $I_{dc\_\max} = |\Delta I_{g0}| + |\Delta I_g| + |\Delta I_{g1}| = 19 + 95 + 10$ mA $= 124$ mA.

For a 3 kW/220 V grid-connected inverter, the allowed DCC is 3000/220 × 0.5% mA = 68 mA according to China standard GB/T 33,593–2017 in Table 6.1. The accumulated DCC from aforementioned parts is 124 mA, it is almost two times of standard limitation. Therefore, solutions are needed to eliminate DCC injected into the grid.

Based on operating principles, DCC rejection methods can be classified into three major categories, detection & feedback methods, capacitor blocking methods, and intelligent algorithms, as shown in Fig. 6.5. Rest of this chapter will discuss these methods one by one.

Fig. 6.5 Classification of DCC rejection methods

6.2 Detection and Feedback Methods

Detection and feedback methods refer to precisely detect the DC component in grid-in current by using various technical measures first, and then feeding this component back into the control loop, and finally a bias voltage is generated in the output voltage of TLIs to suppress DCC. The detection measures of DCC can be classified into two kinds, i.e., direct detection and indirect detection methods.

6.2.1 *Coupled-Inductor Detection Method*

1. Operating principle

The operating principle sketch of coupled-inductor detection method is shown in Fig. 6.6 [2]. A coupled-inductor and a hall sensor are adopted in the sketch. It is worth noting that the turn numbers of primary and secondary windings in the coupled-inductor are equal, and the secondary winding is short-circuited, as well as the primary and secondary windings go through the hall sensor with opposite directions.

Assuming that the primary side current i_p is equal to the grid-in current $i_g = i_{ac} + I_{dc}$, here, i_{ac} and I_{dc} are the AC component, and DC component of the grid-in current, respectively. Due to the same turn numbers in the current transformer, i_s is only equal to the AC component i_{ac}. Therefore, the output signal of the Hall sensor exactly represents the DC component I_{dc} of the grid-in current. Therefore, it is a good way to use a Hall sensor with small capacity range to detect the DC component with low cost and high accuracy.

2. Simulation results

In order to verify the effectiveness of the coupled current transformer detection, a simulation model was built, and the key parameters are as follows: input DC voltage U_{pv} is 360 V, filter inductance L is 2mH, the grid voltage is 220 V/50 Hz, switching frequency f_s is 20 kHz, and rated power is 3 kW. By setting the grid-in current reference as $i_{ref} = 20\sin\omega t + 1$ (contains a DC bias current of 1 A), the grid-in current

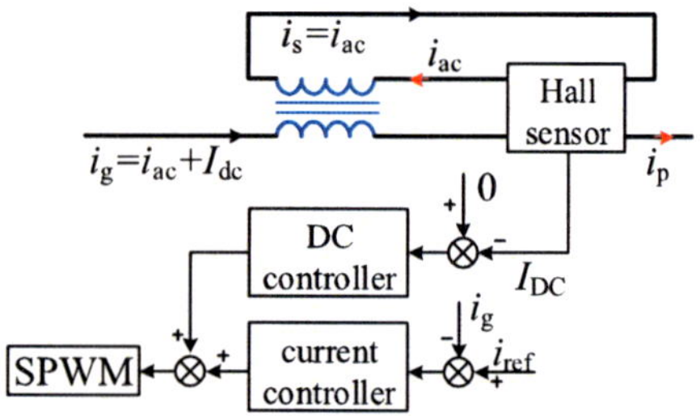

Fig. 6.6 Coupled-inductor detection method

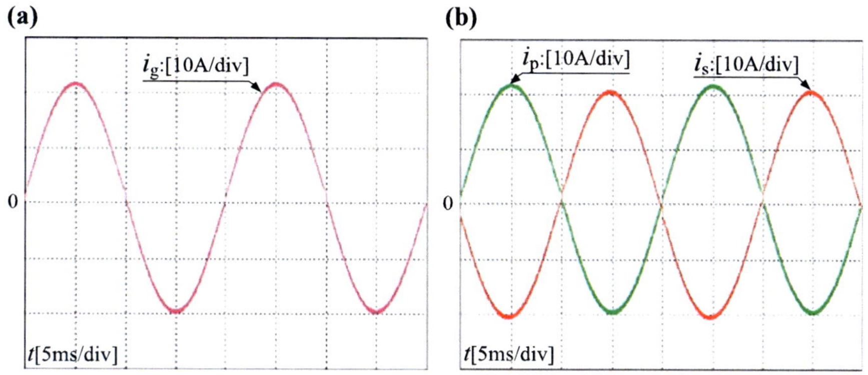

Fig. 6.7 Current waveforms without DCC rejection compensation, **a** Grid-in current, **b** Primary and secondary currents in coupled current transformer

waveform without DCC rejection compensation is shown in Fig. 6.7a, the current waveforms in the primary and secondary windings of the coupled current transform are shown in Fig. 6.7b. It can be seen that the grid-in current contains DCC because of reference set, but the secondary winding current only contains AC component i_{ac} included in grid-in current i_g, it is in agreement with operating principle analysis.

After the DCC rejection strategy is enabled, the grid-in current waveform is shown in Fig. 6.8a, it can be seen that the DCC is mostly disappeared even rough the current reference contains 1A DCC setting. Again, the current waveforms in the primary and secondary windings of the coupled current transform are shown in Fig. 6.8b, they look same because the DC component I_{dc} is restricted. The DCC waveform detected by the Hall sensor is shown in Fig. 6.8c, it can be seen that only about 50 mA of DCC is residual. Therefore, the detection & feedback method is able to make sure the DC components complying with standard for a 3 kW PVPG system.

From the aforementioned analysis, it can be seen that using a small capacity range Hall sensor in the coupled-inductor method can achieve higher detection accuracy and acceptable rejection effect. However, the coupled-inductor needs to be carefully designed to achieve tight coupling between primary and secondary windings. In order to reduce sensor errors and avoid complex design caused by magnetic components, sensor auto-calibrating method and DC-link current sampling method have been proposed in [3, 4], respectively.

6.2.2 Sensor Auto-Calibration Method

1. Operating principle

The operating principle sketch of the sensor auto-calibration method is shown in Fig. 6.9a [3], two current sensors are adopted in the sketch, which is used to measure

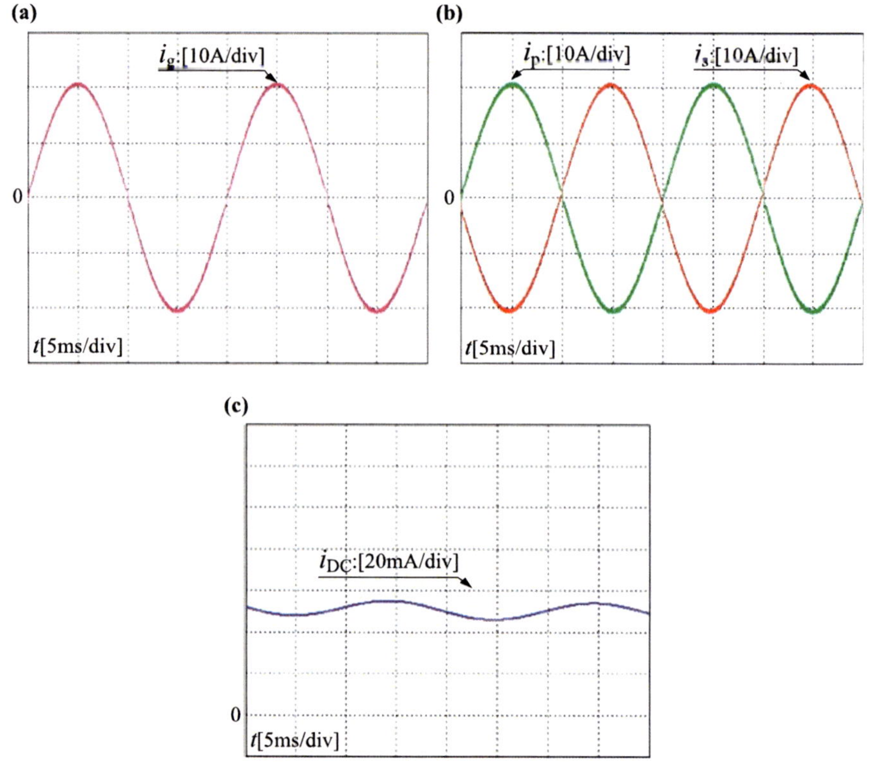

Fig. 6.8 Current waveforms with DCC rejection compensation, **a** Grid-in current, **b** Primary and secondary currents in coupled current transformer, **c** DCC

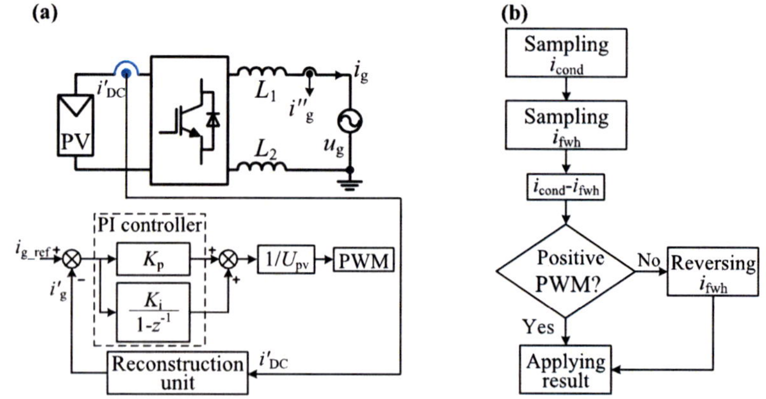

Fig. 6.9 The detection and control structure of the sensor auto-calibration method, **a** Sensor auto-calibration method, **b** Reconstruction process

the DC-link current i'_{DC} and the grid-in current i_g, respectively. Assuming that the full-bridge inverter operates under the unipolar SPWM modulation strategy, the DC-link current i'_{DC} is equal to the absolute value of the grid-in current $|i_g|$ in the energy transfer period, and the DC-link current i'_{DC} is equal to zero in the freewheeling period, ideally. However, the measurement value of i'_{DC} in the freewheeling period is not zero because of the non-linearity and inherent bias of the sensor. Therefore, in order to achieve the auto-calibration of the sensor, the method can be taken that subtracting the i'_{DC} in the freewheeling period from the i'_{DC} in the energy transfer period. The detailed the reconstruction process in Fig. 6.9a is shown in Fig. 6.9b, where i_{cond} is the sampling result of the DC-link current during the energy transfer period, and i_{fwh} is the sampling result of the DC-link current during the freewheeling period. From the aforementioned analysis, i_{fwh} is subtracted from i_{cond} to obtain the corrected result. It is obvious that the corrected result can be directly applied as the grid current reconstruction value if the output PWM command is positive, because the grid voltage is in the positive half cycle at this time. By contrast, if the output PWM instruction is negative, it can be known that it is in the negative half cycle of the grid voltage at this time. Then, the corrected result is reversed to obtain the grid-in current i'_g after eliminating the sensor error. After that, the i'_g is fed back to the grid-in current controller as an accurate measurement result of the grid-in current, and compared with the grid current reference i_{g_ref}. The subtracted current error passes through the grid current regulator and then outputs a PWM wave with dc bias. In the end, this method achieves accurate tracking of the grid current to eliminate the DC component of the grid.

2. Simulation results

In order to verify the effectiveness of the sensor auto-calibration method, a simulation model was built in Matlab/Simulink, and the key parameters are as follows: input DC voltage U_{pv} is 360 V, filter inductance L is 2mH, the grid voltage is 220 V/50 Hz, switching frequency f_s is 20 kHz, and rated power is 3 kW. What is important is that the sampling result is set to contain an offset of 0.1 A which causes that i_{fwh} in the freewheeling period is 0.1 A.

The actual DC-link current is shown in Fig. 6.10a, which is absolutely the high-frequency current pulse. Figure 6.10b shows the i_{cond} waveform obtained after sampling the DC-link current during the energy transfer period. What's more, Fig. 6.10c shows the sampling result of the DC-link current i_{fwh} during the free-wheeling period. It can be seen that there is a slight distortion at the zero-crossing time of i_{cond} in Fig. 6.10b, and i_{fwh} in Fig. 6.10c contains not only a DC component of 0.1 A but also a pulsating component with lower amplitude. The main reason why there are distortions in these waveforms is that there are some sampling points at the time when the DC-link current has not completely returned to zero during the short freewheeling period in the peak region of grid-in current. What's similar is that there are some sampling points at the time when the DC-link current is zero during the short energy transfer period in the zero-cross region of grid-in current.

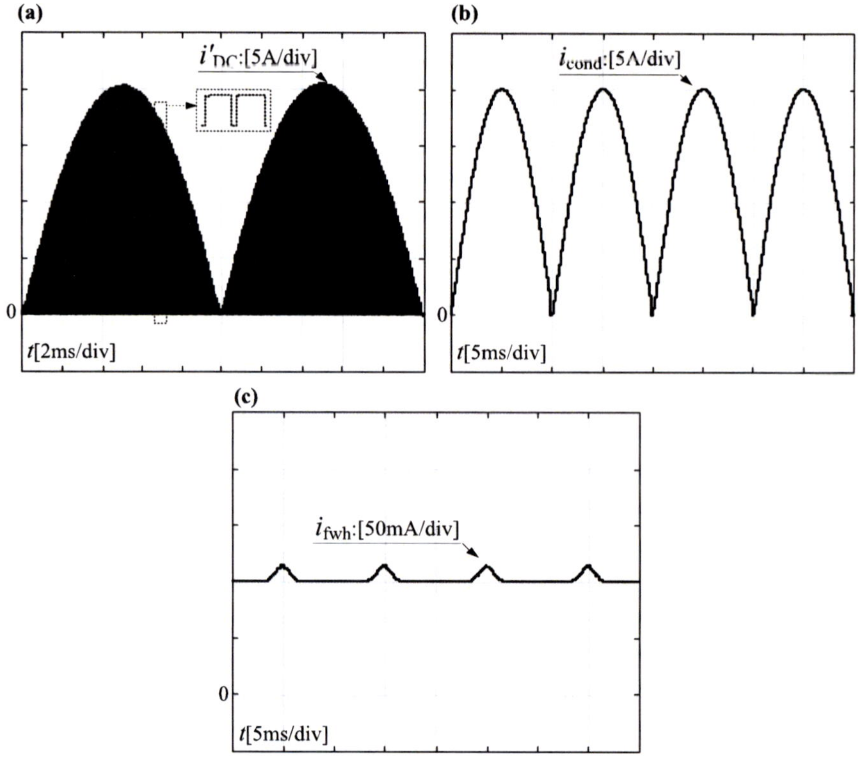

Fig. 6.10 Waveforms of DC-link current, **a** Sampling waveform of DC-link current, **b** Sampling waveform during the energy transfer period, **c** Sampling waveform in the freewheeling period

The i'_g shown in Fig. 6.11a is the grid-in current waveform reconstructed by the process shown in Fig. 6.9b. It can be seen that the i'_g waveform is symmetrical and contains all the information of the grid current. The grid-in current waveform directly measured by the sensor is shown in Fig. 6.11b, which presents that the component caused by the sensor bias error is included.

After the DCC rejection strategy is enabled, it is easy to find that there is almost no DCC in the grid-in current shown in the waveform diagrams of Fig. 6.12a and b. Concretely, it can be seen that the DC component of grid-in current is about 10 mA, which follows the standards to connect the grid. Moreover, Fig. 6.12b and c show the waveform of the grid-in current and the DCC by using the direct current detection method. It can be seen that the DCC in the grid-in current is about 100 mA, which is out of the grid-connection standard range as mentioned above.

From the aforementioned analysis, when applying the auto-calibration method, the sensor has a good effect of eliminating sensor measurement errors. However, if the PWM pulse width is narrow or wide, there is a certain error in the DC-link current sampling. In order to improve the performance and reduce the cost, it is obvious that

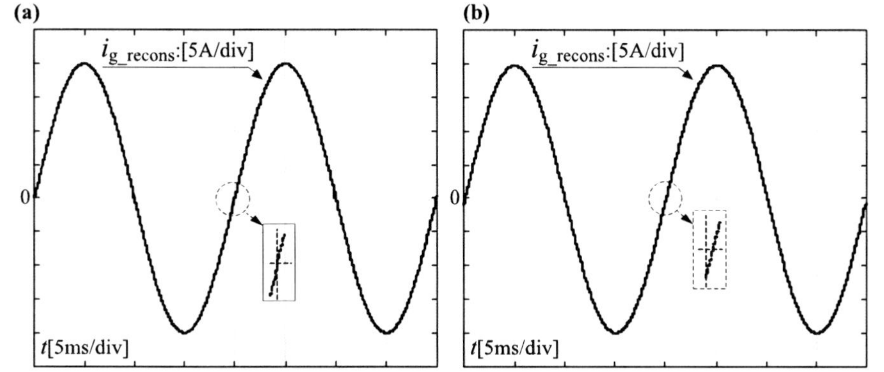

Fig. 6.11 Comparison between reconstructed grid-in current and original grid-in current, **a** Reconstructed grid-in current, **b** Original grid-in current

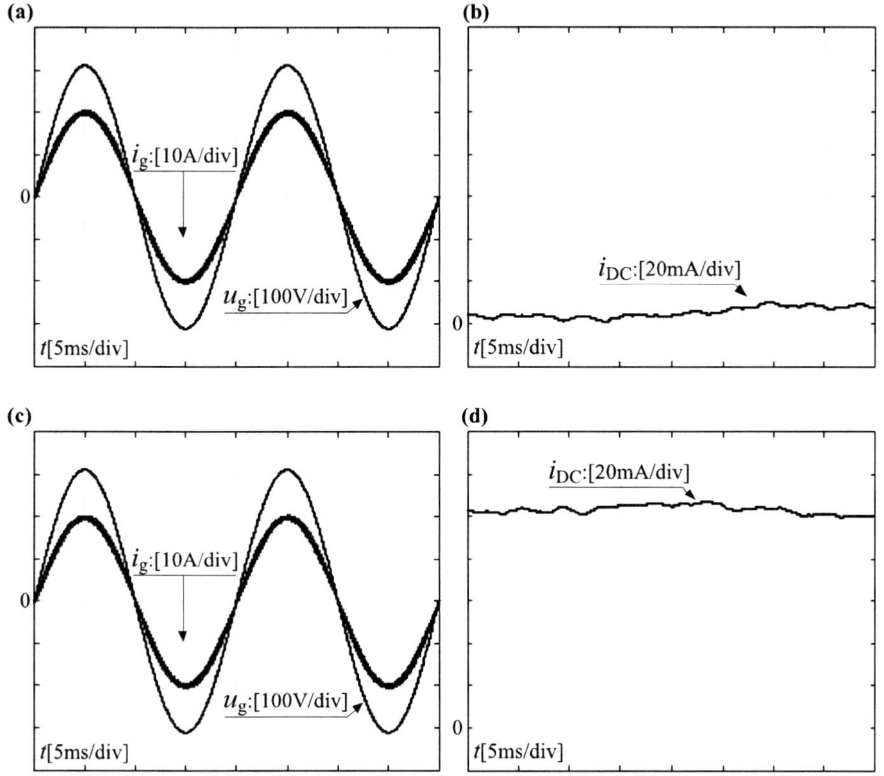

Fig. 6.12 Simulation waveforms of sensor self-calibration method, **a** Grid-in current, **b** DCC of grid-in current using sensor self-calibration method, **c** Grid current (direct measurement), **d** DCC of grid-in current (direct measurement)

the current feedback is the reconstructed grid-in current using this method, so the grid-side current sensor can be omitted. In this way, only the DC-link current sensor is used for sampling, that is, the DC-link current sampling method [4].

6.2.3 DC-Link Current Sampling Method

1. Operating principle

The detection circuit of the DC-link current sampling method is shown in Fig. 6.13a [4], where i'_{DC} is the DC-link sampling current. Assuming that i_{Cond} is the DC-link conduction current, and the grid-in current is i_g, the equation $i_{Cond} = |i_g|$ is satisfied. Considering the DCC of the grid-in current, i_{Cond} divided into two components is shown in (6.13), where I_{AC} represents the output AC amplitude, and I_{DC} represents the injected DCC. The FFT of i_{Cond} is shown in (6.14) in which it can be seen that the even harmonics component of i_{Cond} is related to the fundamental component of the grid-in current, and the DCC of i_{Cond} is determined by odd harmonics of grid-in current. Because of the lower high-harmonic content, 'n' can be replaced by '1' in (6.14) to get (6.15). What's remarkable is that the third term of (6.14) is the line frequency component, whose amplitude is related to I_{DC}. Therefore, in this way, the DCC of the grid-in current can be detected by calculating the power frequency component of i_{Cond}.

The detection process of the DCC is shown in Fig. 6.13b, where the DC-link current i'_{DC} is sampled and fed into the microprocessor through the current sensor and the ADC. What's more, the phase-locked loop is used to obtain the phase information

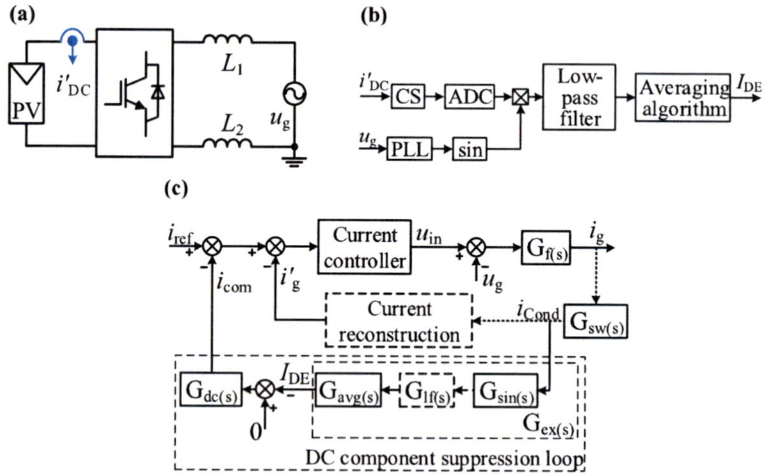

Fig. 6.13 Detection and control structure of DC-link current sampling method, **a** DCC detection circuit, **b** DCC detection structure, **c** Current control structure

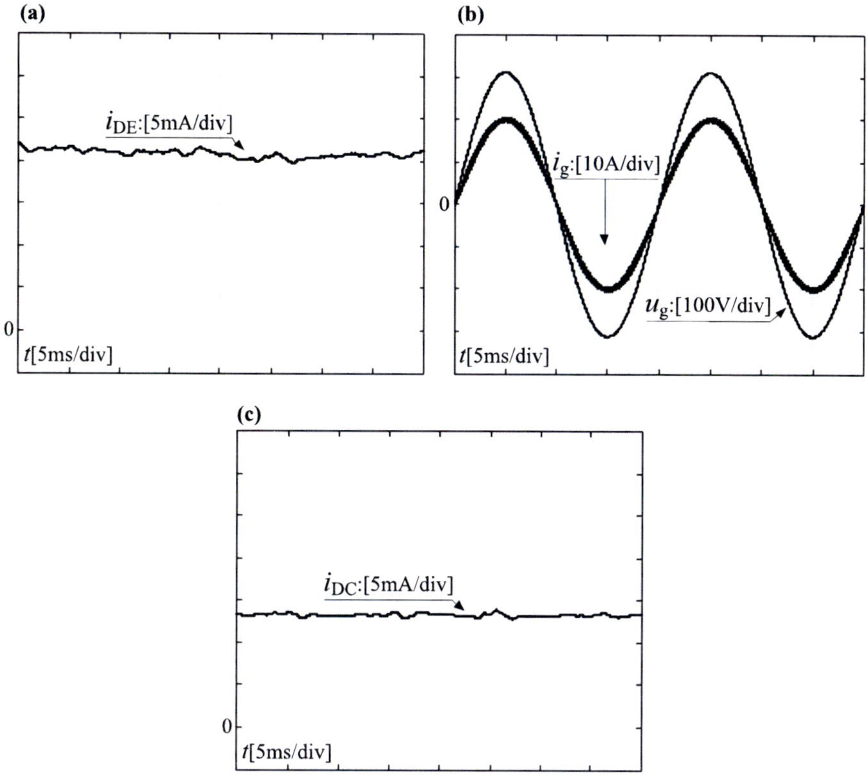

Fig. 6.14 Current and voltage waveforms with DC-link current sampling method, **a** DCC of DC-link detection, **b** Grid-in current, **c** DCC of grid-in current (direct measurement)

of the grid. Based on the FFT result of (6.13), i_{Cond} is multiplied by sinωt to extract its power frequency component, and further obtain I_{DC}, which is shown in (6.14). Substituting (6.13) into (6.14) to obtain (6.15), the DCC detected is the last term of (6.15). Last, in order to eliminate the AC component to get the final DCC I_{DE}, the i_{Ex} is passed the through the low-pass filter and averaging algorithm, as shown in (6.16).

On the whole, the control structure of this rejection strategy is shown in Fig. 6.13c, where $G_f(s)$ is the gain of the output filter, $G_{sw}(s)$ is the switching transfer function, $G_{sin}(s)$ is the transfer function of the sine function, and $G_{lf}(s)$ is The transfer function of the low-pass filter in Fig. 6.13b, $G_{avg}(s)$ is the transfer function of the averaging algorithm, and i_{com} is the output compensation of the DC suppression loop. With the output of DCC regulator $G_{dc}(s)$ fed into the main control loop, the controller finally outputs the switching signal with suppression bias to offset the DCC in the grid-in current.

$$i_{Cond}(t) = \left|i_g(t)\right| = |I_{AC}\sin\omega t + I_{DC}| \tag{6.11}$$

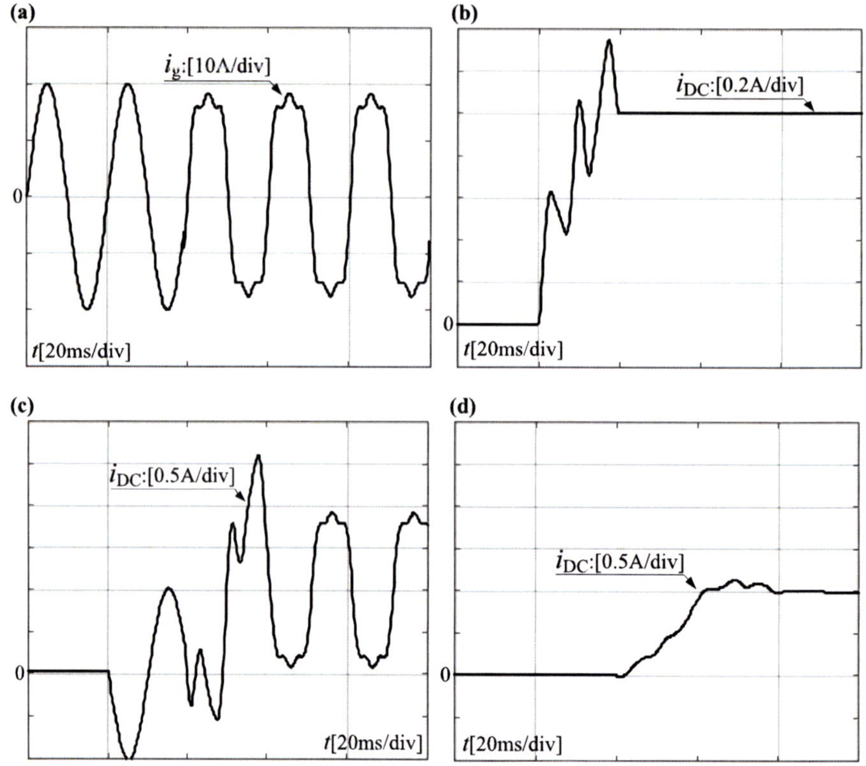

Fig. 6.15 Integration results of the double integration method, **a** Grid-in current, **b** Integration result when using grid period, **c** Single integration, **d** Double integration

Fig. 6.16 Resonance-circuit detection

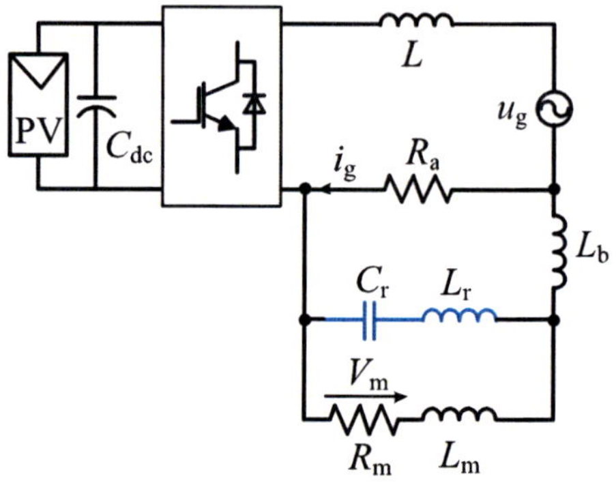

$$i_{\text{Cond}}(t) = \frac{4I_{AC}}{\pi}\left(\frac{1}{2} - \sum_{n=1}^{\infty}\frac{1}{4n^2-1}\cos 2n\omega t\right) + \sum_{n=1}^{\infty}\frac{4I_{DC}}{(2n-1)}\sin[(2n-1)\omega t)], (n = 1, 2, 3...) \tag{6.12}$$

$$i_{\text{Cond}}(t) = \frac{2I_{AC}}{\pi} - \frac{4I_{AC}}{3\pi}\cos 2n\omega t + \frac{4I_{DC}}{\pi}\sin\omega t \tag{6.13}$$

$$i_{Ex}(t) = i_{\text{Cond}}(t)\sin\omega t \tag{6.14}$$

$$i_{Ex}(t) = i_{\text{Cond}}(t)\sin\omega t = \frac{2I_{AC}}{\pi}\sin\omega t - \frac{4I_{AC}}{3\pi}\cos 2\omega t\sin\omega t - \frac{2I_{DC}}{\pi}\cos 2\omega t + \frac{2I_{DC}}{\pi} \tag{6.15}$$

$$I_{DE} = \frac{2I_{DC}}{\pi} \tag{6.16}$$

2. Simulation results

In order to verify the effectiveness of the DC-link current sampling method, here comes the simulation using Matlab/Simulink, and the key parameters are as follows: input DC voltage U_{pv} is 360 V, filter inductance L is 2mH, the grid voltage is 220 V/50 Hz, switching frequency f_s is 20 kHz, and rated power is 3 kW. It is worth noticing that the reference of grid-in current here is $i_{\text{ref}} = 20\sin\omega t + 1$, which contains a DC bias of 1A.

After the DCC rejection strategy is enabled, the steady-state simulation waveforms of the DC-link current sampling method are shown in Fig. 6.14. It can be seen that the sampling of DC-link current and the reconstruction results of grid-in current in this method are consistent with the sensor self-calibration method. The DC-link current i'_{DC} and the reconstructed grid-in current i'_g can be referred to Figs. 6.10a and 6.11a. Figure 6.14a presents the waveforms i_{DE} which is the DCC of grid-in current, and Fig. 6.14b and c show the grid-in current and the DCC by directly measured. It can be seen that the lower DCC of grid-in current is π/2 times of i_{DE} in Fig. 6.14c, which is consistent with the previous analysis.

It can be seen from the aforementioned analysis that this method only needs to use one current sensor to detect the DC-link current, and the calculation of DCC is completed in the microprocessor. However, the DCC extraction process is complicated and is still based on one current sensor. In order to make better use of the resources of the microprocessor, something else methods should be proposed. Considering that the incoming current is a sine form term, the DCC of grid-in current can be obtained by an integration method, that is, double integration method.

6.2.4 Double Integration Method

1. Operating principle

By detecting the grid-in current and integrating it, the double integration method obtains the DCC of grid-in current injected into the grid [5]. This method can be divided into two steps. Firstly, after decomposed using the FFT method, the AC component of grid-in current is integrated. It is obvious that the integration result is nearly zero when the integration period is equal to the grid period. Secondly, the overall grid-in current is integrated during the grid period. Theoretically, the integration result is the DCC of grid-in current, but its actual value can be reduced because of errors. Included the DCC in grid-in current, the result of the FFT can be shown in (6.17). By integrating the first term of the (6.17) which is the DCC of grid-in current and dividing by the integration period, the (6.18) can be obtained. Homoplastically, by integrating the second term of the (6.17) which is the AC component, the (6.19) can be obtained. As seen, if f_g = 1/T, then (6.21) is equal to zero, that is, the DCC

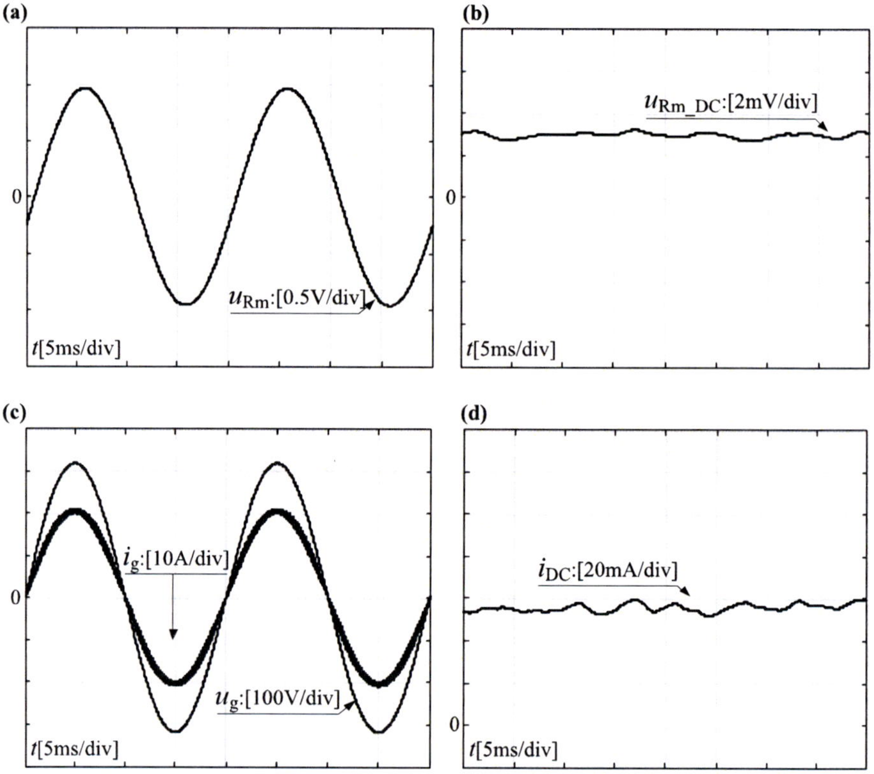

Fig. 6.17 Detection results of resonance-circuit method, **a** Voltage across detection resistor R_m, **b** DCC of voltage across detection resistor R_m, **c** Grid-in current, **d** DCC of grid-in current

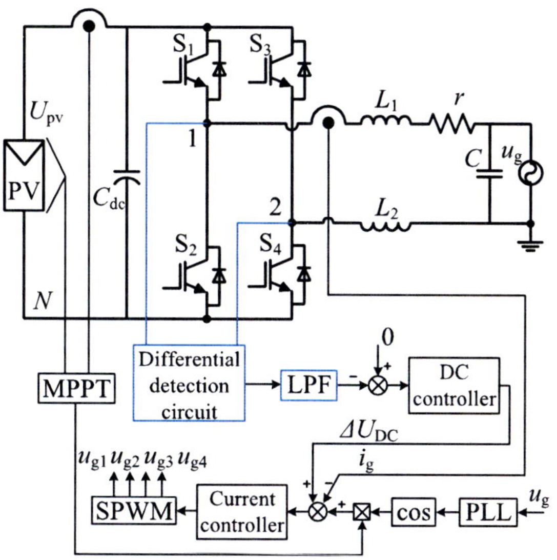

Fig. 6.18 Detection and control structure of bridge-leg-voltage method

of grid-in current can be obtained by integrating the grid current, directly. However, when there is disturbance outside the system such that $f_g \neq 1/T$, the (6.22) can be obtained by integrating the (6.19). It can be seen that if the integration result of grid-in current is directly sampled at this time, a larger detection error will be caused. In order to reduce the error, double integration can be applied on (6.19). The double integration results are shown in (6.23). Comparing (6.22) and (6.23), it is found that the amplitude of AC component in (6.23) is lower and the error is smaller. That is means, the more times integrations are applied, the smaller errors can be got. But too many times integrations will slow down the system response, which is undesired. Taking the right of (6.23) as the detection result of the DCC of grid-in current, the DCC can be eliminated after discretizing and feeding it into the DC rejection loop.

$$i_g(t) = I_{dc} + i_{ac} = I_{dc} + \sum_{n=1,2,3,\ldots} i_n \sin(2\pi n f_g t + \varphi_n) \tag{6.17}$$

$$i_0(t) = \frac{1}{T}\int_{t_0}^{t_0+T} I_{dc}dt = I_{dc} \tag{6.18}$$

$$\begin{aligned} i_1(t) &= \int_{t_0}^{t_0+T} \sum_{n=1,2,3,\ldots} I_n \sin(2\pi n f_i t + \varphi_n)dt \\ &= \sum_{n=1,2,3,\ldots} \frac{-I_n}{2\pi n f_g} \cos(2\pi n f_g t + \varphi_n)\Big|_{t_0}^{t_0+T} \\ &= \sum_{n=1,2,3,\ldots} \frac{I_n}{\pi n f_g} \cdot \sin(2\pi n f_g t_0 + \varphi_n + \pi n f_g T) \cdot \sin(\pi n f_g T) \end{aligned} \tag{6.19}$$

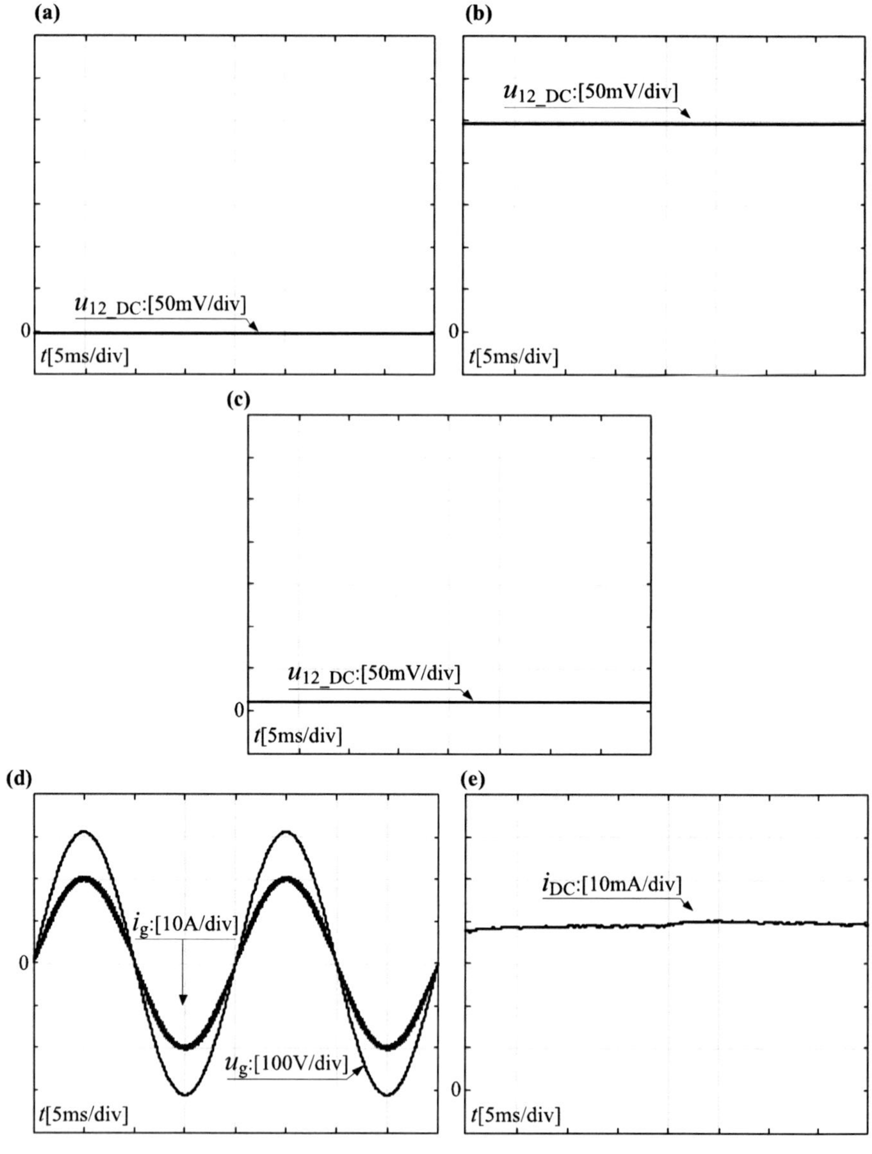

Fig. 6.19 Current and voltage waveforms with Bridge-leg-voltage method enabled, **a** DCC of bridge leg output without bias injected, **b** DCC of bridge leg output with 1 A bias injected, **c** DCC of bridge leg output after compensation, **d** Grid-in current, **e** DCC of grid-in current

$$
\begin{aligned}
i_2(t) &= \frac{1}{T}\int_{t_0}^{t_0+T} i_g(t)dt \\
&= \frac{1}{T}\int_{t_0}^{t_0+T}\left[\int_{t_0}^{t_0+T}\left(I_{dc} + \sum_{n=1,2,3,\ldots} I_n \sin(2\pi n f_g t + \varphi_n)\right)\right]dt \\
&= I_{dc} + \sum_{n=1,2,3,\ldots} \frac{f_i}{\pi n f_g}\sin\left(\frac{\pi n f_g}{f_i}\right)\cdot I_n \sin\left(2\pi n f_g t_0 + \varphi_n + \frac{\pi n f_g}{f_i}\right)
\end{aligned} \tag{6.20}
$$

$$
\begin{aligned}
i_3(t) &= \frac{1}{T^2}\int_{t_0}^{t_0+T}\int_{t_0}^{t_0+T} i_2(t)dt \\
&= I_{dc} + \sum_{n=1,2,3,\ldots}\left[\frac{f_i}{\pi n f_g}\sin(\frac{\pi n f_g}{f_i})\right]^2 \cdot I_n \sin\left(2\pi n f_g t_0 + \varphi_n + \frac{\pi n f_g}{f_i}\right)
\end{aligned} \tag{6.21}
$$

It can be seen from the aforementioned analysis that the double integration method has higher detection accuracy. What is impressive is that only one current should be sampled, but it is higher digital computing requirements that put pressure on microprocessor resources.

2. Simulation results

Assuming that the fundamental component of grid-in current is $i_1 = 20\sin(2\pi f_g t)$, where f_g is the grid frequency equaling to 50 Hz. If the grid current contains low-order harmonics and DCC, it can be expressed as $i_g = 20\sin(2\pi f_g t) + 4\sin(3 \times 2\pi f_g t) + 2\sin(5 \times 2\pi f_g t) + 1$. When f_g is equal to 50 Hz, the waveform of i_g is shown in Fig. 6.15a (lower harmonic and DC components are added after two grid cycles). When the grid period is equal to the integration period, single integration of i_g can obtain the result shown in Fig. 6.15b. In the case of disturbance, if the grid period is not equal to the integration period, single and double integrations are applied on i_g, respectively. And the results are shown in Fig. 6.15c and d. It can be seen from Fig. 6.15 that when the integration period is equal to the grid period, the DCC of grid-in current can be effectively obtained by using a single integration. However, when the integration period is different from the grid period, the result of single integration will fluctuate greatly. At this time, double integration can be applied to get more accurate detection results of DCC.

The method of obtaining the DCC of grid-in current by direct sampling is described above. In order to minimize sampling errors, especially sensor errors, an indirect sampling method is proposed. Namely, the indirect sampling mainly reflects the DCC of grid-in current by sampling the DCC of voltage, including the resonant-circuit method and the bridge-leg-voltage method [7–11].

6.2.5 Resonant-Circuit Method

1. Operating principle

The operating principle sketch of the resonant-circuit detection is shown in Fig. 6.16, in which the resistor R_a is the sampling resistor of grid-in current. In order to reduce the AC component entering the detection branch, a multi-level AC component suppression branch is been applied in the circuit, where the inductor L_b acts as a first stage to block the AC component of the grid-in current from flowing into the detection branch. It is worth noting that the inductor L_r and the capacitor C_r are connected in series to form a resonance circuit as a second stage to bypass the AC component that enters the detection branch. Therefore, this method is called a resonance-circuit detection method. What's more, the inductor L_m is a third-stage to block AC component from flowing into the detection resistor R_m. Finally, the voltage across R_m can reflect the magnitude of the DCC of grid-in current.

It can be seen from the above-mentioned analysis that this method uses a multi-stage circuit for detection. Moreover, there are discrete components used in the circuit and no coupling inductor or transformer, which makes the circuit structure simple. However, in order to minimize the AC component flowing into the detection branch, the parameter design of these components is complicated, because the test results of DCC are highly dependent on the parameters.

From Fig. 6.16, the transfer function of the grid-in current $i_g(s)$ to the detection voltage $v_m(s)$ is shown as

$$G_m(s) = \frac{v_m(s)}{i_g(s)} = \frac{R_m R_a}{\frac{sL_b\left[R_m + s(L_m + L_r) + \frac{1}{sC_r}\right]}{sL_r + \frac{1}{sC_r}} + R_m + sL_m}$$
$$= \frac{R_m R_a s C_r (s^2 L_r C_r + 1)}{s^2 L_b C_r [R_m s C_r + s^2 C_r (L_m + L_r) + 1] + s C_r (R_m + s L_m)(s^2 L_r C_r + 1)} \quad (6.22)$$

2. Simulation results

In order to verify the effectiveness of the resonance-circuit detection, a simulation model was built in Matlab/Simulink, and the key parameters are as follows: input DC voltage U_{pv} is 360 V, filter inductance L is 2 mH, the grid voltage is 220 V/50 Hz, switching frequency f_s is 20 kHz, rated power is 3 kW, inductance L_b is 0.8 mH, and resistance R_m is 3 sΩ. It is important that the reference of the grid-in current i_{ref} is equal to 20sinωt + 0.5, which contains a DC bias of 0.5 A. As shown in Fig. 6.17, the effectiveness of the resonance-circuit detection method can be verified.

Figure 6.17a shows the voltage across the sense resistor R_m, which contains AC component, while Fig. 6.17b shows the DCC in the voltage of sense resistor R_m. What's more, Fig. 6.17c shows the grid current, and the DCC of grid-in current is shown in Fig. 6.17d. From these waveforms in Fig. 6.17, it can be seen that the voltage across the resistor R_m contains only the fundamental frequency component and no high-frequency components. Therefore, this method can indirectly reflect the DCC through the voltage across the resistor R_m. From the grid-in current and its DCC waveform diagram, it is found that the DCC of grid-in current is about 60 mA when the grid-in current reference contains 0.5A DC suppression, which follows the standards to connect the grid.

6.2.6 Bridge-Leg-Voltage Detection Method

1. Operating principle

Bridge-leg-voltage method [9–11] is another indirect detection method that reflects the DCC of grid-in current by detecting the DC voltage in the bridge leg voltage of the inverter. Reference [9] uses a differential amplifier and a second-order low-pass filter to filter out AC components, and the complete control loop is shown in Fig. 6.18 [9–10]. In Fig. 6.18, a differential amplifier circuit is used to detect the TLI output differential mode voltage. By adopting a low-pass filter (LPF) to further eliminate higher harmonic components, a DC voltage component related to the DCC of grid-in current can be obtained. This DCC is fed into the DCC regulator to generate a compensation bias, and this compensation bias is fed into the grid-in current regulator to reject the DCC of grid-in current.

This method has a strong effect on the DCC of grid-in current caused by differences from power devices. However, it has a weak effect on the errors of signal conditioning circuits and the DC voltage component of the grid.

2. Simulation results

In order to verify the effectiveness of the bridge-leg-voltage method, a simulation model was built in Matlab/Simulink, and the key parameters are as follows: input DC voltage U_{pv} is 360 V, filter inductance L is 2mH, the grid voltage is 220 V/50 Hz, switching frequency f_s is 20 kHz, rated power is 3 kW. In addition, the grid-in current reference i_{ref} is equal to $20\sin\omega t + 1$, that is, it contains a DC bias of 1 A.

As shown in Fig. 6.19a, the output DC voltage of the bridge leg is nearly zero when there is no DC bias in grid-in current reference. And Fig. 6.19b shows the DC voltage of the bridge leg when a 1 A DC bias is injected and not fed into DC regulator. What is important is that Fig. 6.19c shows the DC voltage of output from the bridge leg after the 1 A DC bias is injected and the DC regulator is enabled. Grid-in current waveforms and its DCC in steady-state of the bridge leg voltage method are shown in Fig. 6.19d and e, respectively. It is worth noting that when the grid-in current reference is not injected into the DC bias, the DC voltage of bridge leg output is almost equal to zero from these figures. After the DC bias is injected, the DC voltage of bridge leg output increases significantly. Therefore, this method can indirectly detect the DCC of grid-in current. Noticing that the DCC of grid-in current is about 40 mA shown in Fig. 6.19e, it is apparent that the DCC of grid-in current is effectively rejected.

6.3 Capacitor Blocking Methods

Generally speaking, the detection-feedback method feeds the detected DCC of grid-in current into the controller for rejection, so the result of the detection loop and the control parameters of the system will have a greater impact on the suppression effect. Adding a "blocking capacitor" to the loop of the grid-in current is a method for rejecting DCC that does not rely on detection accuracy and has a good rejection effect. According to the position of blocking capacitor placed, it can be divided into DC side and AC side capacitor blocking methods.

6.3.1 *DC-Side Capacitor Blocking Method*

It is obvious that a half-bridge topology is a typical structure with a blocking capacitor on the DC side, as shown in Fig. 6.20a which is a traditional two-level half-bridge TLI circuit. In this circuit, the grid-in current flows through the capacitor during the energy transfer period and the freewheeling period, when the blocking capacitor can effectively eliminate the DCC of grid-in current. But its output is a two-level voltage which means higher harmonic components contained in output. Introduced in Sect. 4.2.1 of this book, the NPC half-bridge grid-connected inverter [12] also has a half-bridge structure. The output of this structure is a three-level voltage with

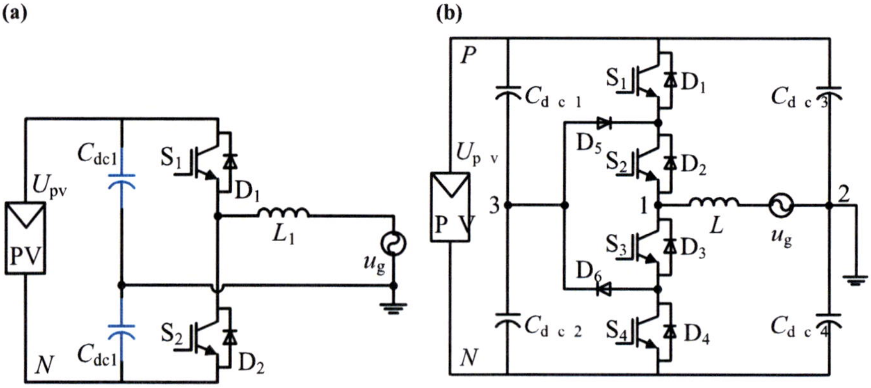

Fig. 6.20 DC-side blocking capacitor method, **a** Two-level half-bridge TLI topology, **b** Three-level half-bridge TLI topology

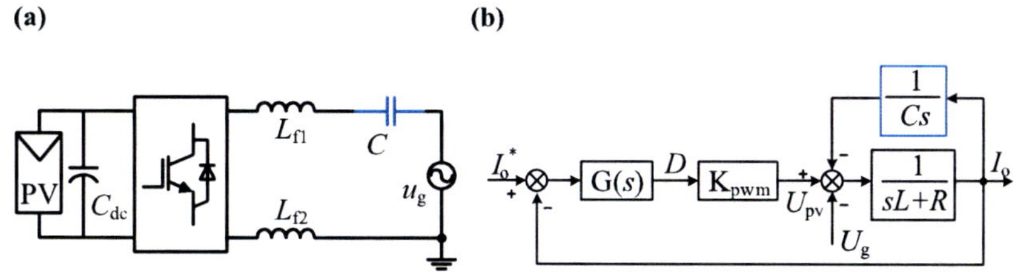

Fig. 6.21 Sketch of actual series capacitor method, **a** Circuit sketch with actual capacitors, **b** Transfer function model

better differential mode characteristics. However, during the freewheeling period, the grid-in current cannot flow through the capacitor, which weakens the blocking effect. The structure of the double-capacitor NPC grid-connected inverter [13] introduced in Sect. 4.2.3 of this book is shown in Fig. 6.20b. This structure allows the grid current to flow through the blocking capacitor during the energy transfer period and the freewheeling period. In this way, the DC blocking effect in the energy transfer period and the freewheeling period can be guaranteed and the three-level differential mode voltage output can be maintained.

As shown in Table 4.8, Sect. 4.2.3 of this book analyzes the DCC of grid-in current for HERIC, NPC, and the circuit shown in Fig. 6.20b.

The DC-side blocking capacitor method generally requires a larger volume to place electrolytic capacitors and higher requirements for the voltage equalization circuit. In other words, the imbalance of the capacitor voltage will affect the DC blocking effect. For this reason, researchers have proposed the AC-side blocking capacitor method [14–18]. The AC side blocking capacitor method places the capacitor on the AC output side in which capacitor can be an actual physical capacitor [14] or a "virtual capacitor" [15–18] method.

6.3.2 AC-Side Capacitor Blocking Method

1. Actual series capacitor method

The circuit sketch of the AC-side blocking capacitor method using actual physical capacitors [14] is shown in Fig. 6.21a, where the equivalent transfer function model is shown in Fig. 6.21b, that is, a blocking capacitor is placed on the grid side. The meanings of the symbols are as follows: U_{pv} is the bridge leg voltage, R is the equivalent series resistor of the inductor, I_o* is the reference of grid-in current, C is the actual physical capacitor, G(s) is the grid-in current controller, and D is the duty ratio, K_{pwm} is the pulse width modulation gain. With clear principles and simple implementation, the method still has many disadvantages. Firstly, the capacitance value is difficult to determine, which means that larger capacitors will slow down

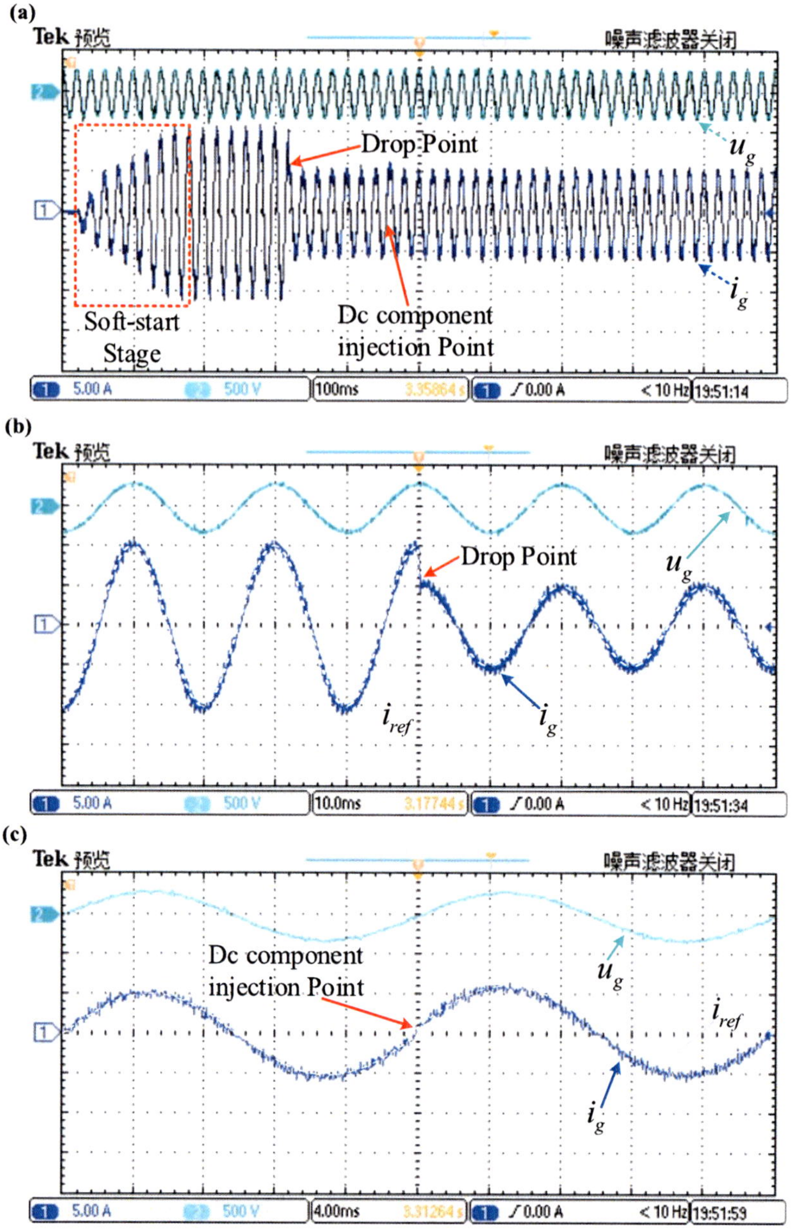

Fig. 6.22 Key waveforms of actual series capacitor method in experiment test, **a** Grid-in current respance waveforms with soft-start, sudden amplitude change and dc component injection, **b** Dynamic details of sudden current amplitude change, **c** Respance details of dc component suppression

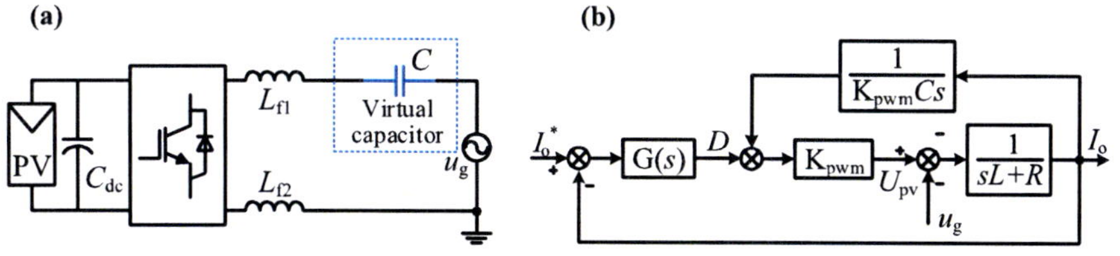

Fig. 6.23 Sketch of virtual capacitor method, **a** Circuit sketch with virtual capacitor, **b** Transfer function model

the dynamic performance of the system, and smaller capacitors will easily cause oscillation. Secondly, the series resistor of capacitor will also reduce the efficiency of the TLI.

2. Experimental results of the actual series capacitor method

In order to verify the effectiveness of the actual physical capacitor method, a hardware platform was built and the key parameters are as follows: input DC voltage U_{pv} is 360 V, grid voltage is 220 V/50 Hz, inductance L is 2.2 mH, and capacitance C is 600 μF. The effectiveness of this method can be evaluated, as shown in Fig. 6.22. In addition, the grid-in current reference i_{ref} is changing according to certain rules [14].

Figure 6.22a shows the different working stages for a grid-connected inverter, and Fig. 6.22b shows the dynamic process waveform of the system with actual capacitor. In addition, the dc component suppression process is shown in Fig. 6.22c. It can be seen from Fig. 6.22a that the grid-connected inverter experiences a soft-start process firstly, and then it keeps running with 10 A current amplitude, next the reference current amplitude suddenly changes to 5 A at the positive peak point of current reference, and finally a 1 A dc component is superimposed on the reference current at the zero crossing point, i.e. the reference current is $5\sin\omega t + 1$. The gray dotted line represents the reference current i_{ref}, and the change points are aligned in the center of the oscilloscope window, as shown in Fig. 6.22b and c. It is worth noting that the dynamic performance of the system is good under this method, and the dc component can be quickly suppressed within one line cycle.

In addition, after the reference current is superimposed with a positive DC bias, the grid-in current will have a larger ripple in the negative half cycle than the positive half cycle, which will increase the harmonic components of the grid-in current. In summary, actual capacitor method can achieve the function of DC component suppression, but considering factors such as volume and cost, virtual capacitor method should be used instead in actual applications.

3. Virtual capacitor method

The circuit sketch of the AC-side blocking capacitor method using the virtual capacitor method [15–18] is shown in Fig. 6.23a and b shows the equivalent transfer function model of this circuit, where the grid-in current is fed forward to the duty cycle. In Fig. 6.23b, $1/(K_{pwm}Cs)$ is the equivalent virtual capacitor which can be realized using a software algorithm. In this way, this method not only realizes the function of suppressing the DCC of grid-in current, but also avoids a series of disadvantages of the actual capacitor.

4. Experimental results of virtual capacitor

The experimental of blocking capacitor method using virtual capacitors and all the key parameters are consistent with the capacitor blocking method using actual physical capacitor.

By observing the key waveforms of virtual capacitor method in experimenting shown in Fig. 6.24, it is found that when using virtual capacitors, the grid-connected inverter has no significant difference in dynamic performance and dc component suppression effect from the actual physical capacitors, and the current positive and negative half-cycle ripples are symmetrical, and there is no overmodulation problem. Therefore, experiment waveforms have confirmed that virtual capacitors can replace actual physical capacitors in practice with low cost.

Suitable control parameters should be decided in the detection & feedback methods, and the capacitor blocking method also needs to adjust its capacitance. The control parameters cannot be adjusted in time when the system reference changes, which may make the system unstable or increase tracking error, because of the fixed parameters used in controller in traditional control methods. It is obvious that the controller parameters have a great influence on both the tracking effect on grid-in current and the suppressing effect on the DCC. However, the precise design of controller parameters is very complicated. With the rapid development of computer hardware resources, suppressing methods using intelligent control algorithms have been gradually adopted, such as iterative PI control algorithms [19], fuzzy control algorithms [20], and neural network ABP (adaptive back propagation) control algorithms [21] and so on.

6.4 Intelligent Algorithms

6.4.1 *Fuzzy Iterative PI Method*

A control block diagram of the fuzzy iterative PI method proposed in [19] is shown in Fig. 6.25 for a three-phase grid-connected inverter. Using a PI controller as the main controller of the circuit, the improved sliding average filtering algorithm is used to obtain the high-precision DCC of grid-in current in real-time. What's more, an

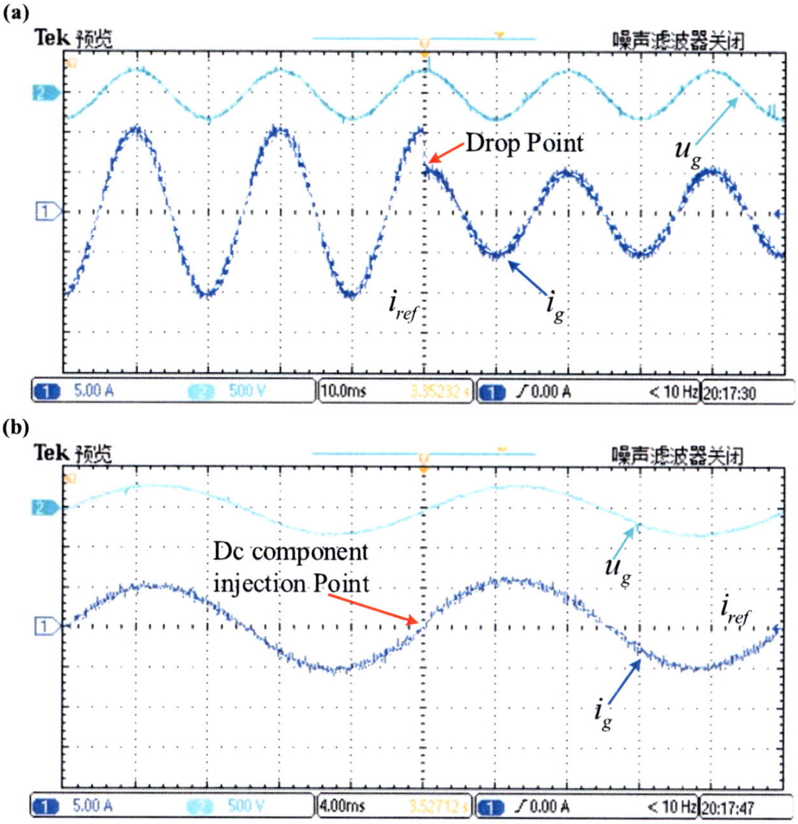

Fig. 6.24 Key waveforms of virtual capacitor method in experiment test, **a** Dynamic details of sudden current amplitude change, **b** Respance details of dc component suppression

iterative PI controller is applied to the DCC suppression loop whose parameters are generated in real-time by fuzzy computing. When combined with the parameter self-adjusting strategy of fuzzy control, it is worth noticing that iterative PI control can eliminate steady-state errors and improve the dynamic performance of the system. With iterative PI control enable, the compensation bias is fed into the main controller loop to eliminate the DCC of grid-in current.

6.4.2 BP Neural Network Method

As shown in Fig. 6.26, a BP neural network algorithm for three-phase grid-connected inverters is proposed in [21], where the main controller of current loop is the QPR controller. In order to accelerate the calculating speed in the microprocessor, an adaptive learning speed regulator is used to select a suitable speed learning factor for

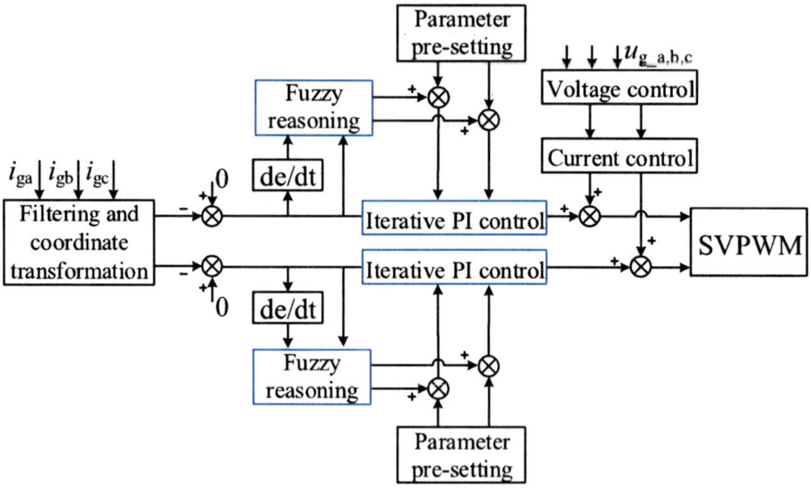

Fig. 6.25 Control strategy of fuzzy iterative PI method

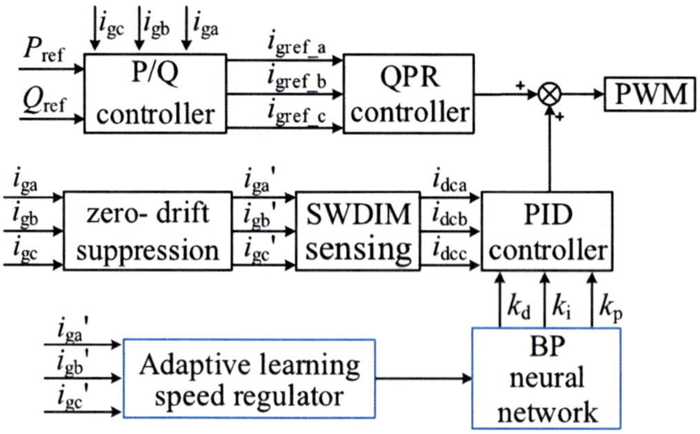

Fig. 6.26 Control strategy of BP neural network method

the subsequent operations of BP neural network. After compensated by the zero drift suppression units, the sampled grid-in current is fed to the sliding window double integration method (SWDIM) to extract the DCC of grid-in current, which is fed into the PI controller of the DCC suppression loop next. Analogously, the control parameters of PI controller are generated in real-time by the adaptive BP algorithm, which can achieve accurate suppression of the DCC of grid-in current.

6.5 Summary

To summarize, the method of eliminating the DCC of grid-in current in a TLI system are discussed in this chapter, which mainly includes detection-feedback method, blocking capacitor method, and intelligent control algorithm. Detection results of DCC fed back to the main controller to generate a modulation strategy with DC bias, the detection-feedback method can eliminate the DCC of grid-in current. Meanwhile, the blocking capacitor method does not need to detect the DCC of grid-in current, because the DC blocking property of capacitor just coincides with the purpose of eliminating the DCC. What's more, unlike detection-feedback method which affected by control parameters of the system easily, the intelligent control algorithm can achieve the purpose of eliminating DCC of grid-in current by precise control with real-time parameters.

References

1. Tang J (2007) Error analysis of inverse proportional arithmetic circuit. Sci Technol Inform 2:41–43 (in Chinese)
2. Abdelhakim A, Mattavelli P, Yang D et al (2018) Coupled-inductor-based DC current measurement technique for transformerless grid-tied inverters. IEEE Trans Power Electron 33(1):18–23
3. Armstrong M, Atkinson DJ, Johnson CM et al (2006) Auto-calibrating DC link current sensing technique for transformerless grid connected H-bridge inverter systems. IEEE Trans Power Electron 21(5):1385–1393
4. Zhang W, Armstrong M, Elgendy MA (2019a) DC injection suppression in transformer-less grid-connected inverter using a DC-link current sensing and active control approach. IEEE Trans Energy Convers 34(1):396–404
5. Wang BC, Guo XQ, Gu HR et al. Real-time DC injection measurement technique for transformerless PV systems. In: 2010 IEEE international symposium on power electronics for distributed generation systems, Hefei, China, pp 980–983
6. Sharma R. Removal of DC offset current from transformerless PV inverters connected to utility. In: 2005 international universities power engineering conference, Cork, Ireland
7. Ahmed A, Li R (2012) Precise detection and elimination of grid injected DC from single phase inverters. Int J Precis Eng Manuf 13(8):1341–1347
8. Bowtell L, Ahfock A (2010) Direct current offset controller for transformerless single-phase photovoltaic grid-connected inverters. IET Renew Power Gener 4(5):428–437
9. He GF, Xu DH, Chen M (2015) A novel control strategy of suppressing DC current injection to the grid for single-phase PV inverter. IEEE Trans Power Electron 30(3):1266–1274
10. Chen M, Xu DH, Zhang T et al (2018) A novel DC current injection suppression method for three-phase grid-connected inverter without the isolation transformer. IEEE Trans Industr Electron 6(11):8656–8666
11. Zhang W, Armstrong M, Elgendy MA (2019b) Mitigation of DC current injection in transformer-less grid-connected inverters using a voltage filtering DC extraction approach. IEEE Trans Energy Convers 34(1):426–434
12. Nabae A, Takahashi I, Akagi H (1981) A new neutral-point-clamped PWM inverter. IEEE Trans Ind Appl 17(5):518–523
13. Gonzalez R, Gubia E, Lopez J et al (2008) Transformerless single-phase multilevel photovoltaic inverter. IEEE Trans Industr Electron 55(7):2694–2702

14. Blewitt WM, Atkinson DJ, Kelly J et al (2010) Approach to low-cost prevention of DC injection in transformerless grid connected inverters. IET Power Electronics 3(1):111–119
15. Wang BC, Guo XQ, Mei Q, Sun XF, Wu WY (2009) DC injection control for transformerless PV grid-connected inverters. Proc CSEE 29(36):23–28 (in Chinese)
16. Yan Q, Wu X, Yuan X et al (2015) Minimization of the DC component in transformerless three-phase grid-connected photovoltaic inverters. IEEE Trans Power Electron 30(7):3984–3997
17. Long B (2019) Design and implementation of a virtual capacitor based DC current suppression method for grid-connected inverters. ISA Trans 92:257–272
18. Wang W, Wang P, Bei TZ et al (2015) DC injection control for grid-connected single-phase inverters based on virtual capacitor. J Power Electr 15(5):1338–1347
19. Xu P, Wu L (2015) Suppression strategy of DC component based on fuzzy iteration PI. J Jiangnan Univ (Nat Sci Edn) 14(5):585–589 (in Chinese)
20. Kumarappan AL, Sandhiya S (2016) Fuzzy based minimization of the DC component in transformerless three-phase grid-connected photovoltaic inverters. Imp J Interdisc Res 2(6):30–36
21. Long B, Huang LJ, Sun HB et al (2018) An intelligent dc current minimization method for transformerless grid connected photovoltaic inverters. ISA Trans 88:268–279